STUDY GUIDE

LINEAR ALGEBRA
AND ITS APPLICATIONS

THIRD EDITION UPDATE

David C. Lay

University of Maryland – College Park

PEARSON

Addison
Wesley

Boston   San Francisco   New York
London   Toronto   Sydney   Tokyo   Singapore   Madrid
Mexico City   Munich   Paris   Cape Town   Hong Kong   Montreal

Contents

Introduction

This *Study Guide* is designed to help you succeed in your linear algebra course. It shows you how to study mathematics, to learn new material, and to prepare effective review sheets for tests. **Key Ideas** and **Study Notes** guide you through each section, with summaries of important ideas and tables that connect related ideas. Detailed solutions to hundreds of exercises (usually every third odd exercise) allow you to check your work or help you get started on a difficult problem. Also, complete explanations are provided for each writing exercise whose answer in the text is only a "Hint." **Study Tips** point out important exercises, give hints about what to study, and sometimes highlight potential exam questions. Frequent **Warnings** identify common student errors. Don't ever take an exam without reviewing these warnings!

The most important material in this *Study Guide* is on pages ix and x of this introduction. Students who follow the strategies in *How to Study Linear Algebra* invariably achieve remarkable results in this course. You can be one of those students.

TECHNOLOGY SUPPORT

If you are using technology with your course, you will need this *Study Guide*. Besides its valuable support for the course material, the *Guide* includes "Lab Manuals" for three computer programs and four graphic calculators. Everything you need to know about using this technology with your text is here. New commands are introduced gradually, and detailed instructions are given for their use. Also, data for more than 850 exercises from the text are stored in electronic files for each type of technology. You'll save hours of time and avoid errors in typing. The files also contain special programs that reinforce basic concepts in the course. If your class is using MATLAB, Maple, or Mathematica, your files are probably already loaded on the school computer system or in specified labs. If you are using a TI-83+, TI-86, TI-89, or HP-48G, your instructor may have plans to download the files to your calculator. In any case, you can always download the files yourself from the Web site:

http://www.laylinalgebra.com

A ReadMe file with each data set describes how to incorporate the data into your software or load it into your calculator. The files on the Web will always reflect the latest versions available.

Special **MATLAB** boxes at the ends of many *Study Guide* sections explain how to use MATLAB for your homework, introducing simple commands as they are needed for the exercises. The first appendix at the back of the *Guide* contains a quick introduction, *Getting Started with MATLAB*, and an index of MATLAB commands. I encourage you to try MATLAB—it is easy to learn.

Notes for the **Maple** and **Mathematica** computer algebra systems and for the **TI-83+/86/89** and **HP-48G** graphic calculators are included in the last four appendices to this *Study Guide*. The notes correspond to the MATLAB boxes and translate MATLAB commands into syntax appropriate for the other technologies. Each set of notes has its own index. Whenever you see a MATLAB box in this *Guide*, turn to the appropriate appendix for help with your technology.

The following faculty members wrote the notes and developed the special programs:

Maple	Douglas Meade, University of South Carolina, Columbia, SC
Mathematica	Lyle Cochran, Whitworth College, Spokane, WA
TI-83+/86/89	Michael Miller, Western Baptist College, Salem, OR
HP-48G	Thomas W. Polaski, Winthrop University, Rock Hill, SC

Also, Professor Jeremy Case, of Taylor University, Marion, IN, helped with the data and other materials involving MATLAB. These colleagues have given good advice based on teaching with our text and *Study Guide*. I appreciate their contributions to this revision of the *Study Guide*.

REVIEW MATERIALS ON THE WEB

In addition to the help in this *Study Guide*, I have provided some material on the Web that my students really appreciate—review sheets and practice exams. Please heed the advice below, because using these study aids in the wrong way can lead to a disaster at exam time. I suggest four steps to prepare for each exam.

1. Assemble a review sheet. The Web review sheets reflect what I emphasize in *my* courses. If possible, you need to find out what *your* instructor expects of you. The three courses on the Web material vary somewhat in their content and approach, and they organize the material differently. To construct a review sheet for one of your exams, you may need to combine parts of two sheets from the Web.

2. Try to complete an initial review for an exam a day or two early. Hard to do, but worth the effort. Don't read the sample exams! Studying from an old exam is a big mistake. Instead, study your lecture notes, looking for items that were emphasized in class. Read over your homework and old quizzes. I insist that my students learn key definitions, practically word for word. If they cannot write a definition properly, they don't know what they are writing about.

3. After the review, *take* the sample exam whose subject material most closely fits what your exam will cover. Find a quiet place and time when you can work the entire exam without stopping , and *without looking at the text or your notes*. (Of course, skip any questions that are inappropriate for your course.) Write your answers. Time yourself.

4. Finally, after you have completed the test, look at the solutions. Identify the areas that need further review. If possible, find an example in the text related to an area in which you are weak. Cover up the solution to the example and try to write out the solution yourself. Peek, if necessary. Don't be reluctant to ask for help.

How to Study Linear Algebra

A first course in linear algebra is dramatically different from most mathematics courses that precede it. The focus shifts from learning computational procedures to digesting and mastering basic concepts that underlie the computations. To survive, you may need to learn a new way to study mathematics. That's why I wrote this *Study Guide*—to show you how to succeed in the course and to give you tools to do this.

Because you are likely to use linear algebra later in your career, you need to learn the material at a level that will carry you far beyond the final exam. I believe that the strategies below are crucial to success.

STRATEGIES FOR SUCCESS IN LINEAR ALGEBRA

1. **Study before you start to work on exercises**. Most students don't do this in courses that precede linear algebra. They survive by looking at the examples when they cannot solve an exercise. That simply will not work in linear algebra. If you "copy" an example (with necessary modifications), you may think you understand the problem, but very little true learning has taken place. (You'll find that out on your first exam.) For this course, in addition to knowing *how* to carry out a certain procedure, you must learn *when* that procedure is appropriate and (most importantly) *why* it works.

    For success with homework, read the text section first, perhaps taking a few notes. Then, read the **Key Ideas** or **Study Notes** in the *Study Guide* for that section. Finally, start to work on the assigned exercises. In the long run, this approach will improve your performance *and* save you time. The preparation time spent here will greatly reduce your exam preparation.

2. **Prepare for each class period as you would for a language class**. Mastery of the subject requires that you learn a rich vocabulary. Your goal now is to become so familiar with concepts that you can use them easily (and correctly) in conversation and in writing. For homework, try to write complete sentences, such as you'll find in the *Study Guide* solutions. Pay attention, too, to the warnings here about misuse of terminology.

This course resembles a language course because of the preparation needed between class meetings, to avoid falling behind. Most sections in the text build on preceding sections. Once you are behind, catching up with the class is often difficult. The fact that concepts may seem "simple" does not mean that you can afford to postpone your study until the weekend. The homework may be harder than you expect. The most valuable advice I can give is to keep up with the course.

3. **Concentrate more on learning definitions, facts, and concepts** than on practicing routine computations or algorithms. See connections *between* concepts. Many theorems and boxed "facts" describe such connections. For examples, see Theorem 2 in Section 1.2, and Theorems 3 and 4 in Section 1.4. Your goal is to think in general terms, to *imagine* typical computations without performing any arithmetic, and to focus on the principles behind the computations.

4. **Review frequently**. Review and reflection are key ingredients for success in learning the material. At strategic points in this *Guide*, I have inserted special subsections labeled "Mastering Linear Algebra Concepts." They provide specific help for your review of each main concept. I urge you to prepare the review sheets described as you reach each review point. Later, you may choose to add further notes. Of course, use the sheets to review for exams. A Glossary Checklist at the end of each chapter in the *Guide* may help you learn important definitions.

CAUTION   Because you can find complete solutions here to many exercises, you will be tempted to read the explanations before you really try to write out the solutions yourself. Don't do it! If you merely think a bit about a problem and then check to see if your idea is basically correct, you are likely to overestimate your understanding. Some of my students have done this and miserably failed the first exam. By then the damage was done, and they had great difficulty catching up with the class. Proper use of the *Study Guide*, however, will help you to succeed and enjoy the course at the same time.

A PERSONAL NOTE

Students who have used this material have told me how much it helped them learn linear algebra and prepare for tests. The first time my students used the *Study Guide* notes, they had already taken one exam. Grades on the next exam were substantial. For some students, the improvement was dramatic. I hope the *Study Guide* will encourage you to master linear algebra and to perform at a level higher than you ever dreamed possible.

David C. Lay

1 Linear Equations in Linear Algebra

As you work through this chapter and the next, your experience may resemble several walks through a village at different seasons of the year. The surroundings will be familiar, but the landscape will change. You will examine various mathematical concepts from several points of view, and a major problem will be to learn all the new terminology and the many connections between the concepts. In Chapter 4, you will see these ideas in a more abstract setting. Diligent work now will make the trip through Chapter 4 just another walk through the same village.

1.1   SYSTEMS OF LINEAR EQUATIONS

The fundamental concepts presented in this section and the next must be mastered for they will be used throughout the course.

STUDY NOTES

Please read **How to Study Linear Algebra**, on the preceding two pages, before you continue.

The text uses boldface type to identify important terms the first time they appear. You need to learn them; some students write selected terms on 3×5 cards, for review. At the end of each chapter in this *Study Guide*, a glossary checklist may help you learn definitions.

The text defines the **size** of a matrix. Don't use the term *dimension*, even though that appears in some computer programming languages, because in linear algebra, *dimension* refers to another concept (in Section 4.5).

The first few examples are so simple that they could be solved by a variety of techniques. But it is important to learn the systematic method presented here, because it easily handles more complicated linear systems, and it works in all cases.

The calculations in this section are based on the following important fact:

> *When elementary row operations are applied to a linear system, the new system has exactly the same solution set.*

(See the text.) The steps in the summary below will be modified slightly in Section 1.2.

Summary of the Elimination Method (for This Section)

1. The first equation must contain an x_1. Interchange equations, if necessary. This will create a nonzero entry in the first row, first column, of the augmented matrix.

2. Eliminate x_1 terms in the other equations. That is, use replacement operations to create zeros in the first column of the matrix below the first row.

3. Obtain an x_2 term in the second equation. (Interchange the second equation with one below, if needed, but don't touch the first equation.) You may scale the second equation, if desired, to create a 1 in the second column and second row of the matrix.

4. Eliminate x_2 terms in equations below the second equation, using replacement operations.

5. Continue with x_3 in the third equation, x_4 in the fourth equation, etc., eliminating these variables in the equations below. This will produce a "triangular" system (at least for systems in this section).

6. Check if the system in triangular form is consistent. If it is, a solution is found by starting with the last nonzero equation and working back up to the first equation. Each variable on the "diagonal" is used to eliminate the terms in that variable above it. The solution to the system becomes apparent when the system is finally transformed into "diagonal" form.

7. Check any solutions you find by substituting them into the original system.

The *solution set* of a system of linear equations either is empty, or contains one solution, or contains infinitely many solutions. When asked to "solve" a system, you may write "inconsistent" if the system has no solution.

As you will see later, determining the number of solutions in the solution set is sometimes more important than actually computing the solution or solutions. For that reason, pay close attention to the subsection on existence and uniqueness questions. Key Exercises: 19–22 and 25.

SOLUTIONS TO EXERCISES

Get into the habit now of working the Practice Problems before you start the exercises. Probably, you should attempt all the Practice Problems before checking the solutions at the end of the exercise set, because once you start reading the first solution, you might tend to read on through the other solutions and spoil your chance to benefit from those problems.

For brevity, the symbols R1, R2, . . ., stand for row 1 (or equation 1), row 2 (or equation 2), and so on.

1.
$$\begin{aligned} x_1 + 5x_2 &= 7 \\ -2x_1 - 7x_2 &= -5 \end{aligned} \qquad \begin{bmatrix} 1 & 5 & 7 \\ -2 & -7 & -5 \end{bmatrix}.$$

Replace R2 by R2 + (2)R1:
$$\begin{aligned} x_1 + 5x_2 &= 7 \\ 3x_2 &= 9 \end{aligned} \qquad \begin{bmatrix} 1 & 5 & 7 \\ 0 & 3 & 9 \end{bmatrix}$$

Scale R2 by 1/3:
$$\begin{aligned} x_1 + 5x_2 &= 7 \\ x_2 &= 3 \end{aligned} \qquad \begin{bmatrix} 1 & 5 & 7 \\ 0 & 1 & 3 \end{bmatrix}$$

Replace R1 by R1 + (−5)R2:
$$\begin{aligned} x_1 &= -8 \\ x_2 &= 3 \end{aligned} \qquad \begin{bmatrix} 1 & 0 & -8 \\ 0 & 1 & 3 \end{bmatrix}$$

The solution is $(x_1, x_2) = (-8, 3)$, or simply $(-8, 3)$. Check:
$$\begin{aligned} (-8) + 5(3) &= -8 + 15 = 7 \\ -2(-8) - 7(3) &= 16 - 21 = -5 \end{aligned}$$

7. $\begin{bmatrix} 1 & 7 & 3 & -4 \\ 0 & 1 & -1 & 3 \\ 0 & 0 & 0 & 1 \\ 0 & 0 & 1 & -2 \end{bmatrix}$. Ordinarily, the next step would be to interchange R3 and R4, to put a 1

in the third row and third column. But in this case, the third row of the augmented matrix corresponds to the equation $0x_1 + 0x_2 + 0x_3 = 1$, or simply, $0 = 1$. A system containing this condition has no solution. Further row operations are unnecessary once an equation such as $0 = 1$ is evident.

The solution set is empty.

Study Tip: When writing a coefficient matrix or augmented matrix for a system of linear equations, be sure that the variables appear *in the same order* in each equation. Arrange the variables in columns, as in the text, placing zeros in the matrix whenever a variable is missing from an equation.

13. $\begin{bmatrix} 1 & 0 & -3 & 8 \\ 2 & 2 & 9 & 7 \\ 0 & 1 & 5 & -2 \end{bmatrix} \sim \begin{bmatrix} 1 & 0 & -3 & 8 \\ 0 & 2 & 15 & -9 \\ 0 & 1 & 5 & -2 \end{bmatrix} \sim \begin{bmatrix} 1 & 0 & -3 & 8 \\ 0 & 1 & 5 & -2 \\ 0 & 2 & 15 & -9 \end{bmatrix} \sim \begin{bmatrix} 1 & 0 & -3 & 8 \\ 0 & 1 & 5 & -2 \\ 0 & 0 & 5 & -5 \end{bmatrix}$

$\sim \begin{bmatrix} 1 & 0 & -3 & 8 \\ 0 & 1 & 5 & -2 \\ 0 & 0 & 1 & -1 \end{bmatrix} \sim \begin{bmatrix} 1 & 0 & 0 & 5 \\ 0 & 1 & 0 & 3 \\ 0 & 0 & 1 & -1 \end{bmatrix}$. The solution is $(5, 3, -1)$.

Study Tip: Pay attention to how a problem is worded. If you only need to determine the existence or uniqueness of a solution, stop row operations when you reach a "triangular" form. Exercises 15–18 do not require you to solve the systems of equation.

19. $\begin{bmatrix} 1 & h & 4 \\ 3 & 6 & 8 \end{bmatrix} \sim \begin{bmatrix} 1 & h & 4 \\ 0 & 6-3h & -4 \end{bmatrix}$. Think of $6 - 3h$ as a constant, c. When c is zero, that is,

when $h = 2$, the system has no solution, because $0\,x_2 = -4$ has no solution. Otherwise, when c is nonzero, that is, when $h \neq 2$, the system has a solution.

23. My own students have recommended that I never give the complete answers to the true/false questions. They felt that the temptation to read the answers is too great. After working both with and without answers, they realized how much they benefited from doing the true/false work by themselves. So, all you will see here are the places where you can find the answers.

   a. See the remarks following the box titled "Elementary Row Operations."

   b. The size of a matrix is defined just before the subsection titled "Solving a Linear System."

   c. The solution set of a linear system is the set of all solutions of the system. See page 3.

   d. See the box before Example 2.

25. $\begin{bmatrix} 1 & -4 & 7 & g \\ 0 & 3 & -5 & h \\ -2 & 5 & -9 & k \end{bmatrix} \sim \begin{bmatrix} 1 & -4 & 7 & g \\ 0 & 3 & -5 & h \\ 0 & -3 & 5 & k+2g \end{bmatrix} \sim \begin{bmatrix} 1 & -4 & 7 & g \\ 0 & 3 & -5 & h \\ 0 & 0 & 0 & k+2g+h \end{bmatrix}$

Let b denote the number $k + 2g + h$. Then the third equation represented by the augmented matrix above is $0x_3 = b$. If b is nonzero, this equation has no solution, so the system is inconsistent. The system is consistent if b is zero, that is, if $k + 2g + h = 0$, then the system

$$x_1 - 4x_2 + 7x_3 = g$$
$$3x_2 - 5x_3 = h$$
$$0 = 0$$

has a solution no matter what the values of g and h. The text will explore this situation more in Section 1.2. Briefly, here is why this system, and hence the original system, is consistent. In this case, the third equation can be ignored, and the second equation, $3x_2 - 5x_3 = h$ has many solutions. Imagine choosing any values for x_2 and x_3 that satisfy the second equation, and substituting those values for x_2 and x_3 in the first equation. The resulting first equation can be solved for x_1. These values for x_1, x_2, and x_3 will satisfy all three equations.

31. Look at the first column. The next row operation should replace the 4 in the third row by a 0. To do this, replace R3 by R3 + (–4)R1. To reverse the operation, replace R3 by R3 + (4)R1.

A Mathematical Note: "If . . . , then . . . ."

Many important facts and theorems in the text are written as implication statements, in the form "If P, then Q", where P and Q represent complete sentences. For instance, the statement in the box at the top of page 8 has the form

$$\text{If} \left\{ \begin{array}{c} \text{the augmented matrices} \\ \text{of two linear systems} \\ \text{are row equivalent} \end{array} \right\}, \text{then} \left\{ \begin{array}{c} \text{the two systems} \\ \text{have the same} \\ \text{solution set} \end{array} \right\} \qquad (1)$$

An implication statement "If P, then Q" is itself true provided that statement Q is true *whenever* statement P is true. In mathematical terminology, we say that "P implies Q," and we write $P \Rightarrow Q$.

Be careful to distinguish between an implication statement "P implies Q" and the **converse** or "opposite" implication, "Q implies P". The converse may or may not be true when the original implication is true. For instance, the converse of (1) above is not true, because there exist two linear systems with the same solution set but whose augmented matrices are not row equivalent. For example:

$$\begin{aligned} x_1 + x_2 &= 1 \\ 2x_1 + x_2 &= 2 \end{aligned} \qquad\qquad \begin{aligned} x_1 + x_2 &= 1 \\ 2x_1 + 2x_2 &= 2 \\ 3x_1 + 3x_2 &= 3 \end{aligned}$$

MATLAB Row Operations

To use MATLAB for your homework in this course, the MATLAB program must contain the data for the exercises in this text. At some schools, the campus-wide version of MATLAB already has this data available on some or all computers. (The same may be true for Maple or Mathematica.) Ask your instructor. If you plan to run MATLAB at home, you will need to download the MATLAB Laydata Toolbox from the website

www.laylinalgebra.com

and follow the instructions there. Data files are also available at this site for Maple, Mathematica, and the graphic calculators TI-83+/86/89 and HP-48G. Basic instructions for using these matrix programs in this course are given in appendices at the end of this *Study Guide*. Specific commands for MATLAB will be introduced as needed at the end of some sections. Corresponding commands for other matrix programs can be found in the appendices.

While you are running MATLAB, type the command `c1s1` (which stands for chapter 1 section 1) at the MATLAB prompt. If the data are not installed, you will get a message such as "Undefined function". Otherwise, you should see a list of exercises in Section 1.1 for which data are available. Type the number of the appropriate exercise and press <Enter>.

For Section 1.1, the MATLAB data for each exercise are stored in a matrix called M. You can perform row operations on M with the following commands (which are in the Laydata Toolbox along with the data):

replace(M,r,m,s)	Replaces row r of matrix M by row $r + m \cdot$ row s
swap(M,r,s)	Interchanges rows r and s of M
scale(M,r,c)	Multiplies row r of M by a nonzero scalar c

(Press <Enter> after each MATLAB command, displayed in boldface type.) The name of any matrix in your MATLAB workspace can be inserted in place of M; the letters r, m, s, and c stand for any whole numbers you choose.

If you enter one of these commands, say, **swap(M,1,3)**, then the new matrix, produced from M, is stored in the matrix "ans" (for "answer"). If, instead, you type **M1 = swap(M,1,3)**, then the answer is stored in a new matrix $M1$. If the next operation is **M2 = replace(M1,2,5,1)**, then the result of changing $M1$ is placed in $M2$, and so on.

The advantage of giving a new name to each new matrix is that you can easily go back a step if you don't like what you just did to a matrix. If, instead, you type **M = replace(M, 2,5,1)**, then the result is placed back in M and the "old" M is lost. Of course, the "reverse" operation, **M = replace(M,2,-5,1)** will bring back the old M.

Note: For the simple problems in this section and the next, the multiple m you need in the command **replace(M,r,m,s)** will usually be a small integer or fraction that you can compute in your head. In general, m may not be so easy to compute mentally. The next two paragraphs describe how to handle such a case.

The entry in row r and column c of a matrix M is denoted by $M(r, c)$. If the number stored in $M(r, c)$ is displayed with a decimal point, then the displayed value may be accurate to only about five digits. In this case, use the *symbol $M(r, c)$* instead of the displayed value in calculations.

For instance, if you want to use the entry $M(s, c)$ to change $M(r, c)$ to 0, enter the commands

m = -M(r,c)/M(s,c)	The multiple of row s to be added to row r
M = replace(M,r,m,s)	Adds m times row s to row r

Or, you can use just one command: **M = replace(M,r,-M(r,c)/M(s,c),s)**.

Finally, the command **format compact** will eliminate extra space between displays, so you can see more data on the screen. The command **format** will return the screen to the normal display.

Warning: Using a matrix program such as MATLAB is fun and will save you time, but make sure you can perform row operations rapidly and accurately with pencil and paper. Probably, you should work all the exercises in Section 1.1 by hand and use your matrix program only to check your work.

1.2 ROW REDUCTION AND ECHELON FORMS

Our interest in the row reduction algorithm lies mostly in the echelon forms that are created by the algorithm. For practical work, a computer should perform the calculations. However, you need to understand the algorithm so you can learn how to use it for various tasks. Also, unless you take your exams at a computer or with a matrix programmable calculator, you must be able to perform row reduction quickly and accurately by hand.

STUDY NOTES

The row reduction algorithm applies to any matrix, not just an augmented matrix for a linear system. In many cases, all you need is an echelon form. The reduced echelon form is mainly used when it comes from an augmented matrix and you have to find all the solutions of a linear system.

Strategies for faster and more accurate row reduction:

- Avoid subtraction in a row replacement. It leads to mistakes in arithmetic. Instead, add a negative multiple of one row to another.

- Always enclose each matrix with brackets or large parentheses.

- To save time, combine all row replacement operations that use the same pivot position, and write just one new matrix. Never "clean out" more than one column at a time. (You can combine several scaling operations, or combine several interchanges, if you are careful. But that seldom will be necessary.)

- *Never* combine an interchange with a replacement. In general, don't combine different types of row operations. This will be particularly important when you evaluate determinants, in Chapters 3 and 5.

How to avoid copying errors:

- Practice neat writing, not too small. Develop proper habits in homework so your work on tests will be accurate, complete, and readable.

- Write a matrix row by row. Your eye may be less likely to read from the wrong row if you place the new matrix beside the old one. Arrange your sequence of matrices across the page, rather than down the page. (Some students prefer to place the matrices in columns. Use whichever method seems to work best for you.)

- Try not to let your work flow from one side of a paper to the reverse side.

Study Tips: Theorem 2 is a key result for future work. Also, study the procedure in the box following Theorem 2. Failure to write out the system of equations (step 4) is a common source of errors.

SOLUTIONS TO EXERCISES

1. To check whether a matrix is in echelon form ask the questions:

   (i) Is every nonzero row above the all-zero rows (if any)?

        The matrix (c) fails this test, so it is *not* in echelon form.

   (ii) Are the leading entries in a stair-step pattern, with zeros below each leading entry?

        The matrices (a), (b), and (d) all pass tests (i) and (ii), so they are in echelon form.

   To check whether a matrix in echelon form is actually in *reduced* echelon form, ask two more questions:

   (iii) Is there a 1 in every pivot position?

         Matrix (d) fails this test, so it is only in echelon form. Finally, ask:

   (iv) Is each leading 1 the only nonzero entry in its column?

        Matrices (a) and (b) pass all four tests, so they are in reduced echelon form.

Study Tip: Exercises 5 and 6 ask you to "visualize" echelon forms and write out matrices whose entries are just symbols. Example 2 suggests the form of your "answer," but it does not show you *how to find* the answer. Later, other exercises will ask you to construct other types of examples. If you look at answers from the text, or the *Study Guide* (or another student), before you try to write your own answers, you will lose most of the value of such exercises. The *process* of trying to understand the question and writing an example is important.

7.
$$\begin{bmatrix} 1 & 3 & 4 & 7 \\ 3 & 9 & 7 & 6 \end{bmatrix} \sim \begin{bmatrix} 1 & 3 & 4 & 7 \\ 0 & 0 & -5 & -15 \end{bmatrix} \sim \begin{bmatrix} 1 & 3 & 4 & 7 \\ 0 & 0 & 1 & 3 \end{bmatrix} \sim \begin{bmatrix} 1 & 3 & 0 & -5 \\ 0 & 0 & 1 & 3 \end{bmatrix}$$

Corresponding system of equations:
$$\begin{aligned} x_1 + 3x_2 \quad &= -5 \\ x_3 &= 3 \end{aligned}$$

The basic variables (corresponding to the pivot positions) are x_1 and x_3. The remaining variable x_2 is free. Solve for the basic variables in terms of the free variable. The general solution is
$$\begin{cases} x_1 = -5 - 3x_2 \\ x_2 \text{ is free} \\ x_3 = 3 \end{cases}$$

13.
$$\begin{bmatrix} 1 & -3 & 0 & -1 & 0 & -2 \\ 0 & 1 & 0 & 0 & -4 & 1 \\ 0 & 0 & 0 & 1 & 9 & 4 \\ 0 & 0 & 0 & 0 & 0 & 0 \end{bmatrix} \sim \begin{bmatrix} 1 & -3 & 0 & 0 & 9 & 2 \\ 0 & 1 & 0 & 0 & -4 & 1 \\ 0 & 0 & 0 & 1 & 9 & 4 \\ 0 & 0 & 0 & 0 & 0 & 0 \end{bmatrix} \sim \begin{bmatrix} 1 & 0 & 0 & 0 & -3 & 5 \\ 0 & 1 & 0 & 0 & -4 & 1 \\ 0 & 0 & 0 & 1 & 9 & 4 \\ 0 & 0 & 0 & 0 & 0 & 0 \end{bmatrix}$$

Corresponding system:
$$\begin{aligned} x_1 \qquad\qquad - 3x_5 &= 5 \\ x_2 \qquad\quad - 4x_5 &= 1 \\ x_4 + 9x_5 &= 4 \\ 0 &= 0 \end{aligned}$$

Basic variables: x_1, x_2, x_4; free variables: x_3, x_5. General solution:
$$\begin{cases} x_1 = 5 + 3x_5 \\ x_2 = 1 + 4x_5 \\ x_3 \text{ is free} \\ x_4 = 4 - 9x_5 \\ x_5 \text{ is free} \end{cases}$$

Note: A common error in this exercise is to assume that x_3 is zero. Another common error is to say *nothing* about x_3 and write only x_1, x_2, x_4, and x_5, as above. To avoid these mistakes, identify the basic variables first. Any remaining variables are *free*. (This type of computation will arise in Chapter 5.) See also Exercise 8.

Study Tip: Be sure to work Exercises 17–20. The experience will help you later. These exercises make nice quiz questions, too.

19. $\begin{bmatrix} 1 & h & 2 \\ 4 & 8 & k \end{bmatrix} \sim \begin{bmatrix} 1 & h & 2 \\ 0 & 8-4h & k-8 \end{bmatrix}$. Look first at $8 - 4h$. If this number is not zero, then the system must be consistent. Also, the solution will be unique because there are no free variables. This is case (b), when $h \neq 2$. Now, if $8 - 4h$ is zero, that is, if $h = 2$, there are two possibilities—either k equals 8 or k does not equal 8. If $h = 2$ and $k = 8$, the second equation is $0x_2 = 0$. The system is consistent and has a free variable, so the system has infinitely many solutions. This is case (c). When $h = 2$ and $k \neq 8$, the second equation is $0x_2 = b$, with b nonzero, and the system has no solution. This is case (a).

21. a. See Theorem 1.

    b. See the second paragraph of the section.

    c. Basic variables are defined after equation (4).

    d. See the beginning of the subsection, "Parametric Descriptions of Solution Sets."
Actually, this question does not consider the case of an inconsistent system. A better true/false statement would be: "If a linear system is consistent, then finding a parametric description of the solution set is the same as *solving* the system."

    e. The row shown corresponds to the equation $5x_4 = 0$. Could there also be an equation of the form $0x_4 = b$, with b nonzero?

25. A full solution is in the text answer section.

Study Tip: Notice from Exercise 27 that the question of uniqueness of the solution of a linear system is not influenced by the numbers in the rightmost column of the augmented matrix.

31. Yes, a system of linear equations with more equations than unknowns can be consistent. The answer in the text includes an example.

34. The data for this exercise comes from one of my students who was working part time for a private wind tunnel company near the University of Maryland. You will need a matrix program to solve this problem. The basic instructions for MATLAB were given in the *Study Guide* notes for Section 1.1. For Maple, Mathematica, the TI-calculators, or the HP-48G calculators, see the respective appendices at the end of this *Study Guide*.

A Mathematical Note: "If and only if"

You need to know what the phrase "if and only if" means. It was used above in Exercise 27, and you will see it again in theorems and in boxed facts. The phrase "if and only if" always appears between two complete statements. Look at Theorem 2, for instance:

$$\left\{ \begin{array}{c} \text{A specific} \\ \text{linear system} \\ \text{is consistent} \end{array} \right\} \text{ if and only if } \left\{ \begin{array}{c} \text{the rightmost column of the} \\ \text{augmented matrix is not} \\ \text{a pivot column.} \end{array} \right\} \qquad (1)$$

The entire sentence means that the two statements in parentheses are either both true or both false.

Sentence (1) has the general form

$$P \text{ if and only if } Q \qquad (2)$$

where P denotes the first statement and Q denotes the second statement. This sentence says two things:

If statement P is true, then statement Q is also true.
If statement Q is true, then statement P is also true.

A mathematical shorthand for (2) is "$P \Leftrightarrow Q$".

1.3   VECTOR EQUATIONS

Do not be deceived by the rather simple beginning of Section 1.3. The important material on $\text{Span}\{v_1, \ldots, v_p\}$ will take time to digest. Figures 8, 10, and 11 are important, along with Exercises 11–14, 17, 18, 25, and 26. Each of the exercises involves an *existence* question about whether a certain vector equation has a solution. (You don't have to find the solution.) Notice how the same basic question can be asked in several different ways.

STUDY NOTES

Develop the habit of reading the section carefully once or twice before looking at the *Study Guide* and before starting the exercises. (Don't just look at the pictures and examples! Important comments lurk in between.)

In nearly all of the text, a *scalar* is just a real number. By convention, scalars are usually written to the left of vectors, such as 5**v** or c**v**, rather than **v**5 or **v**c. To identify vectors in your lecture notes and homework, you can write underlined letters for vectors. (Some students write arrows above the letters, but that takes longer.)

Vectors must be the same size to be added or used in a linear combination. For instance, a vector in $\mathbb{R}^3$ cannot be added to a vector in $\mathbb{R}^2$.

SOLUTIONS TO EXERCISES

1. $\mathbf{u} + \mathbf{v} = \begin{bmatrix} -1 \\ 2 \end{bmatrix} + \begin{bmatrix} -3 \\ -1 \end{bmatrix} = \begin{bmatrix} -1 + (-3) \\ 2 + (-1) \end{bmatrix} = \begin{bmatrix} -4 \\ 1 \end{bmatrix}.$

Using the definitions carefully,

$\mathbf{u} - 2\mathbf{v} = \begin{bmatrix} -1 \\ 2 \end{bmatrix} + (-2)\begin{bmatrix} -3 \\ -1 \end{bmatrix} = \begin{bmatrix} -1 \\ 2 \end{bmatrix} + \begin{bmatrix} (-2)(-3) \\ (-2)(-1) \end{bmatrix} = \begin{bmatrix} -1 + 6 \\ 2 + 2 \end{bmatrix} = \begin{bmatrix} 5 \\ 4 \end{bmatrix}$, or, more quickly,

$\mathbf{u} - 2\mathbf{v} = \begin{bmatrix} -1 \\ 2 \end{bmatrix} - 2\begin{bmatrix} -3 \\ -1 \end{bmatrix} = \begin{bmatrix} -1 + 6 \\ 2 + 2 \end{bmatrix} = \begin{bmatrix} 5 \\ 4 \end{bmatrix}.$ The intermediate step is often not written.

7. See the figure below. Since the grid can be extended in every direction, the figure suggests that every vector in $\mathbb{R}^2$ can be written as a linear combination of **u** and **v**. To write a vector **a** as a linear combination of **u** and **v**, imagine walking from the origin to **a** along the grid "streets" and keep track of how many "blocks" you travel in the **u**-direction and how many in the **v**-direction.

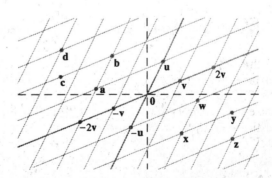

a. To reach **a** from the origin, you might travel 1 unit in the **u**-direction and -2 units in the **v**-direction (that is, 2 units in the negative **v**-direction). Hence $\mathbf{a} = \mathbf{u} - 2\mathbf{v}$.

b. To reach **b** from the origin, travel 2 units in the **u**-direction and -2 units in the **v**-direction. So $\mathbf{b} = 2\mathbf{u} - 2\mathbf{v}$. Or, use the fact that **b** is 1 unit in the **u**-direction from **a**, so that

$$\mathbf{b} = \mathbf{a} + \mathbf{u} = (\mathbf{u} - 2\mathbf{v}) + \mathbf{u} = 2\mathbf{u} - 2\mathbf{v}$$

c. The vector **c** is -1.5 units from **b** in the **v**-direction, so

$$\mathbf{c} = \mathbf{b} - 1.5\mathbf{v} = (2\mathbf{u} - 2\mathbf{v}) - 1.5\mathbf{v} = 2\mathbf{u} - 3.5\mathbf{v}$$

d. The "map" suggests that you can reach **d** if you travel 3 units in the **u**-direction and -4 units in the **v**-direction. If you prefer to stay on the paths displayed on the map, you might travel from the origin to $-3\mathbf{v}$, then move 3 units in the **u**-direction, and finally move -1 unit in the **v**-direction. So

$$\mathbf{d} = -3\mathbf{v} + 3\mathbf{u} - \mathbf{v} = 3\mathbf{u} - 4\mathbf{v}$$

Another solution is

$$\mathbf{d} = \mathbf{b} - 2\mathbf{v} + \mathbf{u} = (2\mathbf{u} - 2\mathbf{v}) - 2\mathbf{v} + \mathbf{u} = 3\mathbf{u} - 4\mathbf{v}$$

9. Here are the intermediate calculations, which usually are not displayed. Check with your instructor whether you need to "show work" on a problem such as this.

$$\begin{array}{rcl} x_2 + 5x_3 &=& 0 \\ 4x_1 + 6x_2 - x_3 &=& 0, \\ -x_1 + 3x_2 - 8x_3 &=& 0 \end{array} \qquad \begin{bmatrix} x_2 + 5x_3 \\ 4x_1 + 6x_2 - x_3 \\ -x_1 + 3x_2 - 8x_3 \end{bmatrix} = \begin{bmatrix} 0 \\ 0 \\ 0 \end{bmatrix},$$

$$\begin{bmatrix} 0 \\ 4x_1 \\ -x_1 \end{bmatrix} + \begin{bmatrix} x_2 \\ 6x_2 \\ 3x_2 \end{bmatrix} + \begin{bmatrix} 5x_3 \\ -x_3 \\ -8x_3 \end{bmatrix} = \begin{bmatrix} 0 \\ 0 \\ 0 \end{bmatrix}, \qquad x_1 \begin{bmatrix} 0 \\ 4 \\ -1 \end{bmatrix} + x_2 \begin{bmatrix} 1 \\ 6 \\ 3 \end{bmatrix} + x_3 \begin{bmatrix} 5 \\ -1 \\ -8 \end{bmatrix} = \begin{bmatrix} 0 \\ 0 \\ 0 \end{bmatrix}$$

Helpful Hint: As you work Exercises 11–14, circle the pivots in an echelon form of an appropriate matrix. This will help you visualize the cases when a vector either is or is not a linear combination of other vectors.

13. Denote the columns of A by a_1, a_2, a_3. To determine if b is a linear combination of these columns, use the boxed fact on page 34. Row reduce the augmented matrix until you reach echelon form:

$$\begin{bmatrix} 1 & -4 & 2 & 3 \\ 0 & 3 & 5 & -7 \\ -2 & 8 & -4 & -3 \end{bmatrix} \sim \begin{bmatrix} ① & -4 & 2 & 3 \\ 0 & ③ & 5 & -7 \\ 0 & 0 & 0 & ③ \end{bmatrix}$$

The system for this augmented matrix is inconsistent, so b is *not* a linear combination of the columns of A.

19. By inspection, $v_2 = (3/2)v_1$. Any linear combination of v_1 and v_2 is actually just a multiple of v_1. For instance,

$$av_1 + bv_2 = av_1 + b(3/2)v_1 = (a + 3b/2)v_1$$

So Span$\{v_1, v_2\}$ is the set of points on the line through v_1 and 0.

Warning: Although Figures 8 and 11 provide the most common ways to view Span$\{u, v\}$, don't forget Exercise 19, which shows that in a special case, Span$\{u, v\}$ can be just a line through the origin. In fact, Span$\{u, v\}$ can also be just the origin itself. How?

21. Let $y = \begin{bmatrix} h \\ k \end{bmatrix}$. Then $[u \quad v \quad y] = \begin{bmatrix} 2 & 2 & h \\ -1 & 1 & k \end{bmatrix} \sim \begin{bmatrix} ② & 2 & h \\ 0 & ② & k+h/2 \end{bmatrix}$. This augmented matrix corresponds to a consistent system for all h and k. So y is in Span$\{u, v\}$ for all h and k.

23. a. The alternative notation for a (column) vector is discussed after Example 1.

   b. Plot the points to check the assertion. Or, see the statement preceding Example 3.

   c. See the line displayed just before Example 4.

   d. See the box that discusses the matrix in (5).

   e. Read the geometric description of Span$\{u, v\}$ very carefully.

Study Tip: I urge my own students to work by themselves on the true/false questions and then meet together in groups of two or three, to compare and discuss their answers.

25. a. There are only three vectors in the set $\{a_1, a_2, a_3\}$, and b is not one of them.

   b. There are infinitely many vectors in $W = $ Span$\{a_1, a_2, a_3\}$. To determine if b is in W, use the method of Exercise 13.

$$\begin{bmatrix} 1 & 0 & -4 & 4 \\ 0 & 3 & -2 & 1 \\ -2 & 6 & 3 & -4 \end{bmatrix} \sim \begin{bmatrix} 1 & 0 & -4 & 4 \\ 0 & 3 & -2 & 1 \\ 0 & 6 & -5 & 4 \end{bmatrix} \sim \begin{bmatrix} ① & 0 & -4 & 4 \\ 0 & ③ & -2 & 1 \\ 0 & 0 & ⊖① & 2 \end{bmatrix}$$
$$\;\;\uparrow \quad\; \uparrow \quad\; \uparrow \quad\; \uparrow$$
$$\;\;a_1 \quad a_2 \quad a_3 \quad b$$

The system for this augmented matrix is consistent, so $\mathbf{b}$ is in W.

c. $\mathbf{a}_1 = 1\mathbf{a}_1 + 0\mathbf{a}_2 + 0\mathbf{a}_3$. See the discussion following the definition of Span$\{\mathbf{v}_1, \ldots, \mathbf{v}_p\}$.

31. a. The center of mass is $\dfrac{1}{3}\left(1 \cdot \begin{bmatrix} 0 \\ 1 \end{bmatrix} + 1 \cdot \begin{bmatrix} 8 \\ 1 \end{bmatrix} + 1 \cdot \begin{bmatrix} 2 \\ 4 \end{bmatrix}\right) = \begin{bmatrix} 10/3 \\ 2 \end{bmatrix}$.

b. The total mass of the new system is 9 grams. The three masses added, w_1, w_2, and w_3, satisfy the equation

$$\frac{1}{9}\left((w_1 + 1) \cdot \begin{bmatrix} 0 \\ 1 \end{bmatrix} + (w_2 + 1) \cdot \begin{bmatrix} 8 \\ 1 \end{bmatrix} + (w_3 + 1) \cdot \begin{bmatrix} 2 \\ 4 \end{bmatrix}\right) = \begin{bmatrix} 2 \\ 2 \end{bmatrix}$$

which can be rearranged to

$$(w_1 + 1) \cdot \begin{bmatrix} 0 \\ 1 \end{bmatrix} + (w_2 + 1) \cdot \begin{bmatrix} 8 \\ 1 \end{bmatrix} + (w_3 + 1) \cdot \begin{bmatrix} 2 \\ 4 \end{bmatrix} = \begin{bmatrix} 18 \\ 18 \end{bmatrix}$$

and

$$w_1 \cdot \begin{bmatrix} 0 \\ 1 \end{bmatrix} + w_2 \cdot \begin{bmatrix} 8 \\ 1 \end{bmatrix} + w_3 \cdot \begin{bmatrix} 2 \\ 4 \end{bmatrix} = \begin{bmatrix} 8 \\ 12 \end{bmatrix}$$

The condition $w_1 + w_2 + w_3 = 6$ and the vector equation above combine to produce a system of three equations whose augmented matrix is shown below, along with a sequence of row operations:

$$\begin{bmatrix} 1 & 1 & 1 & 6 \\ 0 & 8 & 2 & 8 \\ 1 & 1 & 4 & 12 \end{bmatrix} \sim \begin{bmatrix} 1 & 1 & 1 & 6 \\ 0 & 8 & 2 & 8 \\ 0 & 0 & 3 & 6 \end{bmatrix} \sim \begin{bmatrix} 1 & 1 & 1 & 6 \\ 0 & 8 & 2 & 8 \\ 0 & 0 & 1 & 2 \end{bmatrix}$$

$$\sim \begin{bmatrix} 1 & 1 & 0 & 4 \\ 0 & 8 & 0 & 4 \\ 0 & 0 & 1 & 2 \end{bmatrix} \sim \begin{bmatrix} 1 & 0 & 0 & 3.5 \\ 0 & 8 & 0 & 4 \\ 0 & 0 & 1 & 2 \end{bmatrix} \sim \begin{bmatrix} ① & 0 & 0 & 3.5 \\ 0 & ① & 0 & .5 \\ 0 & 0 & ① & 2 \end{bmatrix}$$

Answer: Add 3.5 g at (0, 1), add .5 g at (8, 1), and add 2 g at (2, 4).

33. a. For $j = 1, \ldots, n$, the jth entry of $(\mathbf{u} + \mathbf{v}) + \mathbf{w}$ is $(u_j + v_j) + w_j$. By associativity of addition in $\mathbb{R}$, this entry equals $u_j + (v_j + w_j)$, which is the jth entry of $\mathbf{u} + (\mathbf{v} + \mathbf{w})$. By definition of equality of vectors, $(\mathbf{u} + \mathbf{v}) + \mathbf{w} = \mathbf{u} + (\mathbf{v} + \mathbf{w})$.

b. For any scalar c, the jth entry of $c(\mathbf{u} + \mathbf{v})$ is $c(u_j + v_j)$, and the jth entry of $c\mathbf{u} + c\mathbf{v}$ is $cu_j + cv_j$ (by definition of scalar multiplication and vector addition). These entries are equal, by a distributive law in $\mathbb{R}$. So $c(\mathbf{u} + \mathbf{v}) = c\mathbf{u} + c\mathbf{v}$.

MATLAB   Constructing a Matrix

To access the data for Section 1.3, give the command **cls3**. The data for Exercise 25, for example, consists of a matrix A, its columns **a1**, **a2**, **a3**, and the vector **b**. The command $\mathbf{M} = [\mathbf{a1} \quad \mathbf{a2} \quad \mathbf{a3} \quad \mathbf{b}]$ creates a matrix using the vectors as its columns. The same matrix is created by the command $\mathbf{M} = [\mathbf{A} \quad \mathbf{b}]$.

Each time you want data for a new exercise in Section 1.3, you need the command **cls3**. After the first exercise, you can use the up-arrow ($\uparrow$). This will make MATLAB scroll back through your old commands. You may be able to find "c2sl" faster than you can retype it. Press <Enter> to reuse the command.

Exercises 11–14, 25–28, and 31 can be solved using the commands **replace**, **swap**, and (occasionally) **scale**, described on page 1-6.

1.4   THE MATRIX EQUATION $A\mathbf{x} = \mathbf{b}$

The ideas, boxed statements, and theorems in this section are absolutely fundamental for the rest of the text, so you should read the section extremely carefully.

KEY IDEAS

The definition of $A\mathbf{x}$ as a linear combination of the columns of A will be used often. You should learn the definition in *words* as well as symbols. *Note*: It is not wrong to write a scalar on the *right* side of a vector and write $A\mathbf{x}$ as $\mathbf{a}_1 x_1 + \cdots + \mathbf{a}_n x_n$, but the text follows the usual practice of writing a scalar on the *left* side of a vector.

You need to understand *why* Theorem 4 is true. That may take some time and effort. Example 3 should help, along with the proof. Theorem 4(d) can be restated as "The reduced echelon form of A has no row of zeros."

The phrase *logically equivalent* is explained in the statement of Theorem 4. This phrase is used with several statements in the same way that *if and only if* (or the symbol $\Leftrightarrow$) is used between two statements. (See the Mathematical Note at the end of Section 1.2 in this *Guide*.)

Saying that statements (a), (b), (c), and (d) are logically equivalent means the same thing as saying that (a) ⇔ (b), (b) ⇔ (c), and (c) ⇔ (d).

Key exercises are 1–20, 27, 28, 31 and 32. Think about 31 and 32, even if they are not assigned, because they introduce ideas you will need soon. (Don't check the solution of Exercise 31 until you have written your own answer.)

Checkpoint 1: True or False? If an augmented matrix $[A \quad \mathbf{b}]$ has a pivot position in every row, then the equation $A\mathbf{x} = \mathbf{b}$ is consistent.

Note: You should work a checkpoint problem when you first see it, provided that you have already read the text at least once. Always *write* your answer before comparing it with the one I have written. The checkpoint answer will be at the end of the solutions for this section.

SOLUTIONS TO EXERCISES

1. The text has the solution. Exercises 1–12 are designed to help you learn Theorem 3 and the definition of $A\mathbf{x}$. If a problem involves vectors—say, $\mathbf{v}_1, \mathbf{v}_2, \mathbf{v}_3$ — you can place the vectors into a matrix $[\mathbf{v}_1 \quad \mathbf{v}_2 \quad \mathbf{v}_3]$, if that is helpful. If a problem involves a matrix A, you can give names to the columns of A—say, $\mathbf{a}_1, \mathbf{a}_2, \mathbf{a}_3$—and reformulate a matrix equation as a vector equation. If a problem leads to a system of linear equations, you may regard it as either a vector equation or a matrix equation, whichever is most useful.

7. The left side of the equation is a linear combination of three vectors. Write the matrix A whose columns are those three vectors, and create a variable vector $\mathbf{x}$ with three entries:

$$A = \begin{bmatrix} \begin{bmatrix} 4 \\ -1 \\ 7 \\ -4 \end{bmatrix} & \begin{bmatrix} -5 \\ 3 \\ -5 \\ 1 \end{bmatrix} & \begin{bmatrix} 7 \\ -8 \\ 0 \\ 2 \end{bmatrix} \end{bmatrix} = \begin{bmatrix} 4 & -5 & 7 \\ -1 & 3 & -8 \\ 7 & -5 & 0 \\ -4 & 1 & 2 \end{bmatrix} \text{ and } \mathbf{x} = \begin{bmatrix} x_1 \\ x_2 \\ x_3 \end{bmatrix}.$$

Thus the equation $A\mathbf{x} = \mathbf{b}$ is $\begin{bmatrix} 4 & -5 & 7 \\ -1 & 3 & -8 \\ 7 & -5 & 0 \\ -4 & 1 & 2 \end{bmatrix} \begin{bmatrix} x_1 \\ x_2 \\ x_3 \end{bmatrix} = \begin{bmatrix} 6 \\ -8 \\ 0 \\ -7 \end{bmatrix}$

Warning: Be careful to distinguish between the *matrix equation* $Ax = b$ and the *augmented matrix* $[\mathbf{a}_1 \cdots \mathbf{a}_p \ \mathbf{b}]$, which is used in Theorem 3 to refer to a system of linear equations having this augmented matrix. Thus, the answer to Exercise 7 is *not* the augmented matrix at the right:

$$\begin{bmatrix} 4 & -5 & 7 & 6 \\ -1 & 3 & -8 & -8 \\ 7 & -5 & 0 & 0 \\ -4 & 1 & 2 & -7 \end{bmatrix}$$

13. The vector **u** is in the plane spanned by the columns of A if and only if **u** is a linear combination of the columns of A. This happens if and only if the equation $Ax = \mathbf{u}$ has a solution. (See the box preceding Example 3 in Section 1.4.) To study this equation, reduce the augmented matrix $[A \quad \mathbf{u}]$:

$$\begin{bmatrix} 3 & -5 & 0 \\ -2 & 6 & 4 \\ 1 & 1 & 4 \end{bmatrix} \sim \begin{bmatrix} 1 & 1 & 4 \\ -2 & 6 & 4 \\ 3 & -5 & 0 \end{bmatrix} \sim \begin{bmatrix} 1 & 1 & 4 \\ 0 & 8 & 12 \\ 0 & -8 & -12 \end{bmatrix} \sim \begin{bmatrix} ① & 1 & 4 \\ 0 & ⑧ & 12 \\ 0 & 0 & 0 \end{bmatrix}$$

The equation $Ax = \mathbf{u}$ has a solution, so **u** is in the plane spanned by the columns of A.

Study Tip: Exercises 17–20 require written explanations as well as calculations. For instance, your calculation for Exercise 17 might show the row reduction

$$A = \begin{bmatrix} 1 & 3 & 0 & 3 \\ -1 & -1 & -1 & 1 \\ 0 & -4 & 2 & -8 \\ 2 & 0 & 3 & -1 \end{bmatrix} \sim \begin{bmatrix} 1 & 3 & 0 & 3 \\ 0 & 2 & -1 & 4 \\ 0 & -4 & 2 & -8 \\ 0 & -6 & 3 & -7 \end{bmatrix} \sim \begin{bmatrix} 1 & 3 & 0 & 3 \\ 0 & 2 & -1 & 4 \\ 0 & 0 & 0 & 0 \\ 0 & 0 & 0 & 5 \end{bmatrix} \sim \begin{bmatrix} ① & 3 & 0 & 3 \\ 0 & ② & -1 & 4 \\ 0 & 0 & 0 & ⑤ \\ 0 & 0 & 0 & 0 \end{bmatrix}$$

After this, it is not enough to write "No, by Theorem 4." Instead, you should show that you know *why* Theorem 4 is relevant. For instance, you might write:

> The matrix A does *not* have a pivot in every row. By Theorem 4, the equation $Ax = b$ does *not* have a solution for each **b** in $\mathbb{R}^4$.

On a test, you probably would not have to know the theorem number. It might be enough to say "By a theorem," instead of "By Theorem 4." (Check with your instructor.)

19. The work in Exercise 17 shows that the equation $Ax = b$ does not have a solution for each **b**. That is, statement (d) in Theorem 4 is false. So all four statements in Theorem 4 are false. Since statement (b) is false, not all vectors in $\mathbb{R}^4$ can be written as a linear combination of the columns of A. Since statement (c) is false, the columns of A do *not* span $\mathbb{R}^4$.

Checkpoint 2: Given v_1, v_2, v_3 as in Exercise 21, find a specific vector in $\mathbb{R}^4$ that is not in Span$\{v_1, v_2, v_3\}$. (If necessary, reread Example 3.)

23. **a.** See the paragraph following equation (3). **b.** See the box before Example 3.

   c. See the warning following Theorem 4.   **d.** See Example 4.

   e. See Theorem 4.   **f.** See Theorem 4.

25. By definition, the matrix-vector product on the left is a linear combination of the columns of the matrix, in this case using weights –3, –1, and 2. So $c_1 = -3$, $c_2 = -1$, and $c_3 = 2$.

29. Start with any 3×3 matrix B in echelon form that has three pivot positions. Perform a row operation (a row interchange or a row replacement) that creates a matrix A that is *not* in echelon form. Then A has the desired property. The justification is given by row reducing A to B, in order to display the pivot positions. Since A has a pivot position in every row, the columns of A span $\mathbb{R}^3$, by Theorem 4.

31. A 3×2 matrix has three rows and two columns. With only two columns, A can have at most two pivot columns, and so A has at most two pivot positions, which is not enough to fill all three rows. By Theorem 4, the equation $A\mathbf{x} = \mathbf{b}$ cannot be consistent for all $\mathbf{b}$ in $\mathbb{R}^3$.

33. If the equation $A\mathbf{x} = \mathbf{b}$ has a unique solution, then the associated system of equations does not have any free variables. If every variable is a basic variable, then each column of A is a pivot column. So the reduced echelon form of A must be $\begin{bmatrix} 1 & 0 & 0 \\ 0 & 1 & 0 \\ 0 & 0 & 1 \\ 0 & 0 & 0 \end{bmatrix}$.

37. **[M]** The original matrix has no pivot in the fourth row, so its columns do not span $\mathbb{R}^4$, by Theorem 4.

Helpful Hint: For Exercises 41 and 42, use a matrix program to obtain an echelon form of the matrix. Try covering various columns of this matrix, one at a time, and ask yourself if the columns of the resulting matrix span $\mathbb{R}^4$. If you can delete one column, can you delete a second column? Why or why not?

   The analysis here depends on the following idea, which is fairly obvious but is not explicitly mentioned in the text. When a row operation is performed on a matrix A, the calculations for each new entry depend only on the other entries in the *same column*. If a column of A is removed, forming a new matrix, the absence of this column has no affect on any row-operation calculations for entries in the other columns of A. (The absence of a column might affect the particular *choice* of row operations performed for some purpose, but that is not relevant.)

Answers to Checkpoints:

1. False. See the Warning after Theorem 4. If you missed this, you are not studying the text properly. You should read the text thoroughly *before* you look at the *Study Guide* and before you work on the exercises.

2. Let $A = [\mathbf{v}_1 \quad \mathbf{v}_2 \quad \mathbf{v}_3] = \begin{bmatrix} 1 & 0 & 1 \\ 0 & 1 & 0 \\ -1 & 0 & 0 \\ 0 & -1 & -1 \end{bmatrix}$ and $\mathbf{b} = \begin{bmatrix} b_1 \\ b_2 \\ b_3 \\ b_4 \end{bmatrix}$. Row reduce the augmented matrix for $A\mathbf{x} = \mathbf{b}$ to determine values of $b_1, \ldots, b_4$ that make the equation *inconsistent*.

$$\begin{bmatrix} 1 & 0 & 1 & b_1 \\ 0 & 1 & 0 & b_2 \\ -1 & 0 & 0 & b_3 \\ 0 & -1 & -1 & b_4 \end{bmatrix} \sim \begin{bmatrix} 1 & 0 & 1 & b_1 \\ 0 & 1 & 0 & b_2 \\ 0 & 0 & 1 & b_3 + b_1 \\ 0 & -1 & -1 & b_4 \end{bmatrix} \sim \begin{bmatrix} 1 & 0 & 1 & b_1 \\ 0 & 1 & 0 & b_2 \\ 0 & 0 & 1 & b_3 + b_1 \\ 0 & 0 & -1 & b_4 + b_2 \end{bmatrix}$$

$$\begin{bmatrix} 1 & 0 & 1 & b_1 \\ 0 & 1 & 0 & b_2 \\ 0 & 0 & 1 & b_3 + b_1 \\ 0 & 0 & 0 & b_4 + b_2 + b_3 + b_1 \end{bmatrix}$$

Take $\mathbf{b} = (1, 1, 0, 0)$, for example, or any other choice of $b_1, \ldots, b_4$ whose sum is *not* zero.

Mastering Linear Algebra Concepts: Span

Please begin by reviewing "How to Study Linear Algebra," at the beginning of this *Study Guide*.

To really understand a key concept, you need to form an image in your mind that consists of the basic definition(s) together with many related ideas. Your goal at this point is to collect various ideas associated with the set Span$\{\mathbf{v}_1, \ldots, \mathbf{v}_p\}$ and the concept of a set that "spans" $\mathbb{R}^n$. Here are specific things to do now as you prepare a sheet (or sheets) for review and reference.

•   Write the **definition** of Span$\{\mathbf{v}_1, \ldots, \mathbf{v}_p\}$. (Learn it word for word.)

•   Write the **definition** of the phrase: $\{\mathbf{v}_1, \ldots, \mathbf{v}_p\}$ spans $\mathbb{R}^n$. (See page 43.) Here *span* is a verb rather than a noun as in Span$\{\mathbf{v}_1, \ldots, \mathbf{v}_p\}$.

•   Add the **equivalent description** (not definition) of what is meant for a vector **b** to be in Span$\{\mathbf{v}_1, \ldots, \mathbf{v}_p\}$. (See page 35.)

•   Copy **Theorem 4** word for word. (If you try to rephrase or summarize it in your own words, you are likely to change the meaning.)

- Sketch some **geometric interpretations** of Span$\{v_1, \ldots, v_p\}$. (Select some of Figs. 8, 10, 11, and Exercises 19, 20 in Section 1.3.)

- Identify **special cases**. (Describe Span$\{u\}$ and Span$\{u, v\}$.)

- Summarize **algorithms** or **typical computations** (such as Example 6 and Exercises 11–14, 17, 18, 25, and 26 in Section 1.3, or Example 3 and Exercises 13–22 in Section 1.4.

- Describe connections with other concepts. (See pages 42–43.)

Whenever you encounter new examples or situations that help you understand the concept of a spanning set, add them to this review sheet.

MATLAB   gauss **and** bgauss

To solve **Ax** = **b**, row reduce the matrix **M** = [**A** **b**]. The command **x** = [5;3; – 7] creates a column vector **x** with entries 5, 3, –7. Matrix vector multiplication is **A*x**.

To speed up row reduction of **M** = [**A** **b**], the command **gauss(M,r)** will use the leading entry in row r of M as a pivot, and use row replacements to create zeros in the pivot column below this pivot entry. The result is stored in the default matrix "ans", unless you assign the result to some other variable, such as M itself.

For the backward phase of row reduction, use **bgauss(M,r)**, which selects the leading entry in row r of M as the pivot, and creates zeros in the column *above* the pivot. Use **scale** to create leading 1's in the pivot positions. The commands **gauss**, **bgauss**, and **scale** are in the Laydata Toolbox, which you can download from the web.

1.5   SOLUTION SETS OF LINEAR SYSTEMS _____

Many of the concepts and computations in linear algebra involve sets of vectors which are visualized geometrically as lines and planes. The most important examples of such sets are the solution sets of linear systems.

KEY IDEAS

Visualize the solution set of a homogeneous equation $Ax = 0$ as:

- the single point **0**, when $Ax = 0$ has only the trivial solution,

- a line through **0**, when $Ax = 0$ has one free variable,

- a plane through **0**, when $Ax = 0$ has two free variables.
  (For more than two free variables, also use a plane through **0**.)

For $\mathbf{b} \neq \mathbf{0}$, visualize the solution set of $A\mathbf{x} = \mathbf{b}$ as:

- empty, if $\mathbf{b}$ is not a linear combination of the columns of A,

- one nonzero point (vector), when $A\mathbf{x} = \mathbf{b}$ has a unique solution,

- a line not through $\mathbf{0}$, when $A\mathbf{x} = \mathbf{b}$ is consistent and has one free variable,

- a plane not through $\mathbf{0}$, when $A\mathbf{x} = \mathbf{b}$ is consistent and has two or more free variables.

The solution set of $A\mathbf{x} = \mathbf{b}$ is said to be described *implicitly*, because the equation is a condition an $\mathbf{x}$ must satisfy in order to be in the set, yet the equation does not show how to find such an $\mathbf{x}$. When the solution set of $A\mathbf{x} = \mathbf{0}$ is written as Span$\{\mathbf{v}_1, \ldots, \mathbf{v}_p\}$, the set is said to be described *explicitly*; each element in the set is produced by forming a linear combination of $\mathbf{v}_1, \ldots, \mathbf{v}_p$.

A common explicit description of a set is an equation in *parametric vector form*. Examples are:

$\mathbf{x} = t\mathbf{v},$	a line through $\mathbf{0}$ in the direction of $\mathbf{v}$,
$\mathbf{x} = \mathbf{p} + t\mathbf{v},$	a line through $\mathbf{p}$ in the direction of $\mathbf{v}$,
$\mathbf{x} = x_2\mathbf{u} + x_3\mathbf{v},$	a plane through $\mathbf{0}$, $\mathbf{u}$, and $\mathbf{v}$,
$\mathbf{x} = \mathbf{p} + x_2\mathbf{u} + x_3\mathbf{v},$	a plane through $\mathbf{p}$ parallel to the plane whose equation is $\mathbf{x} = x_2\mathbf{u} + x_3\mathbf{v}$.

An equation in parametric vector form describes a set explicitly because the equation shows how to produce each $\mathbf{x}$ in the set.

To *solve* an equation $A\mathbf{x} = \mathbf{b}$ means to find an explicit description of the solution set. If the system is inconsistent, the solution set is empty. Otherwise, the description of all solutions can be written in parametric vector form, in which the parameters are the free variables from the system. *Important*: The number of free variables in $A\mathbf{x} = \mathbf{b}$ depends only on A, not on $\mathbf{b}$.

Theorem 6 and the paragraph following it are important. They describe how the solutions of $A\mathbf{x} = \mathbf{0}$ and $A\mathbf{x} = \mathbf{b}$ are related when the solution set of $A\mathbf{x} = \mathbf{b}$ is nonempty. See Figs. 5 and 6. Key exercises: 5–16, 29–32, 37.

SOLUTIONS TO EXERCISES

1. Reduce the augmented matrix to echelon form and circle the pivot positions. If a column of the *coefficient* matrix is not a pivot column, the corresponding variable is free and the system of equations has a nontrivial solution. Otherwise, the system has *only* the trivial solution.

$$\begin{bmatrix} 2 & -5 & 8 & 0 \\ -2 & -7 & 1 & 0 \\ 4 & 2 & 7 & 0 \end{bmatrix} \sim \begin{bmatrix} 2 & -5 & 8 & 0 \\ 0 & -12 & 9 & 0 \\ 0 & 12 & -9 & 0 \end{bmatrix} \sim \begin{bmatrix} ② & -5 & 8 & 0 \\ 0 & ㊱12 & 9 & 0 \\ 0 & 0 & 0 & 0 \end{bmatrix}$$

The variable x_3 is free, so the system has a nontrivial solution.

7. Always use the *reduced* echelon form of an augmented matrix to find the solutions of a system. See the text's discussion of back substitution on pages 22–23.

$$\begin{bmatrix} 1 & 3 & -3 & 7 \\ 0 & 1 & -4 & 5 \end{bmatrix} \sim \begin{bmatrix} \boxed{1} & 0 & 9 & -8 \\ 0 & \boxed{1} & -4 & 5 \end{bmatrix}, \quad \begin{array}{r} \boxed{x_1} \quad + 9x_3 = -8 \\ \boxed{x_2} - 4x_3 = 5 \end{array}$$

If you wrote something like the system above, then you made a common mistake. The matrix in the text problem is a coefficient matrix, not an augmented matrix. You should row reduce $[A \quad \mathbf{0}]$. The correct system of equations is

$$\begin{array}{r} \boxed{x_1} \quad + 9x_3 - 8x_4 = 0 \\ \boxed{x_2} - 4x_3 + 5x_4 = 0 \end{array}$$

The basic variables are x_1 and x_2, with x_3 and x_4 free. Next, $x_1 = -9x_3 + 8x_4$, and $x_2 = 4x_3 - 5x_4$. The general solution is

$$\mathbf{x} = \begin{bmatrix} x_1 \\ x_2 \\ x_3 \\ x_4 \end{bmatrix} = \begin{bmatrix} -9x_3 + 8x_4 \\ 4x_3 - 5x_4 \\ x_3 \\ x_4 \end{bmatrix} = \begin{bmatrix} -9x_3 \\ 4x_3 \\ x_3 \\ 0 \end{bmatrix} + \begin{bmatrix} 8x_4 \\ -5x_4 \\ 0 \\ x_4 \end{bmatrix} = x_3 \begin{bmatrix} -9 \\ 4 \\ 1 \\ 0 \end{bmatrix} + x_4 \begin{bmatrix} 8 \\ -5 \\ 0 \\ 1 \end{bmatrix}$$

The solution set is the same as Span$\{\mathbf{u}, \mathbf{v}\}$, where $\mathbf{u} = (-9, 4, 1, 0)$ and $\mathbf{v} = (8, -5, 0, 1)$. Originally, the solution set was described implicitly, by a set of equations. Now the solution set is described explicitly, in parametric vector form.

11.

$$\begin{bmatrix} 1 & -4 & -2 & 0 & 3 & -5 & 0 \\ 0 & 0 & 1 & 0 & 0 & -1 & 0 \\ 0 & 0 & 0 & 0 & 1 & -4 & 0 \\ 0 & 0 & 0 & 0 & 0 & 0 & 0 \end{bmatrix} \sim \begin{bmatrix} 1 & -4 & -2 & 0 & 0 & 7 & 0 \\ 0 & 0 & 1 & 0 & 0 & -1 & 0 \\ 0 & 0 & 0 & 0 & 1 & -4 & 0 \\ 0 & 0 & 0 & 0 & 0 & 0 & 0 \end{bmatrix} \sim \begin{bmatrix} \boxed{1} & -4 & 0 & 0 & 0 & 5 & 0 \\ 0 & 0 & \boxed{1} & 0 & 0 & -1 & 0 \\ 0 & 0 & 0 & 0 & \boxed{1} & -4 & 0 \\ 0 & 0 & 0 & 0 & 0 & 0 & 0 \end{bmatrix}$$

$$\begin{array}{r} \boxed{x_1} - 4x_2 \qquad\qquad + 5x_6 = 0 \\ \boxed{x_3} \qquad - x_6 = 0 \\ \boxed{x_5} - 4x_6 = 0 \\ 0 = 0 \end{array}$$

Some students are not sure what to do with x_4. Some ignore it; others set it equal to zero. In fact, x_4 is free; there is no constraint on x_4 at all. The basic variables are x_1, x_3, and x_5. The remaining variables are free. So, $x_1 = 4x_2 - 5x_6$, $x_3 = x_6$, and $x_5 = 4x_6$, with x_2, x_4, and x_6 free. In parametric vector form,

$$
\mathbf{x} = \begin{bmatrix} x_1 \\ x_2 \\ x_3 \\ x_4 \\ x_5 \\ x_6 \end{bmatrix} = \begin{bmatrix} 4x_2 - 5x_6 \\ x_2 \\ x_6 \\ x_4 \\ 4x_6 \\ x_6 \end{bmatrix} = \begin{bmatrix} 4x_2 \\ x_2 \\ 0 \\ 0 \\ 0 \\ 0 \end{bmatrix} + \begin{bmatrix} 0 \\ 0 \\ 0 \\ x_4 \\ 0 \\ 0 \end{bmatrix} + \begin{bmatrix} -5x_6 \\ 0 \\ x_6 \\ 0 \\ 4x_6 \\ x_6 \end{bmatrix} = x_2 \begin{bmatrix} 4 \\ 1 \\ 0 \\ 0 \\ 0 \\ 0 \end{bmatrix} + x_4 \begin{bmatrix} 0 \\ 0 \\ 0 \\ 1 \\ 0 \\ 0 \end{bmatrix} + x_6 \begin{bmatrix} -5 \\ 0 \\ 1 \\ 0 \\ 4 \\ 1 \end{bmatrix}
$$

$$
 \uparrow \qquad \uparrow \qquad \uparrow
$$
$$
 \mathbf{u} \qquad \mathbf{v} \qquad \mathbf{w}
$$

The solution set is the same as Span$\{\mathbf{u}, \mathbf{v}, \mathbf{w}\}$.

Study Tip: When solving a system, identify (and perhaps circle) the basic variables. All other variables are free.

13. To write the general solution in parametric vector form, pull out the constant terms that do not involve the free variable:

$$
\mathbf{x} = \begin{bmatrix} x_1 \\ x_2 \\ x_3 \end{bmatrix} = \begin{bmatrix} 5 + 4x_3 \\ -2 - 7x_3 \\ x_3 \end{bmatrix} = \begin{bmatrix} 5 \\ -2 \\ 0 \end{bmatrix} + \begin{bmatrix} 4x_3 \\ -7x_3 \\ x_3 \end{bmatrix} = \begin{bmatrix} 5 \\ -2 \\ 0 \end{bmatrix} + x_3 \begin{bmatrix} 4 \\ -7 \\ 1 \end{bmatrix}
$$

$$
 \uparrow \qquad\qquad \uparrow \qquad = \mathbf{p} + x_3 \mathbf{q}
$$
$$
 \mathbf{p} \qquad\qquad \mathbf{q}
$$

Geometrically, the solution set is the line through $\begin{bmatrix} 5 \\ -2 \\ 0 \end{bmatrix}$ parallel to $\begin{bmatrix} 4 \\ -7 \\ 1 \end{bmatrix}$.

Checkpoint: Let A be a 2 $\times$ 2 matrix. Answer True or False: If the solution set of $A\mathbf{x} = \mathbf{0}$ is a line through the origin in $\mathbb{R}^2$ and if $\mathbf{b} \neq \mathbf{0}$, then the solution set of $A\mathbf{x} = \mathbf{b}$ is a line not through the origin.

19. The line through $\mathbf{a}$ parallel to $\mathbf{b}$ can be written as $\mathbf{x} = \mathbf{a} + t\mathbf{b}$, where t represents a parameter:

$$
\mathbf{x} = \begin{bmatrix} x_1 \\ x_2 \end{bmatrix} = \begin{bmatrix} -2 \\ 0 \end{bmatrix} + t \begin{bmatrix} -5 \\ 3 \end{bmatrix}, \text{ or } \begin{cases} x_1 = -2 - 5t \\ x_2 = 3t \end{cases}
$$

23. **a.** See the first paragraph of the subsection titled "Homogeneous Linear Systems."
    b. See the first two sentences of the subsection titled "Parametric Vector Form."

c. See the box before Example 1.

d. See the paragraph that precedes Fig. 5.

e. See Theorem 6.

25. Suppose $\mathbf{p}$ satisfies $A\mathbf{x} = \mathbf{b}$. Then $A\mathbf{p} = \mathbf{b}$. Theorem 6 says that the solution set of $A\mathbf{x} = \mathbf{b}$ equals the set $S = \{\mathbf{w} : \mathbf{w} = \mathbf{p} + \mathbf{v}_h$ for some $\mathbf{v}_h$ such that $A\mathbf{v}_h = \mathbf{0}\}$. There are two things to prove: (a) every vector in S satisfies $A\mathbf{x} = \mathbf{b}$, (b) every vector that satisfies $A\mathbf{x} = \mathbf{b}$ is in S.

a. Let $\mathbf{w}$ have the form $\mathbf{w} = \mathbf{p} + \mathbf{v}_h$,   where $A\mathbf{v}_h = \mathbf{0}$. Then

$$A\mathbf{w} = A(\mathbf{p} + \mathbf{v}_h) = A\mathbf{p} + A\mathbf{v}_h \qquad \text{By Theorem 5(a) in Section 1.4}$$
$$= \mathbf{b} + \mathbf{0} = \mathbf{b}$$

So every vector of the form $\mathbf{p} + \mathbf{v}_h$ satisfies $A\mathbf{x} = \mathbf{b}$.

b. Now let $\mathbf{w}$ be any solution of $A\mathbf{x} = \mathbf{b}$, and set $\mathbf{v}_h = \mathbf{w} - \mathbf{p}$. Then

$$A\mathbf{v}_h = A(\mathbf{w} - \mathbf{p}) = A\mathbf{w} - A\mathbf{p} = \mathbf{b} - \mathbf{b} = \mathbf{0}$$

So $\mathbf{v}_h$ satisfies $A\mathbf{x} = \mathbf{0}$. Thus every solution of $A\mathbf{x} = \mathbf{b}$ has the form

$$\mathbf{w} = \mathbf{p} + \mathbf{v}_h.$$

31. A is a 3×2 matrix with two pivot positions.

a. Since A has a pivot position in each column, each variable in $A\mathbf{x} = \mathbf{0}$ is a basic variable. So the equation $A\mathbf{x} = \mathbf{0}$ has no free variables and hence no nontrivial solution.

b. With two pivot positions and three rows, A cannot have a pivot in every row. So the equation $A\mathbf{x} = \mathbf{b}$ cannot have a solution for every possible $\mathbf{b}$ (in $\mathbb{R}^3$), by Theorem 4 in Section 1.4.

37. If you worked on the Checkpoint when you first saw it, you should be ready for this exercise. Since the solution set of $A\mathbf{x} = \mathbf{0}$ contains the point $(4, 1)$, the vector $\mathbf{x} = (4, 1)$ satisfies $A\mathbf{x} = \mathbf{0}$. Write this equation as a vector equation, using $\mathbf{a}_1$ and $\mathbf{a}_2$ for the columns of A:

$$4 \cdot \mathbf{a}_1 + 1 \cdot \mathbf{a}_2 = \mathbf{0}$$

Then $\mathbf{a}_2 = -4\mathbf{a}_1$. So choose any nonzero vector for the first column of A and multiply that column by -4 to get the second column of A. For example, set $A = \begin{bmatrix} 1 & -4 \\ 1 & -4 \end{bmatrix}$.

Finally, the only way the solution set of $A\mathbf{x} = \mathbf{b}$ could *not* be parallel to the line through $(4, 1)$ and the origin is for the solution set of $A\mathbf{x} = \mathbf{b}$ to be *empty*. (Theorem 6 applies only to the case when the equation $A\mathbf{x} = \mathbf{b}$ has a nonempty solution set.) For $\mathbf{b}$, take any vector that is *not* a multiple of the columns of A.

Answer to Checkpoint: False. The solution set could be empty. In this case, the solution set of $Ax = b$ is not produced by translating the (nonempty) solution set of $Ax = 0$. See the Warning after Theorem 6.

MATLAB   Zero Matrices

The command `zeros(m,n)` creates an $m \times n$ matrix of zeros. When solving an equation $Ax = 0$, create an augmented matrix:

   `M = [A zeros(m, 1)]`   m is the number of rows in A.

Then use `gauss`, `bgauss`, and `scale` to row reduce M completely.

1.6   APPLICATIONS OF LINEAR SYSTEMS _____

All of the examples and exercises in this system involve linear systems that have multiple solutions. In each case, make a note of *why* you should expect the system to have many solutions.

STUDY NOTES

The Leontief exchange model concerns the dollar value (called the *price*) of the annual output of each sector of a nation's economy. An equilibrium price vector **p** provides a list of prices, one for each section, such that each sector's expenses and income are in balance. Example 1 shows that there are many equilibrium price vectors; each one is a multiple of a fixed equilibrium price vector. This means that once the prices are all in balance, multiplying all the prices by a fixed constant does not affect the balance. For instance, if all prices are doubled, then each sector's expenses and income are doubled at the same time and hence they remain in balance.

A solution of a chemical equation-balance problem is a list of coefficients that appear on the various terms in the chemical equation. When a chemical equation is balanced, the number of atoms of each type on the left side of the equation matches the number of corresponding atoms on the right side. If the coefficients in the equation are each multiplied by a fixed positive integer, the equation will remain balanced. So, there are many solutions to a chemical equation-balance problem.

The problems here in network flow have multiple solutions for the simple reason that there are more variables than there are constraint equations. The equations for network flow are mostly nonhomogeneous. In contrast, the Leontief model and the chemical equation-balance problem both lead to systems of homogeneous equations.

SOLUTIONS TO EXERCISES

1. Fill in the exchange table one column at a time. The entries in a column describe where a sector's output goes. The decimal fractions in each column sum to 1.

Distribution of
Output From:

	Goods	Services		Purchased by:
output	↓	↓	input	
	.2	.7	→	Goods
	.8	.3	→	Services

Denote the total annual output (in dollars) of the sectors by p_G and p_S. From the first row, the total input to the Goods sector is $.2\,p_G + .7\,p_S$. The Goods sector must pay for that. So the equilibrium prices must satisfy

income   expenses

$$p_G \;=\; .2p_G + .7p_S$$

From the second row, the input (that is, the expense) of the Services sector is $.8\,p_G + .3\,p_S$. The equilibrium equation for the Services sector is

income   expenses

$$p_S \;=\; .8p_G + .3p_S$$

Move all variables to the left side and combine like terms:

$$.8p_G - .7p_S = 0$$
$$-.8p_G + .7p_S = 0$$

Row reduce the augmented matrix:

$$\begin{bmatrix} .8 & -.7 & 0 \\ -.8 & .7 & 0 \end{bmatrix} \sim \begin{bmatrix} .8 & -.7 & 0 \\ 0 & 0 & 0 \end{bmatrix} \sim \begin{bmatrix} 1 & -.875 & 0 \\ 0 & 0 & 0 \end{bmatrix}$$

The general solution is $p_G = .875\,p_S$, with p_S free. One equilibrium solution is $p_S = 1000$ and $p_G = 875$. If one uses fractions instead of decimals in the calculations, the general solution would be written $p_G = (7/8)\,p_S$, and a natural choice of prices might be $p_S = 80$ and $p_G = 70$. Only the *ratio* of the prices is important: $p_G = .875\,p_S$. The economic equilibrium is unaffected by a proportional change in prices.

7. The following vectors list the numbers of atoms of sodium (Na), hydrogen (H), carbon (C), and oxygen (O):

$$NaHCO_3: \begin{bmatrix} 1 \\ 1 \\ 1 \\ 3 \end{bmatrix}, \ H_3C_6H_5O_7: \begin{bmatrix} 0 \\ 8 \\ 6 \\ 7 \end{bmatrix}, \ Na_3C_6H_5O_7: \begin{bmatrix} 3 \\ 5 \\ 6 \\ 7 \end{bmatrix}, \ H_2O: \begin{bmatrix} 0 \\ 2 \\ 0 \\ 1 \end{bmatrix}, \ CO_2: \begin{bmatrix} 0 \\ 0 \\ 1 \\ 2 \end{bmatrix} \begin{matrix} \text{sodium} \\ \text{hydrogen} \\ \text{carbon} \\ \text{oxygen} \end{matrix}$$

The order of the various atoms is not important. The list here was selected by writing the elements in the order in which they first appear in the chemical equation, reading left to right:

$$x_1 \cdot NaHCO_3 + x_2 \cdot H_3C_6H_5O_7 \rightarrow x_3 \cdot Na_3C_6H_5O_7 + x_4 \cdot H_2O + x_5 \cdot CO_2$$

The coefficients $x_1, \ldots, x_5$ satisfy the vector equation

$$x_1 \begin{bmatrix} 1 \\ 1 \\ 1 \\ 3 \end{bmatrix} + x_2 \begin{bmatrix} 0 \\ 8 \\ 6 \\ 7 \end{bmatrix} = x_3 \begin{bmatrix} 3 \\ 5 \\ 6 \\ 7 \end{bmatrix} + x_4 \begin{bmatrix} 0 \\ 2 \\ 0 \\ 1 \end{bmatrix} + x_5 \begin{bmatrix} 0 \\ 0 \\ 1 \\ 2 \end{bmatrix}$$

Move all terms to the left side (changing the sign of each entry in the third, fourth, and fifth vectors) and reduce the augmented matrix:

$$\begin{bmatrix} 1 & 0 & -3 & 0 & 0 & 0 \\ 1 & 8 & -5 & -2 & 0 & 0 \\ 1 & 6 & -6 & 0 & -1 & 0 \\ 3 & 7 & -7 & -1 & -2 & 0 \end{bmatrix} \sim \cdots \sim \begin{bmatrix} 1 & 0 & 0 & 0 & -1 & 0 \\ 0 & 1 & 0 & 0 & -1/3 & 0 \\ 0 & 0 & 1 & 0 & -1/3 & 0 \\ 0 & 0 & 0 & 1 & -1 & 0 \end{bmatrix}$$

The general solution is $x_1 = x_5$, $x_2 = (1/3)x_5$, $x_3 = (1/3)x_5$, $x_4 = x_5$, and x_5 is free. Take $x_5 = 3$. Then $x_1 = x_4 = 3$, and $x_2 = x_3 = 1$. The balanced equation is

$$3NaHCO_3 + H_3C_6H_5O_7 \rightarrow Na_3C_6H_5O_7 + 3H_2O + 3CO_2$$

13. Write the equations for each intersection (see the diagram for the intersection labels):

Intersection	Flow in		Flow out	Rearrange the equations:
A	$x_2 + 30$	=	$x_1 + 80$	$x_1 - x_2 \qquad\qquad\qquad\qquad = -50$
B	$x_3 + x_5$	=	$x_2 + x_4$	$x_2 - x_3 + x_4 - x_5 \qquad = 0$
C	$x_6 + 100$	=	$x_5 + 40$	$x_5 - x_6 = 60$
D	$x_4 + 40$	=	$x_6 + 90$	$x_4 \qquad\quad - x_6 = 50$
E	$x_1 + 60$	=	$x_3 + 20$	$x_1 \qquad\quad - x_3 \qquad\qquad = -40$
Total flow:	230	=	230	

Completely reduce the augmented matrix:

$$
\begin{bmatrix}
1 & -1 & 0 & 0 & 0 & 0 & -50 \\
0 & 1 & -1 & 1 & -1 & 0 & 0 \\
0 & 0 & 0 & 0 & 1 & -1 & 60 \\
0 & 0 & 0 & 1 & 0 & -1 & 50 \\
1 & 0 & -1 & 0 & 0 & 0 & -40
\end{bmatrix}
\sim \cdots \sim
\begin{bmatrix}
1 & -1 & 0 & 0 & 0 & 0 & -50 \\
0 & 1 & -1 & 1 & -1 & 0 & 0 \\
0 & 0 & 0 & 1 & 0 & -1 & 50 \\
0 & 0 & 0 & 0 & 1 & -1 & 60 \\
0 & 0 & 0 & 0 & 0 & 0 & 0
\end{bmatrix}
$$

$$
\sim \cdots \sim
\begin{bmatrix}
1 & 0 & -1 & 0 & 0 & 0 & -40 \\
0 & 1 & -1 & 0 & 0 & 0 & 10 \\
0 & 0 & 0 & 1 & 0 & -1 & 50 \\
0 & 0 & 0 & 0 & 1 & -1 & 60 \\
0 & 0 & 0 & 0 & 0 & 0 & 0
\end{bmatrix}
$$

a. The general solution is
$$
\begin{cases}
x_1 = x_3 - 40 \\
x_2 = x_3 + 10 \\
x_3 \text{ is free} \\
x_4 = x_6 + 50 \\
x_5 = x_6 + 60 \\
x_6 \text{ is free}
\end{cases}
$$

b. To find minimum flows, note that since x_1 cannot be negative, $x_3 \geq 40$. This implies that $x_2 \geq 50$. Also, since x_6 cannot be negative, $x_4 \geq 50$ and $x_5 \geq 60$. The minimum flows are $x_2 = 50$, $x_3 = 40$, $x_4 = 50$, $x_5 = 60$ (when $x_1 = 0$ and $x_6 = 0$).

MATLAB   Rational Format

Chemical equation-balance problems are studied best using exact or symbolic arithmetic, because the balance variables must be whole numbers (with no round-off allowed). In MATLAB, a simple approach is to execute the command `format rat`, which will make MATLAB display matrix or vector entries as rational numbers. In general, the rational number displayed might be only an approximation for a floating-point number. But since the chemical equations studied here have integer coefficients, `format rat` will make MATLAB display the exact (rational) value of every entry during row reduction. Use `format` or `format short` to return to the standard MATLAB display of numbers.

Once you find a rational solution of a chemical equation-balance problem, you can multiply the entries in the solution vector by a suitable integer to produce a solution that involves only whole numbers.

1.7   LINEAR INDEPENDENCE

This section is as important as Section 1.4 and should be studied just as carefully. Full understanding of the concepts will take time, so get started on the section now.

KEY IDEAS

Figures 1 and 2, along with Theorem 7, will help you understand the nature of a linearly dependent set. (Fig. 2 applies only when **u** and **v** are independent.) But you must also learn the *definitions* of linear dependence and linear independence, word for word! Many theoretical problems involving a linearly dependent set are treated by the definition, because it provides an equation (the dependence equation) with which to work. (See the proof of Theorem 7.)

The box before Example 2 contains a very useful fact. Any time you need to study the linear independence of a set of p vectors in $\mathbb{R}^n$, you can always form an $n \times p$ matrix A with those vectors as columns and then study the matrix equation $A\mathbf{x} = \mathbf{0}$. This is not the only method, however. Stay alert for three special situations:

- A set of two vectors. Always check this by inspection; don't waste time on row reduction of [A **0**]. The set is linearly independent if neither of the vectors is a multiple of the other. (For brevity, I sometimes say that "the vectors are not multiples.") See Example 3.

- A set that contains too many vectors, that is, more vectors than entries in the vectors; the columns of a short, fat matrix. Theorem 8.

- A set that contains the zero vector. Theorem 9.

The most common mistake students make when checking a set of three or more vectors for independence is to think they only have to verify that no vector is a multiple of one of the other vectors. Wrong! Study Example 5 and Figure 4.

Key exercises are 9–20 and 23–28, and 30. Try Exercise 35, even if it is not assigned. Think carefully, and write your answer before checking the answer section.

SOLUTIONS TO EXERCISES

1. Use an augmented matrix to study the solution set of $x_1\mathbf{u} + x_2\mathbf{v} + x_3\mathbf{w} = \mathbf{0}$ (*), where $\mathbf{u}$, $\mathbf{v}$, and

$\mathbf{w}$ are the three given vectors. Since $\begin{bmatrix} 5 & 7 & 9 & 0 \\ 0 & 2 & 4 & 0 \\ 0 & -6 & -8 & 0 \end{bmatrix} \sim \begin{bmatrix} ⑤ & 7 & 9 & 0 \\ 0 & ② & 4 & 0 \\ 0 & 0 & ④ & 0 \end{bmatrix}$, there are no free

variables. So the homogeneous equation (*) has only the trivial solution. The vectors are linearly independent.

Warning: Whenever you study a homogeneous equation, you may be tempted to omit the augmented column of zeros because it never changes under row operations. I urge you to keep the zeros, to avoid possibly misinterpreting your own calculations. In Exercise 1, if you wrote

$\begin{bmatrix} 5 & 7 & 9 \\ 0 & 2 & 4 \\ 0 & -6 & -8 \end{bmatrix} \sim \begin{bmatrix} ⑤ & 7 & 9 \\ 0 & ② & 4 \\ 0 & 0 & ④ \end{bmatrix}$

you might conclude that "the system is inconsistent" and then go on to make some crazy statement about linear dependence or independence. Don't laugh. I have seen this happen on exams. A more common error occurs in a problem like Exercise 7. In that exercise, if you write

$\begin{bmatrix} 1 & 4 & -3 & 0 \\ -2 & -7 & 5 & 1 \\ -4 & -5 & 7 & -5 \end{bmatrix} \sim \begin{bmatrix} 1 & 4 & -3 & 0 \\ 0 & 1 & -1 & 1 \\ 0 & 11 & -5 & 5 \end{bmatrix} \sim \begin{bmatrix} ① & 4 & -3 & 0 \\ 0 & ① & -1 & 1 \\ 0 & 0 & ⑥ & -6 \end{bmatrix}$

you might conclude that "the system has a unique solution" and the vectors are linearly independent. However, the four columns are actually linearly dependent. In both cases, the error is to misinterpret your matrix as an augmented matrix.

7. Study the equation $Ax = 0$. Some people may start with the method of Example 2:

$$\begin{bmatrix} 1 & 4 & -3 & 0 & 0 \\ -2 & -7 & 5 & 1 & 0 \\ -4 & -5 & 7 & -5 & 0 \end{bmatrix} \sim \begin{bmatrix} 1 & 4 & -3 & 0 & 0 \\ 0 & 1 & -1 & 1 & 0 \\ 0 & 11 & -5 & 5 & 0 \end{bmatrix} \sim \begin{bmatrix} ① & 4 & -3 & 0 & 0 \\ 0 & ① & -1 & 1 & 0 \\ 0 & 0 & ⑥ & -6 & 0 \end{bmatrix}$$

But this is a waste of time. There are only 3 rows, so there are at most three pivot positions. Hence, at least one of the four variables must be free. So the equation $Ax = 0$ has a nontrivial solution and the columns of A are linearly dependent.

Warning: Exercise 9 and Practice Problem 3 emphasize that to check whether a set such as $\{v_1, v_2, v_3\}$ is linearly dependent, it is *not* wise to check instead whether v_3 is a linear combination of v_1 and v_2.

13. To study the linear dependence of three vectors, say v_1, v_2, v_3, row reduce the augmented matrix $[v_1 \quad v_2 \quad v_3 \quad 0]$:

$$\begin{bmatrix} 1 & -2 & 3 & ⓪ \\ 5 & -9 & h & 0 \\ -3 & 6 & -9 & 0 \end{bmatrix} \sim \begin{bmatrix} 1 & -2 & 3 & 0 \\ ⓪ & 1 & h-15 & 0 \\ 0 & 0 & 0 & 0 \end{bmatrix}$$

The equation $x_1v_1 + x_2v_2 + x_3v_3 = 0$ has a free variable and hence a nontrivial solution no matter what the value of h. So the vectors are linearly dependent for all values of h.

Checkpoint: What is wrong with the following statement?

The vectors $\begin{bmatrix} 3 \\ -1 \end{bmatrix}, \begin{bmatrix} 2 \\ 8 \end{bmatrix}, \begin{bmatrix} -5 \\ 3 \end{bmatrix}, \begin{bmatrix} 7 \\ -4 \end{bmatrix}$ are linearly dependent "because there is a free variable," or "because there are more variables than equations."

15. The set $\left\{ \begin{bmatrix} 5 \\ 1 \end{bmatrix}, \begin{bmatrix} 2 \\ 8 \end{bmatrix}, \begin{bmatrix} 1 \\ 3 \end{bmatrix}, \begin{bmatrix} -1 \\ 7 \end{bmatrix} \right\}$ is obviously linearly dependent, by Theorem 8, because there are more vectors (4) than entries in the vectors. On a test, you probably will not have to know the theorem number. Check with your instructor.

19. The set is linearly independent because neither vector is a multiple of the other vector. [Two of the entries in the first vector are −4 times the corresponding entry in the second vector. But this multiple does not work for the third entries.]

21. a. See the box before Example 2.

    b. See the warning after Theorem 7.

    c. See Fig. 3, after Theorem 8.

    d. See the remark following Example 4.

25.
$$\begin{bmatrix} \blacksquare & * \\ 0 & \blacksquare \\ 0 & 0 \\ 0 & 0 \end{bmatrix} \text{ and } \begin{bmatrix} 0 & \blacksquare \\ 0 & 0 \\ 0 & 0 \\ 0 & 0 \end{bmatrix}$$

31. Think of $A = [\mathbf{a}_1 \quad \mathbf{a}_2 \quad \mathbf{a}_3]$. The text points out that $\mathbf{a}_3 = \mathbf{a}_1 + \mathbf{a}_2$. Rewrite this as $\mathbf{a}_1 + \mathbf{a}_2 - \mathbf{a}_3 = \mathbf{0}$. As a matrix equation, $A\mathbf{x} = \mathbf{0}$ for $\mathbf{x} = (1, 1, -1)$.

33. The text uses Theorem 7 to conclude that $\{\mathbf{v}_1, \ldots, \mathbf{v}_4\}$ is linearly dependent. Another argument is to rewrite the equation $\mathbf{v}_3 = 2\mathbf{v}_1 + \mathbf{v}_2$ as $2\mathbf{v}_1 + 1\mathbf{v}_2 + (-1)\mathbf{v}_3 + 0\mathbf{v}_4 = \mathbf{0}$. This is a linear dependence relation. Some students think of this argument rather than Theorem 7. Did you? (I hope you did not read the answer before trying this problem.)

37. True. The text gives a complete answer.

39. If for all $\mathbf{b}$ the equation $A\mathbf{x} = \mathbf{b}$ has at most one solution, then take $\mathbf{b} = \mathbf{0}$, and conclude that the equation $A\mathbf{x} = \mathbf{0}$ has at most one solution. Then the trivial solution is the only solution, and so the columns of A are linearly independent.

43. [M] Make $\mathbf{v}$ any one of the columns of A that is not in B and row reduce the augmented matrix $[B \quad \mathbf{v}]$. The calculations will show that the equation $B\mathbf{x} = \mathbf{v}$ is consistent, which means that $\mathbf{v}$ is a linear combination of the columns of B. Thus, each column of A that is not a column of B is in the set spanned by the columns of B.

Answer to Checkpoint: The set of four vectors contains only vectors, no variables of any kind, and no equations. It makes no sense to talk about the variables in a set of vectors. Variables appear in an equation. One cannot assume that the writer of the statement has any idea of the appropriate equation. If you want to give an explanation involving variables, then you must specify the equation. One correct answer is: the vectors are linearly dependent because the

equation $x_1 \begin{bmatrix} 3 \\ -1 \end{bmatrix} + x_2 \begin{bmatrix} 2 \\ 8 \end{bmatrix} + x_3 \begin{bmatrix} -5 \\ 3 \end{bmatrix} + x_4 \begin{bmatrix} 7 \\ -4 \end{bmatrix} = \begin{bmatrix} 0 \\ 0 \end{bmatrix}$ necessarily has a free variable.

Mastering Linear Algebra Concepts: Linear Independence

In Section 1.4 of this *Guide*, I described how to begin forming a mental image of the concept of a spanning set. The same technique works for linear independence. The goal is to merge all the ideas you find regarding linear independence into a single mental image, with each part immediately available in your mind for use as needed. Start now to organize on paper your understanding of linear independence/dependence, using the following list as a guide. In each case, write information that you think will be helpful. (Definitions and theorems should be copied word-for-word.)

- definitions of linear independence and dependence
- equivalent descriptions   Theorem 7
- geometric interpretations   Figs. 1, 2, 4
- special cases   Theorems 8, 9, box on p. 67, Examples 3, 5, 6
- examples and "counterexamples"   Figs. 1, 2, 3, 4, Exercises 9–20, 33–38
- algorithms or typical computations   Examples 1, 2, Exercises 1–8
- connections with other concepts   Box on p. 66, Examples 2, 4, Exercises 27, 30, 39

As you work on your notes, be careful to use terminology correctly. For instance, the term "linearly independent" may be applied to a set of vectors, but it *never* is applied to a matrix or to an equation. The *columns* of a matrix may be linearly independent, but it is meaningless to refer to a linearly independent matrix. Similarly, *solutions* of a system of linear equations may be linearly independent, but the term "linearly independent equations" has never been defined. Finally, a set of vectors or a matrix cannot have a "nontrivial solution". Only equations have solutions.

1.8   INTRODUCTION TO LINEAR TRANSFORMATIONS _____

Linear transformations are important for both the theory and the applications of linear algebra. You will see both uses in a variety of settings throughout the text. The graphical descriptions in this section will be augmented in Section 1.9 and in a later section on computer graphics.

STUDY NOTES

Viewing the correspondence from a vector **x** to a vector $A\mathbf{x}$ as a mapping provides a dynamic interpretation of matrix-vector multiplication and a new way to understand the equation $A\mathbf{x} = \mathbf{b}$. Using the language of computer science, we can describe a matrix in two ways—as a data structure (a rectangular array of numbers) and as a program (a prescription for transforming vectors). Strictly speaking, however, the actual linear transformation is the function or mapping $\mathbf{x} \mapsto A\mathbf{x}$ rather than just A itself.

Here is a way to visualize a matrix acting as a linear transformation. The entries in the input vector $\mathbf{x}$ are assigned as weights that multiply the corresponding columns of A, then the resulting weighted columns are added together to produce the output vector $\mathbf{b}$.

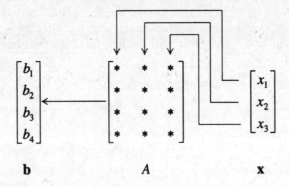

$$\mathbf{b} \qquad\qquad A \qquad\qquad \mathbf{x}$$

As you learn the definition of a linear transformation T, don't forget the crucial phrases "for all $\mathbf{u}$ and $\mathbf{v}$ in the domain of T" and "for all $\mathbf{u}$ and all scalars c." The mapping T defined by $T(x_1, x_2) = (|x_2|, |x_1|)$ is *not* a linear mapping, and yet T satisfies the linearity properties for *some* vectors in its domain and *some* scalars.

The key exercises are 17–20, 25 and 31.

SOLUTIONS TO EXERCISES

1. $T(\mathbf{u}) = A\mathbf{u} = \begin{bmatrix} 2 & 0 \\ 0 & 2 \end{bmatrix}\begin{bmatrix} 1 \\ -3 \end{bmatrix} = \begin{bmatrix} 2 \\ -6 \end{bmatrix}$, $T(\mathbf{v}) = \begin{bmatrix} 2 & 0 \\ 0 & 2 \end{bmatrix}\begin{bmatrix} a \\ b \end{bmatrix} = \begin{bmatrix} 2a \\ 2b \end{bmatrix}$

5. $[A \quad \mathbf{b}] = \begin{bmatrix} 1 & -5 & -7 & -2 \\ -3 & 7 & 5 & -2 \end{bmatrix} \sim \begin{bmatrix} 1 & -5 & -7 & -2 \\ 0 & 1 & 2 & 1 \end{bmatrix} \sim \begin{bmatrix} 1 & 0 & 3 & 3 \\ 0 & 1 & 2 & 1 \end{bmatrix}$

Note that a solution is *not* $\begin{bmatrix} 3 \\ 1 \end{bmatrix}$. To avoid this common error, write the equations:

$$\begin{aligned} x_1 \quad + 3x_3 &= 3 \\ x_2 + 2x_3 &= 1 \end{aligned}$$ and solve for the basic variables: $\begin{cases} x_1 = 3 - 3x_3 \\ x_2 = 1 - 2x_3 \\ x_3 \text{ is free} \end{cases}$

General solution: $\mathbf{x} = \begin{bmatrix} x_1 \\ x_2 \\ x_3 \end{bmatrix} = \begin{bmatrix} 3 - 3x_3 \\ 1 - 2x_3 \\ x_3 \end{bmatrix} = \begin{bmatrix} 3 \\ 1 \\ 0 \end{bmatrix} + x_3 \begin{bmatrix} -3 \\ -2 \\ 1 \end{bmatrix}$. For a particular solution, one might

choose $x_3 = 0$ and $\mathbf{x} = \begin{bmatrix} 3 \\ 1 \\ 0 \end{bmatrix}$.

7. $a = 5$; the domain of T is $\mathbb{R}^5$, because a 6×5 matrix has 5 columns and for $A\mathbf{x}$ to be defined, $\mathbf{x}$ must be in $\mathbb{R}^5$. $b = 6$; the codomain of T is $\mathbb{R}^6$, because $A\mathbf{x}$ is a linear combination of the columns of A, and each column of A is in $\mathbb{R}^6$.

13. The transformation may be described geometrically as a reflection through the origin. Two other correct descriptions are a rotation of π radians about the origin and a rotation of $-\pi$ radians about the origin. See the figure.

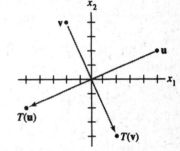

18. *Additional Hint*: Draw a line through $\mathbf{w}$ parallel to $\mathbf{v}$, and draw a line through $\mathbf{w}$ parallel to $\mathbf{u}$. This will help you write $\mathbf{w}$ as a linear combination of $\mathbf{u}$ and $\mathbf{v}$.

19. All you know are the images of $\mathbf{e}_1$ and $\mathbf{e}_2$ and the fact that T is linear. The key idea is to write

$$\mathbf{x} = \begin{bmatrix} 5 \\ -3 \end{bmatrix} = 5\begin{bmatrix} 1 \\ 0 \end{bmatrix} - 3\begin{bmatrix} 0 \\ 1 \end{bmatrix} = 5\mathbf{e}_1 - 3\mathbf{e}_2.$$ Then, from the linearity of T, write

$$T(\mathbf{x}) = T(5\mathbf{e}_1 - 3\mathbf{e}_2) = 5T(\mathbf{e}_1) - 3T(\mathbf{e}_2) = 5\mathbf{y}_1 - 3\mathbf{y}_2 = 5\begin{bmatrix} 2 \\ 5 \end{bmatrix} - 3\begin{bmatrix} -1 \\ 6 \end{bmatrix} = \begin{bmatrix} 13 \\ 7 \end{bmatrix}$$

To find the image of $\begin{bmatrix} x_1 \\ x_2 \end{bmatrix}$, observe that $\mathbf{x} = \begin{bmatrix} x_1 \\ x_2 \end{bmatrix} = x_1 \begin{bmatrix} 1 \\ 0 \end{bmatrix} + x_2 \begin{bmatrix} 0 \\ 1 \end{bmatrix} = x_1\mathbf{e}_1 + x_2\mathbf{e}_2$. Then

$$T(\mathbf{x}) = T(x_1\mathbf{e}_1 + x_2\mathbf{e}_2) = x_1T(\mathbf{e}_1) + x_2T(\mathbf{e}_2) = x_1\begin{bmatrix} 2 \\ 5 \end{bmatrix} + x_2\begin{bmatrix} -1 \\ 6 \end{bmatrix} = \begin{bmatrix} 2x_1 - x_2 \\ 5x_1 + 6x_2 \end{bmatrix}$$

21. a. A function is another word for transformation or mapping.

    b. See the paragraph before Example 1.

    c. See Figure 2. Or, see the paragraph before Example 1.

    d. See the paragraph after the definition of a linear transformation.

    e. See the paragraph following the box that contains equation (4).

25. Any point $\mathbf{x}$ on the line through $\mathbf{p}$ in the direction of $\mathbf{v}$ satisfies the parametric equation $\mathbf{x} = \mathbf{p} + t\mathbf{v}$ for some value of t. By linearity, the image $T(\mathbf{x})$ satisfies the parametric equation

$$T(\mathbf{x}) = T(\mathbf{p} + t\mathbf{v}) = T(\mathbf{p}) + tT(\mathbf{v}) \qquad (*)$$

If $T(\mathbf{v}) = \mathbf{0}$, then $T(\mathbf{x}) = T(\mathbf{p})$ for all values of t, and the image of the original line is just a single point. Otherwise, (*) is the parametric equation of a line through $T(\mathbf{p})$ in the direction of $T(\mathbf{v})$.

Study Tip: Exercise 31 is important, because it will help you to connect the concepts of linear dependence and linear transformation. Be sure to try the exercise first, before looking in the answer section of the text. Don't feel badly if you need to peek at the hint there. Only my best students can do this problem unaided. Once you have seen the hint, try hard to construct the desired explanation without consulting the solution I have written below. Don't give up too soon. Reread the definitions of linear dependence and linear transformation, if necessary.

    After you have written your best attempt at an explanation, check it against the *Study Guide* solution. Also, study the strategy there of how I found the solution. Even if your attempt is quite unsatisfactory, the time spent on this problem is worthwhile, because you will learn more from the solution here.

31. To help you use this *Study Guide* properly, I have hidden the solution at the end of the solutions for Section 1.9. *Do not look there until you have followed the instructions above.* (I may not "hide" a solution again, but I wanted this one time to emphasize the importance of working seriously on a problem before checking the solution.)

Mastering Linear Algebra Concepts: Linear Transformation

Start to form a robust mental image of a linear transformation by preparing a review sheet that covers the following categories:

• definition	Page 77
• equivalent descriptions	Equations (4) and (5)
• geometric interpretations	Figs. 1 or 2
• special cases	Matrix transformation: page 77
• examples and "counterexamples"	Superposition, Examples 2-6 Paragraph before Exercise 1, in this *Guide* Exercises 29, 30, 33
• connections with other concepts	Existence and uniqueness: page 75 Linear dependence: Exercise 31

Note: Exercise 31 should enrich your mental image of linear dependence, so add a note about it to your list for "linear independence". If your course does not emphasize the next section, turn now to the end of the *Study Guide* material for Section 1.9 and read the box on *Existence and Uniqueness*.

1.9   MATRIX OF A LINEAR TRANSFORMATION _____

Every matrix transformation is a linear transformation. This section shows that every linear transformation from $\mathbb{R}^n$ to $\mathbb{R}^m$ is a matrix transformation. Chapters 4 and 5 will discuss other examples of linear transformations.

KEY IDEAS

A linear transformation $T: \mathbb{R}^n \rightarrow \mathbb{R}^m$ is completely determined by what it does to the columns of the identity matrix I_n. The jth column of the standard matrix for T is $T(\mathbf{e}_j)$, where $\mathbf{e}_j$ is the jth column of I_n.

There are two ways to compute the standard matrix A. Either compute $T(\mathbf{e}_1), \ldots, T(\mathbf{e}_n)$, which is easy to do when T is described geometrically, as in Exercises 1–14, or fill in the entries of A by inspection, which is easy to do when T is described by a formula, as in Exercises 15–22.

Existence and uniqueness questions about the mapping $x \mapsto Ax$ are determined by properties of A. You should know how this works. The proof of Theorem 11 also applies to linear transformations on the general vector spaces in Chapter 4. Here is another way to understand Theorem 11 [and Theorem 12(b)] using the language of matrix transformations:

Let A be the standard matrix of T. Then T is one-to-one if and only if the equation $Ax = b$ has at most one solution for each b. This happens if and only if every column of A is a pivot column, which happens if and only if $Ax = 0$ has only the trivial solution.

The "if and only if" phrase in Theorem 11 (and in the proof above) was discussed in *A Mathematical Note*, in Section 1.2 of this *Guide*.

SOLUTIONS TO EXERCISES

1. The columns of the standard matrix A of T are the images of e_1 and e_2. Write these images

$$\text{vertically: } T(e_1) = \begin{bmatrix} 3 \\ 1 \\ 3 \\ 1 \end{bmatrix} \text{ and } T(e_2) = \begin{bmatrix} -5 \\ 2 \\ 0 \\ 0 \end{bmatrix}. \text{ Then } A = [T(e_1)\ T(e_2)] = \begin{bmatrix} 3 & -5 \\ 1 & 2 \\ 3 & 0 \\ 1 & 0 \end{bmatrix}.$$

7. Follow what happens to e_1 and e_2. Since e_1 is on the unit circle in the plane, it rotates through $-3\pi/4$ radians into a point on the unit circle that lies in the third quadrant and on the line $x_2 = x_1$ (that is, $y = x$ in more familiar notation). The point $(-1, -1)$ is on the line $x_2 = x_1$, but its distance from the origin is $\sqrt{2}$. So the rotational image of e_1 is $(-1/\sqrt{2}, -1/\sqrt{2})$. Then this image reflects in the horizontal axis to $(-1/\sqrt{2}, 1/\sqrt{2})$.

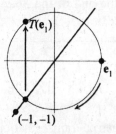

Similarly, e_2 rotates into a point on the unit circle that lies in the second quadrant and on the line $x_2 = -x_1$, namely, $(1/\sqrt{2}, -1/\sqrt{2})$. Then this image reflects in the horizontal axis to $(1/\sqrt{2}, 1/\sqrt{2})$.

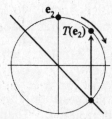

When the two calculations described above are written in vertical vector notation, the transformation's standard matrix $[T(\mathbf{e}_1) \ T(\mathbf{e}_2)]$ is easily seen:

$$\mathbf{e}_1 \to \begin{bmatrix} -1/\sqrt{2} \\ -1/\sqrt{2} \end{bmatrix} \to \begin{bmatrix} -1/\sqrt{2} \\ 1/\sqrt{2} \end{bmatrix}, \quad \mathbf{e}_2 \to \begin{bmatrix} 1/\sqrt{2} \\ -1/\sqrt{2} \end{bmatrix} \to \begin{bmatrix} 1/\sqrt{2} \\ 1/\sqrt{2} \end{bmatrix}, \quad A = \begin{bmatrix} -1/\sqrt{2} & 1/\sqrt{2} \\ 1/\sqrt{2} & 1/\sqrt{2} \end{bmatrix}$$

Checkpoint: Use an idea from this section to explain why the linear transformation T that reflects points through the origin, $T(x_1, x_2) = (-x_1, -x_2)$, is the same as the linear transformation R that rotates points about the origin in $\mathbb{R}^2$ through π radians.

13. Since $(2, 1) = 2\mathbf{e}_1 + \mathbf{e}_2$, the image of $(2, 1)$ under T is $2T(\mathbf{e}_1) + T(\mathbf{e}_2)$, by linearity of T. On the figure in the exercise, locate $2T(\mathbf{e}_1)$ and use it with $T(\mathbf{e}_2)$ to form the parallelogram shown in the text's answers.

19. The matrix A that changes (x_1, x_2, x_3) into $(x_1 - 5x_2 + 4x_3, x_2 - 6x_3)$ can be found by inspection when vectors are written in column formation. Write a blank matrix A to the left of the column vector $\mathbf{x}$ and fill in the entries of A. Since $T(\mathbf{x})$ has 2 entries, A has 2 rows. Since $\mathbf{x}$ has 3 entries, A must have 3 columns.

$$\begin{bmatrix} x_1 - 5x_2 + 4x_3 \\ x_2 - 6x_3 \end{bmatrix} = \begin{bmatrix} & A & \end{bmatrix}\begin{bmatrix} x_1 \\ x_2 \\ x_3 \end{bmatrix} = \begin{bmatrix} 1 & -5 & 4 \\ 0 & 1 & -6 \end{bmatrix}\begin{bmatrix} x_1 \\ x_2 \\ x_3 \end{bmatrix}$$

21. $T(\mathbf{x}) = \begin{bmatrix} x_1 + x_2 \\ 4x_1 + 5x_2 \end{bmatrix} = \begin{bmatrix} & A & \end{bmatrix}\begin{bmatrix} x_1 \\ x_2 \end{bmatrix} = \begin{bmatrix} 1 & 1 \\ 4 & 5 \end{bmatrix}\begin{bmatrix} x_1 \\ x_2 \end{bmatrix}$. To solve $T(\mathbf{x}) = \begin{bmatrix} 3 \\ 8 \end{bmatrix}$, row reduce the

augmented matrix: $\begin{bmatrix} 1 & 1 & 3 \\ 4 & 5 & 8 \end{bmatrix} \sim \begin{bmatrix} 1 & 1 & 3 \\ 0 & 1 & -4 \end{bmatrix} \sim \begin{bmatrix} 1 & 0 & 7 \\ 0 & 1 & -4 \end{bmatrix}$, $\mathbf{x} = \begin{bmatrix} 7 \\ -4 \end{bmatrix}$.

Study Tip: When T is described by a formula, as in Exercises 15–22, you can use the method of Exercise 19 to find an A such that $T(\mathbf{x}) = A\mathbf{x}$, *provided* that T is a linear transformation. (Finding A *proves* that T is linear.) If you can't find the matrix, T is probably *not* a linear transformation. To show that such a T is not linear, you have either to find two vectors $\mathbf{u}$ and $\mathbf{v}$ such that $T(\mathbf{u} + \mathbf{v})$ is not equal to $T(\mathbf{u}) + T(\mathbf{v})$ or to find a vector $\mathbf{u}$ and scalar c such that $T(c\mathbf{u}) \neq cT(\mathbf{u})$.

The text does not give you practice determining whether a transformation is linear because the time needed to develop this skill would have to be taken away from some other topic. If you are expected to have this skill, you will need some exercises (besides Exercises 32 and 33 in Section 1.8). Check with your instructor.

23. a. See Theorem 10.

    b. See Example 3.

    c. See the paragraph before Table 1.

    d. See the definition of *onto*. *Any* function from $\mathbb{R}^n$ to $\mathbb{R}^m$ maps each vector onto another vector.

    e. See Example 5.

25. Three row interchanges on the standard matrix A of the transformation T in Exercise 17

produce $\begin{bmatrix} \textcircled{1} & 1 & 0 & 0 \\ 0 & \textcircled{1} & 1 & 0 \\ 0 & 0 & \textcircled{1} & 1 \\ 0 & 0 & 0 & 0 \end{bmatrix}$. This matrix shows that A has only three pivot positions, so the

equation $A\mathbf{x} = \mathbf{0}$ has a nontrivial solution. By Theorem 11, the transformation T is *not* one-to-one. Also, since A does not have a pivot in each row, the columns of A do not span $\mathbb{R}^4$. By Theorem 12, T does *not* map $\mathbb{R}^4$ onto $\mathbb{R}^4$.

31. *T is one-to-one if and only if* A *has* n *pivot columns.* This statement follows by combining Theorem 12(b) with the statement in Exercise 30 of Section 1.7.

A Mathematical Note: One-to-one

Many students have difficulty with the concept of a one-to-one mapping. Figure 4 should help. The transformation T on the left appears to map three (or even more) points to one image point. In contrast, the transformation T on the right maps three points to one image point. In contrast, the transformation T on the right maps three points to three points. You could say that T is three-to-three (or six-to-six), but the standard terminology is one-to-one.

---

33. Define $T: \mathbb{R}^n \rightarrow \mathbb{R}^m$ by $T(\mathbf{x}) = B\mathbf{x}$ for some $m \times n$ matrix B, and let A be the standard matrix for T. By definition, $A = [T(\mathbf{e}_1) \cdots T(\mathbf{e}_n)]$, where $\mathbf{e}_j$ is the jth column of I_n. However, by matrix-vector multiplication,

    $T(\mathbf{e}_j) = B\mathbf{e}_j = \mathbf{b}_j$, the jth column of B. *So* $A = [\mathbf{b}_1 \cdots \mathbf{b}_n] = B$.

35. If $T: \mathbb{R}^n \rightarrow \mathbb{R}^m$ maps $\mathbb{R}^n$ onto $\mathbb{R}^m$, then its standard matrix A has a pivot in each row, by Theorem 12 and by Theorem 4 in Section 1.4. So A must have at least as many columns as rows, so $m \leq n$.

   When T is one-to-one, A must have a pivot in each column, by Theorem 12, so $m \geq n$.

37. **[M]** There is no pivot in the fourth column, so the columns of the matrix are not linearly independent and hence the linear transformation is not one-to-one (Theorem 12). (Or, use the result of Exercise 31.)

39. **[M]** Row reduction of the matrix shows that columns 1, 2, 3, and 5 contain pivots, but there is no pivot in the fifth row, so the columns of the matrix do not span $\mathbb{R}^5$. By Theorem 12, the linear transformation is not onto.

31. *(This solution is for Section 1.8.) To construct the proof, first write in mathematical terms what is given.*

   Since $\{\mathbf{v}_1, \mathbf{v}_2, \mathbf{v}_3\}$ is linearly dependent, there exist scalars c_1, c_2, c_3, not all zero, such that

$$c_1 \mathbf{v}_1 + c_2 \mathbf{v}_2 + c_3 \mathbf{v}_3 = 0 \qquad\qquad (*)$$

Next, think about what you must prove. In this problem, to prove that the image points are linearly dependent, you need a dependence relation among $T(\mathbf{v}_1)$, $T(\mathbf{v}_2)$, and $T(\mathbf{v}_3)$. That fact suggests the next step.

   Apply T to both sides of (*) and use linearity of T, obtaining

$$T(c_1 \mathbf{v}_1 + c_2 \mathbf{v}_2 + c_3 \mathbf{v}_3) = T(0)$$

and

$$c_1 T(\mathbf{v}_1) + c_2 T(\mathbf{v}_2) + c_3 T(\mathbf{v}_3) = 0$$

Since not all the weights are zero, $\{T(\mathbf{v}_1), T(\mathbf{v}_2), T(\mathbf{v}_3)\}$ is a linearly dependent set. This completes the proof.

Study Tip: Analyze the strategy above for solving Exercise 31 (in Section 1.8). This approach will work later in a variety of situations.

Answer to Checkpoint: The reflection T has the property that $T(\mathbf{e}_1) = -\mathbf{e}_1$ and $T(\mathbf{e}_2) = -\mathbf{e}_2$, while the rotation R has the property that $R(\mathbf{e}_1) = -\mathbf{e}_1$ and $R(\mathbf{e}_2) = -\mathbf{e}_2$. Since a linear transformation is completely determined by what it does to the columns $\mathbf{e}_1$ and $\mathbf{e}_2$ of the identity matrix, T and R must be the same transformation. (You could also explain this by observing that T and R have the same standard matrix, namely, $[-\mathbf{e}_1 \quad -\mathbf{e}_2]$.)

Mastering Linear Algebra Concepts: Existence and Uniqueness

It's time to review and organize what you have learned about existence and uniqueness concepts, if you have not already done so. The review will help to prepare you for an exam on the chapter material.

Search through the chapter and collect all the various ways to express existence and uniqueness statements. Most of them can be found in boxes (and theorems) with an "if and only if" statement. Also, check the exercises. For existence, make two lists—one that concerns the equation $A\mathbf{x} = \mathbf{b}$ for some fixed $\mathbf{b}$ (but not always phrased as a matrix equation), and one that concerns the existence of solutions of $A\mathbf{x} = \mathbf{b}$ for all $\mathbf{b}$.

1.10   LINEAR MODELS IN BUSINESS, SCIENCE, AND ENGINEERING

This is the second of twelve sections devoted to uses of linear algebra. The applications in the text were selected to give you an impression of the power of linear algebra. You are likely to encounter some of these topics again—in school or in your career—and the discussions in your text will be valuable references.

The main point of this section is to present several interesting applications in which "linearity" arises naturally.

STUDY NOTES

Nutrition Problem: In some applied problems such as the nutrition problem considered here, the data are already organized naturally in a manner that leads to a vector equation of the type we have discussed. The steps to the solution in this case may be diagrammed as follows:

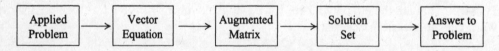

The nutrition model is linear because the nutrients supplied by each foodstuff are *proportional* to the amount of the foodstuff added to the diet mixture, and each nutrient in the mixture is the *sum* of the amounts from each foodstuff. Study equations (1) and (2) on page 94.

The nutrition problem leads naturally into linear programming, a subject that uses linear algebra and has applications in agriculture, business, engineering, and other areas. In the 1950's and 1960's, one of the most common applications of linear algebra (measured in millions of dollars per year for computer time) was to linear programming problems. Such problems are still of great importance in operations research and management science. The following reference gives an entertaining introduction to linear programming. Matrix notation is used in its appendix (pp. 127–152).

Gass, Saul I., *An Illustrated Guide to Linear Programming,* New York: McGraw-Hill, 1970. Republished by Dover Publications, 1990.

Electrical Networks: The linearity of this model, which is evident from the matrix equation $Ri = v$, comes from the linearity of Ohm's law and Kirchhoff's voltage law. (Kirchhoff's current law, which is also linear, is needed when studying another model that involves branch currents.)

Population Movement: The entries in each *column* of the migration matrix must sum to one because the (decimal) fractions in a column account for the entire population in one region. A certain fraction of the population in a region remains in (or moves within) that region, and other fractions move elsewhere.

SOLUTIONS TO EXERCISES

1. a. If x_1 is the number of servings of Cheerios and x_2 is the number of servings of 100% Natural Cereal, then x_1 and x_2 should satisfy

$$x_1 \begin{bmatrix} \text{nutrients} \\ \text{per serving} \\ \text{of Cheerios} \end{bmatrix} + x_2 \begin{bmatrix} \text{nutrients} \\ \text{per serving of} \\ \text{100\% Natural} \end{bmatrix} = \begin{bmatrix} \text{quantities} \\ \text{of nutrients} \\ \text{required} \end{bmatrix}$$

That is,

$$x_1 \begin{bmatrix} 110 \\ 4 \\ 20 \\ 2 \end{bmatrix} + x_2 \begin{bmatrix} 130 \\ 3 \\ 18 \\ 5 \end{bmatrix} = \begin{bmatrix} 295 \\ 9 \\ 48 \\ 8 \end{bmatrix}$$

b. The equivalent matrix equation is $\begin{bmatrix} 110 & 130 \\ 4 & 3 \\ 20 & 18 \\ 2 & 5 \end{bmatrix} \begin{bmatrix} x_1 \\ x_2 \end{bmatrix} = \begin{bmatrix} 295 \\ 9 \\ 48 \\ 8 \end{bmatrix}$. To solve this, row reduce the augmented matrix for this equation.

$$
\begin{bmatrix} 110 & 130 & 295 \\ 4 & 3 & 9 \\ 20 & 18 & 48 \\ 2 & 5 & 8 \end{bmatrix} \sim \begin{bmatrix} 2 & 5 & 8 \\ 4 & 3 & 9 \\ 20 & 18 & 48 \\ 110 & 130 & 295 \end{bmatrix} \sim \begin{bmatrix} 1 & 2.5 & 4 \\ 4 & 3 & 9 \\ 10 & 9 & 24 \\ 110 & 130 & 295 \end{bmatrix}
$$

$$
\sim \begin{bmatrix} 1 & 2.5 & 4 \\ 0 & -7 & -7 \\ 0 & -16 & -16 \\ 0 & -145 & -145 \end{bmatrix} \sim \begin{bmatrix} 1 & 2.5 & 4 \\ 0 & 1 & 1 \\ 0 & 0 & 0 \\ 0 & 0 & 0 \end{bmatrix} \sim \begin{bmatrix} 1 & 0 & 1.5 \\ 0 & 1 & 1 \\ 0 & 0 & 0 \\ 0 & 0 & 0 \end{bmatrix}
$$

The desired nutrients are provided by 1.5 servings of Cheerios together with 1 serving of 100% Natural Cereal.

Study Tip: Be sure to distinguish between (i) the vector equation, (ii) the matrix equation (which has the form $A\mathbf{x} = \mathbf{b}$), and (iii) the augmented matrix (which has the form $[A \quad \mathbf{b}]$) that represents a system of linear equations.

7. Loop 1: The resistance vector is

$$
\mathbf{r}_1 = \begin{bmatrix} 12 \\ -7 \\ 0 \\ -4 \end{bmatrix}
$$

Total of three RI voltage drops for current I_1

Voltage drop for I_2 is negative; I_2 flows in opposite direction

Current I_3 does not flow in loop 1

Voltage drop for I_4 is negative; I_4 flows in opposite direction

Loop 2: The resistance vector is

$$
\mathbf{r}_2 = \begin{bmatrix} -7 \\ 15 \\ -6 \\ 0 \end{bmatrix}
$$

Voltage drop for I_1 is negative; I_1 flows in opposite direction

Total of three RI voltage drops for current I_2

Voltage drop for I_3 is negative; I_3 flows in opposite direction

Current I_4 does not flow in loop 2

Also, $\mathbf{r}_3 = \begin{bmatrix} 0 \\ -6 \\ 14 \\ -5 \end{bmatrix}$, $\mathbf{r}_4 = \begin{bmatrix} -4 \\ 0 \\ -5 \\ 13 \end{bmatrix}$, and $R = [\mathbf{r}_1 \quad \mathbf{r}_2 \quad \mathbf{r}_3 \quad \mathbf{r}_4] = \begin{bmatrix} 12 & -7 & 0 & -4 \\ -7 & 15 & -6 & 0 \\ 0 & -6 & 14 & -5 \\ -4 & 0 & -5 & 13 \end{bmatrix}$.

Note that each off-diagonal entry of R is negative (or zero). This happens because the loop current directions are all chosen in the same direction on the figure. (For each loop j, this choice forces the currents in other loops adjacent to loop j to flow in the direction opposite to current I_j.)

Next, set $\mathbf{v} = \begin{bmatrix} 40 \\ 30 \\ 20 \\ -10 \end{bmatrix}$. Note the negative voltage in loop 4. The current direction chosen in

loop 4 is opposed by the orientation of the voltage source in that loop. Thus $R\mathbf{i} = \mathbf{v}$ becomes

$$\begin{bmatrix} 12 & -7 & 0 & -4 \\ -7 & 15 & -6 & 0 \\ 0 & -6 & 14 & -5 \\ -4 & 0 & -5 & 13 \end{bmatrix} \begin{bmatrix} I_1 \\ I_2 \\ I_3 \\ I_4 \end{bmatrix} = \begin{bmatrix} 40 \\ 30 \\ 20 \\ -10 \end{bmatrix}.$$ [M]: The solution is $\mathbf{i} = \begin{bmatrix} I_1 \\ I_2 \\ I_3 \\ I_4 \end{bmatrix} = \begin{bmatrix} 11.43 \\ 10.55 \\ 8.04 \\ 5.84 \end{bmatrix}.$

13. **[M]** The order of entries in a column of a migration matrix must match the order of the columns. For instance, if the first column concerns the population in the city, then the first entry in *each* column of the matrix must be the fraction of the population that moves to (or remains in) the city. In this case, the data in the exercise leads to

$$M = \begin{bmatrix} .95 & .03 \\ .05 & .97 \end{bmatrix} \text{ and } \mathbf{x}_0 = \begin{bmatrix} 600,000 \\ 400,000 \end{bmatrix}.$$

a. Some of the population vectors are

$$\mathbf{x}_5 = \begin{bmatrix} 523,293 \\ 476,707 \end{bmatrix}, \ \mathbf{x}_{10} = \begin{bmatrix} 472,737 \\ 527,263 \end{bmatrix}, \ \mathbf{x}_{15} = \begin{bmatrix} 439,417 \\ 560,583 \end{bmatrix}, \ \mathbf{x}_{20} = \begin{bmatrix} 417,456 \\ 582,544 \end{bmatrix}$$

The data here shows that the city population is declining and the suburban population is increasing, but the changes in population each year seem to grow smaller.

b. When $\mathbf{x}_0 = \begin{bmatrix} 350,000 \\ 650,000 \end{bmatrix}$, the situation is different. Now

$$\mathbf{x}_5 = \begin{bmatrix} 358,523 \\ 641,477 \end{bmatrix}, \ \mathbf{x}_{10} = \begin{bmatrix} 364,140 \\ 635,860 \end{bmatrix}, \ \mathbf{x}_{15} = \begin{bmatrix} 367,843 \\ 632,157 \end{bmatrix}, \ \mathbf{x}_{20} = \begin{bmatrix} 370,283 \\ 629,717 \end{bmatrix}$$

The city population is increasing slowly and the suburban population is decreasing. No other conclusions are expected. (This example will be analyzed in greater detail later in the text.)

MATLAB   Generating a Sequence

The m-file (in the Laydata Toolbox) for Exercises 9–13 in Section 1.9 stores initial vectors in **x0**. Set **x = x0** to put the initial data into **x**. Then use the command **x = M*x** repeatedly to generate the sequence **x1, x2, ...**   You only type the command once. After that, use the up-arrow (↑) key to recall the command, and press <Enter>.

In Exercise 11, you need 6 decimal places to get four significant figures in $M(1, 2)$. Use the command **format long** and then **M** to see more decimal places in M. The command **format short** will return MATLAB to the standard four decimal place display. (The display format does not affect MATLAB's accuracy in computations.)

Numbers are entered in MATLAB without commas. The number 600,000 in MATLAB scientific notation is 6e5. A small number such as .00000012 is 1.2e-7.

Chapter 1   SUPPLEMENTARY EXERCISES ———————————

The supplementary exercises at the end of each chapter review material from the chapter, synthesize concepts from several chapters, or supplement the chapter material in some way. The text has solutions for most of the odd-numbered exercises. The *Study Guide* provides solutions for selected odd-numbered exercises that have only an answer or a *Hint*.

In each chapter, Exercise 1 consists of many true/false questions, whose level of difficulty varies. Some are similar to the ones that appear in many sections of the text, in which a word or phrase is sometimes missing or slightly misstated. Some follow fairly easily from a theorem; others may need careful reasoning. A few may require an argument that uses several ideas. In each case, think carefully about the statement and attempt to write a solution. The text provides the true/false answer, but you must supply the justification or counterexample. Careful work on Exercise 1 will help you prepare for an exam over the chapter material.

7. a. Set $\mathbf{v}_1 = \begin{bmatrix} 2 \\ -5 \\ 7 \end{bmatrix}$, $\mathbf{v}_2 = \begin{bmatrix} -4 \\ 1 \\ -5 \end{bmatrix}$, $\mathbf{v}_3 = \begin{bmatrix} -2 \\ 1 \\ -3 \end{bmatrix}$ and $\mathbf{b} = \begin{bmatrix} b_1 \\ b_2 \\ b_3 \end{bmatrix}$. "Determine if $\mathbf{v}_1, \mathbf{v}_2, \mathbf{v}_3$ span $\mathbb{R}^3$."

To do this, row reduce $[\mathbf{v}_1 \quad \mathbf{v}_2 \quad \mathbf{v}_3]$:

$$\begin{bmatrix} 1 & -2 & -1 \\ -5 & 1 & 1 \\ 7 & -5 & -3 \end{bmatrix} \sim \begin{bmatrix} 1 & -2 & -1 \\ -5 & 1 & 1 \\ 7 & -5 & -3 \end{bmatrix} \sim \begin{bmatrix} 1 & -2 & -1 \\ 0 & -9 & -4 \\ 0 & 9 & 4 \end{bmatrix} \sim \begin{bmatrix} 1 & -2 & -1 \\ 0 & -9 & -4 \\ 0 & 0 & 0 \end{bmatrix}.$$ The matrix does not have

a pivot in each row, so its columns do not span $\mathbb{R}^3$, by Theorem 4 in Section 1.4.

13. The reduced echelon form of A looks like $E = \begin{bmatrix} 1 & 0 & * \\ 0 & 1 & * \\ 0 & 0 & 0 \end{bmatrix}$. Since E is row equivalent to A,

the equation $E\mathbf{x} = \mathbf{0}$ has the same solutions as $A\mathbf{x} = \mathbf{0}$. Thus $\begin{bmatrix} 1 & 0 & * \\ 0 & 1 & * \\ 0 & 0 & 0 \end{bmatrix} \begin{bmatrix} 3 \\ -2 \\ 1 \end{bmatrix} = \begin{bmatrix} 0 \\ 0 \\ 0 \end{bmatrix}$. By

inspection, $E = \begin{bmatrix} 1 & 0 & -3 \\ 0 & 1 & 2 \\ 0 & 0 & 0 \end{bmatrix}$.

17. Here are two arguments. The first is a "direct" proof. The second is called a "proof by contradiction."

   i. Since $\{\mathbf{v}_1, \mathbf{v}_2, \mathbf{v}_3\}$ is a linearly independent set, $\mathbf{v}_1 \neq \mathbf{0}$. Also, Theorem 7 shows that $\mathbf{v}_2$ cannot be a multiple of $\mathbf{v}_1$, and $\mathbf{v}_3$ cannot be a linear combination of $\mathbf{v}_1$ and $\mathbf{v}_2$. By hypothesis, $\mathbf{v}_4$ is not a linear combination of $\mathbf{v}_1$, $\mathbf{v}_2$, and $\mathbf{v}_3$. Thus, by Theorem 7, $\{\mathbf{v}_1, \mathbf{v}_2, \mathbf{v}_3, \mathbf{v}_4\}$ cannot be a linearly dependent set and so must be linearly independent.

   ii. Since $\{\mathbf{v}_1, \mathbf{v}_2, \mathbf{v}_3\}$ is a linearly independent set, $\mathbf{v}_1 \neq \mathbf{0}$. Suppose that $\{\mathbf{v}_1, \mathbf{v}_2, \mathbf{v}_3, \mathbf{v}_4\}$ is linearly dependent. Then, by Theorem 7, one of the vectors in the set is a linear combination of the preceding vectors. This vector cannot be $\mathbf{v}_4$ because $\mathbf{v}_4$ is *not* in Span$\{\mathbf{v}_1, \mathbf{v}_2, \mathbf{v}_3\}$. Also, none of the vectors in $\{\mathbf{v}_1, \mathbf{v}_2, \mathbf{v}_3\}$ is a linear combination of the preceding vectors, by Theorem 7. So the linear dependence of $\{\mathbf{v}_1, \mathbf{v}_2, \mathbf{v}_3, \mathbf{v}_4\}$ is impossible, and $\{\mathbf{v}_1, \mathbf{v}_2, \mathbf{v}_3, \mathbf{v}_4\}$ is linearly independent.

Chapter 1   GLOSSARY CHECKLIST _____

Check your knowledge by attempting to write definitions of the terms below. Then compare your work with the definitions given in the text's Glossary. Ask your instructor which definitions, if any, might appear on a test.

affine transformation: A mapping $T : \mathbb{R}^n \rightarrow \mathbb{R}^m$ of the form $T(x) = \dots$.

augmented matrix: A matrix made up of a . . . .

back-substitution (with matrix notation):   The . . . phase of row reduction of an . . . .

basic variable: A variable in a linear system that . . . .

codomain (of $T : \mathbb{R}^n \rightarrow \mathbb{R}^m$):   The set . . . that contains . . . .

coefficient matrix: A matrix whose entries are . . . .

consistent linear system: A linear system with . . . .

contraction: A mapping $\mathbf{x} \mapsto$ . . . .

difference equation (or **linear recurrence relation**):   An equation of the form . . . whose solution is . . . .

dilation: A mapping $\mathbf{x} \mapsto$ . . . .

domain (of a transformation T):   The set of . . . .

echelon form (or **row echelon form**, of a matrix):   An echelon matrix that . . . .

echelon matrix (or **row echelon matrix**):   A rectangular matrix that has three properties: (1) . . . (2) . . . (3) . . . .

elementary row operations: (1) . . . (2) . . . (3) . . . .

equal vectors: Vectors in $\mathbb{R}^n$ whose . . . .

equivalent (linear) systems: Linear systems with the . . . .

existence question: Asks, "Does . . . exist?" or "Is . . .?" Also, "Does . . . exist for . . .?"

floating point arithmetic: Arithmetic with numbers represented as . . . .

flop: One arithmetic operation . . . .

free variable: Any variable in a linear system that . . . .

Gaussian elimination: *See* row reduction algorithm.

general solution (of a linear system):   A . . . description of a solution set that expresses . . .

homogenous equation: An equation of the . . . .

identity matrix (denoted by I or I_n):   A square matrix . . . .

image (of a vector $\mathbf{x}$ under a transformation T):   The vector . . . . (Use symbols)

inconsistent linear system: A linear system with . . . .

leading entry: The . . . entry in a row of a matrix.

linear combination: A sum of . . . .

linear dependence relation: A . . . equation where . . . .

linear equation (in the variables $x_1, \ldots, x_n$):   An equation that can be written in the form .

linearly dependent (vectors):   An indexed set $\{\mathbf{v}_1, \ldots, \mathbf{v}_p\}$ with the property that . . . .

linearly independent (vectors):   An indexed set $\{\mathbf{v}_1, \ldots, \mathbf{v}_p\}$ with the property . . . .

linear system: A collection of one or more . . . equations involving . . . .

linear transformation: A transformation $T: \mathbb{R}^n \to \mathbb{R}^m$ is linear if (i) . . . , and (ii) . . . .

line through p parallel to v: The set . . . . (Use symbols)

matrix: A rectangular . . . .

matrix equation: An equation that . . . .

matrix transformation: A mapping $\mathbf{x} \mapsto$ . . . .

migration matrix: A matrix that gives the . . . movement between different locations, from . . . .

$m \times n$ matrix: A matrix with . . . .

nontrivial solution: A nonzero solution of . . . .

one-to-one (mapping):   A mapping $T : \mathbb{R}^n \to \mathbb{R}^m$ such that . . . .

onto (mapping):   A mapping $T : \mathbb{R}^n \to \mathbb{R}^m$ such that . . . .

overdetermined system: A system of equations with . . . .

parallelogram rule for addition: A geometric interpretation of . . . .

parametric equation of a line: An equation of the form . . . .

parametric equation of a plane: An equation of the form . . . .

pivot: A . . . number that either is used . . . or is . . . .

pivot column: A column that . . . .

pivot position: A position in a matrix A that corresponds . . . .

plane through u, v, and the origin: A set whose parametric equation is . . . .

product Ax: . . . .

range (of a linear transformation T):   The set of . . . .

reduced echelon form (or **reduced row echelon form,** of a matrix) :   A rectangular matrix in echelon format that has these additional properties . . . .

roundoff error: Error in floating point arithmetic caused when . . . .

row-column rule for computing Ax: . . . .

row equivalent (matrices):   Two matrices for which there exists . . . .

row reduction algorithm: A systematic method using . . . .

row replacement: An elementary row operation that . . . .

scalar: . . . .

scalar multiple of u by c: The vector . . . .

set spanned by $\{v_1, \ldots, v_p\}$:   . . . .

size (of a matrix):   Two numbers . . . .

solution (of a linear system):   . . . .

solution set: The set of . . . .

Span {v_1, ..., v_p}:   The set . . . .

standard matrix (for a linear transformation T):   The matrix . . . .

system of linear equations (or a linear system):   A collection of . . . .

transformation (or **function** or **mapping**) T **from** $\mathbb{R}^n$ **to** $\mathbb{R}^m$:   A rule that assigns to each vector x in $\mathbb{R}^n$ a . . . . Notation: $T : \mathbb{R}^n \rightarrow \mathbb{R}^m$.

translation (by a vector **p**):   The operation of . . . .

trivial solution: The solution . . . of a . . . .

underdetermined system: A system of equations with . . . .

uniqueness question: Asks, "If a solution of a system . . . ?"

vector: . . . .

vector equation: An equation involving . . . .

weights: . . . .

2 | Matrix Algebra

2.1   MATRIX OPERATIONS

Most of this chapter is an outgrowth of the idea in Section 1.7 that a matrix can transform data. This dynamic role of matrices suggests that we study the *combined effect* of several matrices on data (that is, on a vector or a set of vectors). Sections 2.1 to 2.5 describe this *matrix algebra*.

KEY IDEA

Matrix multiplication corresponds to composition of linear transformations. The definition of *AB*, using the columns of *B*, is critical for the development of both the theory and some of the applications in the text.

STUDY NOTES

Double-subscript notation: The subscripts tell the location of an entry in the matrix—the first subscript identifies the row and the second subscript the column. (Remember: *Row* is shorter than *column*, so *row* goes first.) This convention is opposite to the way a spreadsheet identifies the location of an entry.

In the product *AB*, left-multiplication (that is, multiplication on the left) by *A* acts on the columns of *B*, by definition, while right-multiplication by *B* acts on the rows of *A* (see page 112). That is,

$$\begin{bmatrix} \text{column } j \\ \text{of } AB \end{bmatrix} = A \begin{bmatrix} \text{column } j \\ \text{of } B \end{bmatrix} \quad \text{and} \quad [\text{row } i \text{ of } AB] = [\text{row } i \text{ of } A]B$$

To compute a specific matrix product by hand, use the Row-Column Rule. If A is $m \times n$, then the (i, j)-entry of AB is written with sigma notation as

$$(AB)_{ij} = \sum_{k=1}^{n} a_{ik} b_{kj}$$

Remember that if you change the *order* (position) of the factors in a matrix product, the new product may be different, or it may not even be defined. For instance, $(A + C)B$ and $AB + BC$ are probably *not* equal! Also, see the warning box on page 114.

Notes: Key exercises 13 and 17–22 emphasize the definition of a matrix product. Work at least five of these exercises, for practice.

SOLUTIONS TO EXERCISES

1. $-2A = (-2)\begin{bmatrix} 2 & 0 & -1 \\ 4 & -5 & 2 \end{bmatrix} = \begin{bmatrix} -4 & 0 & 2 \\ -8 & 10 & -4 \end{bmatrix}$. Next, use $B - 2A = B + (-2A)$:

$$B - 2A = \begin{bmatrix} 7 & -5 & 1 \\ 1 & -4 & -3 \end{bmatrix} + \begin{bmatrix} -4 & 0 & 2 \\ -8 & 10 & -4 \end{bmatrix} = \begin{bmatrix} 3 & -5 & 3 \\ -7 & 6 & -7 \end{bmatrix}$$

The product AC is not defined because the number of columns of A does not match the number of rows of C.

$$CD = \begin{bmatrix} 1 & 2 \\ -2 & 1 \end{bmatrix}\begin{bmatrix} 3 & 5 \\ -1 & 4 \end{bmatrix} = \begin{bmatrix} 1 \cdot 3 + 2(-1) & 1 \cdot 5 + 2 \cdot 4 \\ -2 \cdot 3 + 1(-1) & -2 \cdot 5 + 1 \cdot 4 \end{bmatrix} = \begin{bmatrix} 1 & 13 \\ -7 & -6 \end{bmatrix}$$. For mental computation,

the row-column rule is easier to use than the definition.

7. Since A has 3 columns, B must match with 3 rows. Otherwise, AB is undefined. Since AB has 7 columns, so does B. Thus, B is 3×7.

13. If you had difficulty with this problem, read the definition of AB from *right* to *left*. Here is the definition, written in reverse order:

$$[A\mathbf{b}_1 \cdots A\mathbf{b}_p] = A[\mathbf{b}_1 \cdots \mathbf{b}_p] = AB, \text{ when } B = [\mathbf{b}_1 \cdots \mathbf{b}_p].$$

Thus $[Q\mathbf{r}_1 \cdots Q\mathbf{r}_p] = QR$, when $R = [\mathbf{r}_1 \cdots \mathbf{r}_p]$.

15. a. See the definition of AB.   b. See the box after Example 3.

    c. Read Theorem 2(b) from right to left.   d. Read Theorem 3(b) from right to left.

    e. See the box after Theorem 3.

19. A solution is in the text. The main point is that the columns of AB are $A\mathbf{b}_1, \ldots, A\mathbf{b}_p$.

Checkpoint: Show that if $\mathbf{y}$ is a linear combination of the columns of AB, then $\mathbf{y}$ is a linear combination of the columns of A.

21. Let $\mathbf{b}_p$ be the last column of B. By hypothesis, the last column of AB is zero. Thus, $A\mathbf{b}_p = \mathbf{0}$. However, $\mathbf{b}_p$ is not the zero vector, because B has no column of zeros. Thus, the equation $A\mathbf{b}_p = \mathbf{0}$ is a linear dependence relation among the columns of A, and so the columns of A are linearly dependent.

23. If $\mathbf{x}$ satisfies $A\mathbf{x} = \mathbf{0}$, then $CA\mathbf{x} = C\mathbf{0} = \mathbf{0}$ and so $I_n\mathbf{x} = \mathbf{0}$ and $\mathbf{x} = \mathbf{0}$. This shows that the equation $A\mathbf{x} = \mathbf{0}$ has no free variables. So every variable is a basic variable and every column of A is a pivot column. (A variation of this argument could be made using linear independence and Exercise 30 in Section 1.7.) Since each pivot is in a different row, A must have at least as many rows as columns.

25. By Exercise 23, the equation $CA = I_n$ implies that (number of rows in A) $\geq$ (number of columns), that is, $m \geq n$. By Exercise 24, the equation $AD = I_m$ implies that (number of rows in A) $\leq$ (number of columns), that is, $m \leq n$. Thus $m = n$. To prove the second statement, observe that $CAD = C(AD) = CI_m = C$, and also $CAD = (CA)D = I_nD = D$. Thus $C = D$. A shorter calculation is

$$C = CI_n = C(AD) = (CA)D = I_nD = D$$

Study Tip: In Exercises 27 and 28, *inner* products ($\mathbf{u}^T\mathbf{v}$ and $\mathbf{v}^T\mathbf{u}$) have the transpose symbol in the middle. *Outer* products ($\mathbf{u}\mathbf{v}^T$ and $\mathbf{v}\mathbf{u}^T$) have the transpose symbol on the outside.

29. The (i, j)-entry of $A(B + C)$ equals the (i, j)-entry of $AB + AC$, because

$$\sum_{k=1}^{n} a_{ik}(b_{kj} + c_{kj}) = \sum_{k=1}^{n} a_{ik}b_{kj} + \sum_{k=1}^{n} a_{ik}c_{kj}$$

The (i, j)-entry of $(B + C)A$ equals the (i, j)-entry of $BA + CA$, because

$$\sum_{k=1}^{n} (b_{ik} + c_{ik})a_{kj} = \sum_{k=1}^{n} b_{ik}a_{kj} + \sum_{k=1}^{n} c_{ik}a_{kj}$$

31. Use the definition of the product I_mA and the fact that $I_m\mathbf{x} = \mathbf{x}$ for $\mathbf{x}$ in $\mathbf{R}^m$.

$$I_mA = I_m[\mathbf{a}_1 \cdots \mathbf{a}_n] = [I_m\mathbf{a}_1 \cdots I_m\mathbf{a}_n] = [\mathbf{a}_1 \cdots \mathbf{a}_n] = A$$

33. The (i, j)-entry of $(AB)^T$ is the (j, i)-entry of AB, which is

$$a_{j1}b_{1i} + \cdots + a_{jn}b_{ni}$$

The entries in row i of B^T are $b_{1i}, \ldots, b_{ni}$, because they come from column i of B. Likewise, the entries in column j of A^T are $a_{j1}, \ldots, a_{jn}$, because they come from row j of A. Thus the (i, j)-entry in B^TA^T is $a_{j1}b_{1i} + \cdots + a_{jn}b_{ni}$, as above.

Answer to Checkpoint: If $\mathbf{y}$ is a linear combination of the columns of AB, then there is a vector $\mathbf{x}$ such that $\mathbf{y} = (AB)\mathbf{x}$. By definition of matrix multiplication, $\mathbf{y} = A(B\mathbf{x})$. This expresses $\mathbf{y}$ as a linear combination of the columns of A using the entries in the vector $B\mathbf{x}$ as weights.

MATLAB   Matrix Notation and Operations

To create a matrix, enter the data row-by-row, with a space between entries and a semicolon between rows. For instance, the command

    A=[1 2 3; 4 5 -6]   Use brackets around the data.

creates a 2×3 matrix A. If A is $m \times n$, then **size(A)** is the row vector $[m \quad n]$. The (i, j)-entry in A is **A(i, j)**. If i or j is replaced by a colon, the result is a column or row of A, respectively. Examples:

    A(:,3)   Column 3 of A
    A(2,:)   Row 2 of A

To specify columns 3, 4 and 5 of A, you can use

    A(:,[3 4 5])   or   A[:,3:5)

The symbols 3:5 (read "3 to 5") stand for the vector [3 4 5]. Similar notation works for selected rows of A.

MATLAB uses +, −, and * to denote matrix addition, subtraction, and multiplication, respectively. If A is square and k is a positive integer, **A^k** denotes the kth power of A. The transpose of A is **A′** (with an apostrophe for the prime symbol). Note: when A has complex entries, the (i, j)-entry of **A′** is the complex conjugate of the (j, i)-entry of A.

Use a single column (or row) matrix for a vector. If **u** and **v** are column vectors of the same size, then **u′*v** is their inner product, and **u*v′** is an outer product.

MATLAB has commands that construct many special matrices. For example,

    M=zeros(5,6)   A 5 × 6 matrix of zeros
    M=ones(3,5)   A 3 × 5 matrix of ones
    M=eye(6)   The 6 × 6 identity matrix
    M=diag([3 5 7 2 4])   A 5 × 5 diagonal matrix
    M=rand(6)   A 6 × 6 matrix with random entries
    M=randomint(6,4)   A 6 × 4 matrix with random integer entries

Place **help** in front of any command to learn all the features of the command. The former name for **randomint** was **randint**, but MATLAB now uses **randint** for a slightly different command in its Communications Toolbox.

2.2   THE INVERSE OF A MATRIX

Matrix inverses are essential for many discussions in linear algebra. This section and the next describe the main properties of invertible matrices.

STUDY NOTES

The inverse formula for a 2×2 matrix will be used frequently in exercises later in the text. (See Theorem 4.) To invert a 2×2 matrix, interchange the diagonal entries, reverse the signs of the off-diagonal entries, and divide each entry by the determinant (assuming $ad - bc \neq 0$).

Theorem 5 and its proof are important. The phrase "has a unique solution" includes the assertion that a solution exists, so the proof has two parts. The equation $AA^{-1} = I$ is used to prove that a solution exists, and the equation $A^{-1}A = I$ is used to show that the solution is unique.

Except when A is 2×2, Theorem 5 is practically never used to solve $Ax = b$. Row reduction of $[A \ \ b]$ is faster. Actually, in practical work, you will seldom need to compute A^{-1}. (However, Example 3 illustrates a case in which the entries of A^{-1} could be useful.)

When using an inverse in matrix algebra, remember that matrix multiplication is not commutative. The phrase "left-multiply B by A^{-1}" means to multiply B on its left side by A^{-1}. *Never* write $\dfrac{B}{A}$ (or B/A) because it could stand for $A^{-1}B$ or BA^{-1}.

Elementary matrices are used in this text mainly to link row reduction to matrix multiplication. Each elementary row operation amounts to left-multiplication by an elementary matrix. So, if A can be row reduced to U, then there is a product F of elementary matrices such that $FA = U$.

Theorem 7 includes an *if and only if* statement, which was discussed in the Appendix to Section 1.2 in this *Study Guide*. The proof of this statement in Theorem 7 has two parts: (1) assume that A is invertible and prove that $A \sim I_n$; and (2) assume that $A \sim I_n$ and prove that A is invertible.

SOLUTIONS TO EXERCISES

1. $\begin{bmatrix} 8 & 6 \\ 5 & 4 \end{bmatrix}^{-1} = \dfrac{1}{32-30}\begin{bmatrix} 4 & -6 \\ -5 & 8 \end{bmatrix} = \begin{bmatrix} 2 & -3 \\ -5/2 & 4 \end{bmatrix}$

7. a. $\begin{bmatrix} 1 & 2 \\ 5 & 12 \end{bmatrix}^{-1} = \dfrac{1}{1\cdot12-2\cdot5}\begin{bmatrix} 12 & -2 \\ -5 & 1 \end{bmatrix} = \dfrac{1}{2}\begin{bmatrix} 12 & -2 \\ -5 & 1 \end{bmatrix}$ or $\begin{bmatrix} 6 & -1 \\ -2.5 & .5 \end{bmatrix}$

$\mathbf{x} = A^{-1}\mathbf{b}_1 = \dfrac{1}{2}\begin{bmatrix} 12 & -2 \\ -5 & 1 \end{bmatrix}\begin{bmatrix} -1 \\ 3 \end{bmatrix} = \dfrac{1}{2}\begin{bmatrix} -18 \\ 8 \end{bmatrix} = \begin{bmatrix} -9 \\ 4 \end{bmatrix}$. Similar calculations give

$A^{-1}\mathbf{b}_2 = \begin{bmatrix} 11 \\ -5 \end{bmatrix}, A^{-1}\mathbf{b}_3 = \begin{bmatrix} 6 \\ -2 \end{bmatrix}, A^{-1}\mathbf{b}_4 = \begin{bmatrix} 13 \\ -5 \end{bmatrix}.$

b. $[A \quad \mathbf{b}_1 \quad \mathbf{b}_2 \quad \mathbf{b}_3 \quad \mathbf{b}_4] = \begin{bmatrix} 1 & 2 & -1 & 1 & 2 & 3 \\ 5 & 12 & 3 & -5 & 6 & 5 \end{bmatrix}$

$\sim \begin{bmatrix} 1 & 2 & -1 & 1 & 2 & 3 \\ 0 & 2 & 8 & -10 & -4 & -10 \end{bmatrix} \sim \begin{bmatrix} 1 & 2 & -1 & 1 & 2 & 3 \\ 0 & 1 & 4 & -5 & -2 & -5 \end{bmatrix}$

$\sim \begin{bmatrix} 1 & 0 & -9 & 11 & 6 & 13 \\ 0 & 1 & 4 & -5 & -2 & -5 \end{bmatrix}$

The solutions are $\begin{bmatrix} -9 \\ 4 \end{bmatrix}, \begin{bmatrix} 11 \\ -5 \end{bmatrix}, \begin{bmatrix} 6 \\ -2 \end{bmatrix}$, and $\begin{bmatrix} 13 \\ -5 \end{bmatrix}$, the same as in part (a).

Note: This exercise was designed to make the arithmetic simple for both methods, but (a) requires more arithmetic than (b). In fact, (a) requires 22 multiplications or divisions and 9 additions or subtractions. In general, the arithmetic for method (b) can be unpleasant for hand calculation. However, when A is larger than 2×2, method (b) is *much* faster than (a).

Study Tip: Notice in Exercise 7(a) how the 1/2 in the formula for A^{-1} was kept outside the matrix $\begin{bmatrix} 12 & -2 \\ -5 & 1 \end{bmatrix}$ when computing $A^{-1}\mathbf{b}$. This trick sometimes simplifies hand calculations (on exams!) by postponing the arithmetic with fractions (or decimals) until the end.

9. **a.** See the definition of *invertible*.   **b.** See Theorem 6(b).

   c. See Theorem 4.   **d.** See Theorem 5.

   e. See the box just before Example 6.

11. (See the proof of Theorem 5.) The $n \times p$ matrix B is given (but is arbitrary). Since A is invertible, the matrix $A^{-1}B$ satisfies $AX = B$, because $A(A^{-1}B) = AA^{-1}B = IB = B$. To show this solution is unique, let X be any solution of $AX = B$. Then, left-multiplication of each side by A^{-1} shows that X must be $A^{-1}B$:

$$A^{-1}(AX) = A^{-1}B, \quad IX = A^{-1}B, \quad \text{and} \quad X = A^{-1}B.$$

Study Tip: Whenever you are told "A is invertible," you know that A^{-1} exists, and you may use A^{-1} to solve an equation or to make appropriate calculations.

13. Left-multiply each side of the equation $AB = AC$ by A^{-1} to obtain

$$A^{-1}AB = A^{-1}AC, \quad IB = IC, \quad \text{and} \quad B = C.$$

This conclusion does not always follow when A is singular. The matrices in Exercise 10 of Section 2.1 provide a counterexample.

Warning: A common mistake in Exercise 16 is to try to use the formula $(AB)^{-1} = B^{-1}A^{-1}$. But, this formula can be used only when you know, in advance, that both A and B are invertible. In Exercise 16, you must *prove* that A is invertible.

19. Unlike Exercise 17, this exercise asks two things, "Does a solution exist?" and "What is the solution?" First, find what the solution must be, if it exists. That is, suppose X satisfies the equation $C^{-1}(A + X)B^{-1} = I$. Left-multiply each side by C, and then right-multiply each side by B:

$$CC^{-1}(A + X)B^{-1} = CI, \quad I(A + X)B^{-1} = C, \quad (A + X)B^{-1}B = CB, \quad (A + X)I = CB$$

Expand the left side and then subtract A from both sides:

$$AI + XI = CB, \quad A + X = CB, \quad X = CB - A$$

If a solution exists, it must be $CB - A$. To *show* that $CB - A$ really *is* a solution, substitute it for X:

$$C^{-1}[A + (CB - A)]B^{-1} = C^{-1}[CB]B^{-1} = C^{-1}CBB^{-1} = II = I.$$

After this section, your instructor may permit you to include fewer details in your calculations. (Check on this.) For instance, after some practice with algebra, an expression such as $CC^{-1}(A + X)B^{-1}$ could be simplified directly to $(A + X)B^{-1}$ without first replacing CC^{-1} by I.

21. Suppose A is invertible. By Theorem 5, the equation $A\mathbf{x} = \mathbf{0}$ has only one solution, namely, the zero solution. This means that the columns of A are linearly independent, by a remark in Section 1.7.

23. Suppose A is $n \times n$ and the equation $A\mathbf{x} = \mathbf{0}$ has only the trivial solution. Then there are no free variables in this equation, and so A has n pivot columns. Since A is *square* and the n pivot positions must be in different rows, the pivots in an echelon form of A must be on the main diagonal. Hence A is row equivalent to the $n \times n$ identity matrix.

25. Suppose $A = \begin{bmatrix} a & b \\ c & d \end{bmatrix}$ and $ad - bc = 0$. If $a = b = 0$, then examine $\begin{bmatrix} 0 & 0 \\ c & d \end{bmatrix}\begin{bmatrix} x_1 \\ x_2 \end{bmatrix} = \begin{bmatrix} 0 \\ 0 \end{bmatrix}$. This

has the solution $\mathbf{x} = \begin{bmatrix} d \\ -c \end{bmatrix}$. This solution is nonzero, except when $c = d = 0$. In that case,

however, A is the zero matrix, and $A\mathbf{x} = \mathbf{0}$ for *every* vector $\mathbf{x}$. Finally, if a and b are not both

zero, set $\mathbf{u} = \begin{bmatrix} -b \\ a \end{bmatrix}$. Then $A\mathbf{u} = \begin{bmatrix} a & b \\ c & d \end{bmatrix}\begin{bmatrix} -b \\ a \end{bmatrix} = \begin{bmatrix} -ab + ba \\ -cb + da \end{bmatrix} = \begin{bmatrix} 0 \\ 0 \end{bmatrix}$, because $-cb + da = 0$.

Thus, $\mathbf{u}$ is a nontrivial solution of $A\mathbf{x} = \mathbf{0}$. So, in all cases, the equation $A\mathbf{x} = \mathbf{0}$ has more than one solution. This is impossible when A is invertible (by Theorem 5), so A is *not* invertible.

27. a. Interchange A and B in equation (1) after Example 6 of Section 2.1: $\text{row}_i(BA) = \text{row}_i(B) \cdot A$. Then replace B by the identity matrix: $\text{row}_i(A) = \text{row}_i(IA) = \text{row}_i(I) \cdot A$.

b. Using part (a), when rows 1 and 2 of A are interchanged, write the result as

$$\begin{bmatrix} \text{row}_2(A) \\ \text{row}_1(A) \\ \text{row}_3(A) \end{bmatrix} = \begin{bmatrix} \text{row}_2(I) \cdot A \\ \text{row}_1(I) \cdot A \\ \text{row}_3(I) \cdot A \end{bmatrix} = \begin{bmatrix} \text{row}_2(I) \\ \text{row}_1(I) \\ \text{row}_3(I) \end{bmatrix} A = EA \qquad (*)$$

Here, E is obtained by interchanging rows 1 and 2 of I. The second equality in (*) is a consequence of the fact that $\text{row}_i(EA) = \text{row}_i(E) \cdot A$.

c. Using part (a), when row 3 of A is multiplied by 5, write the result as

$$\begin{bmatrix} \text{row}_1(A) \\ \text{row}_2(A) \\ 5 \cdot \text{row}_3(A) \end{bmatrix} = \begin{bmatrix} \text{row}_1(I) \cdot A \\ \text{row}_2(I) \cdot A \\ 5 \cdot \text{row}_3(I) \cdot A \end{bmatrix} = \begin{bmatrix} \text{row}_1(I) \\ \text{row}_2(I) \\ 5 \cdot \text{row}_3(I) \end{bmatrix} A = EA$$

Here, E is obtained by multiplying row 3 of I by 5.

31. $\begin{bmatrix} A & I \end{bmatrix} = \begin{bmatrix} 1 & 0 & -2 & 1 & 0 & 0 \\ -3 & 1 & 4 & 0 & 1 & 0 \\ 2 & -3 & 4 & 0 & 0 & 1 \end{bmatrix} \sim \begin{bmatrix} 1 & 0 & -2 & 1 & 0 & 0 \\ 0 & 1 & -2 & 3 & 1 & 0 \\ 0 & -3 & 8 & -2 & 0 & 1 \end{bmatrix}$

$\sim \begin{bmatrix} 1 & 0 & -2 & 1 & 0 & 0 \\ 0 & 1 & -2 & 3 & 1 & 0 \\ 0 & 0 & 2 & 7 & 3 & 1 \end{bmatrix} \sim \begin{bmatrix} 1 & 0 & 0 & 8 & 3 & 1 \\ 0 & 1 & 0 & 10 & 4 & 1 \\ 0 & 0 & 2 & 7 & 3 & 1 \end{bmatrix}$

$\sim \begin{bmatrix} 1 & 0 & 0 & 8 & 3 & 1 \\ 0 & 1 & 0 & 10 & 4 & 1 \\ 0 & 0 & 1 & 7/2 & 3/2 & 1/2 \end{bmatrix}. \quad A^{-1} = \begin{bmatrix} 8 & 3 & 1 \\ 10 & 4 & 1 \\ 7/2 & 3/2 & 1/2 \end{bmatrix}$

33. Let $B = \begin{bmatrix} 1 & 0 & 0 & \cdots & 0 \\ -1 & 1 & 0 & & 0 \\ 0 & -1 & 1 & & \\ \vdots & & \ddots & \ddots & \vdots \\ 0 & 0 & \cdots & -1 & 1 \end{bmatrix}$, and for $j = 1, \ldots, n$, let $\mathbf{a}_j$, $\mathbf{b}_j$, and $\mathbf{e}_j$ denote the jth columns

of A, B, and I, respectively. Note that for $j = 1, \ldots, n-1$, $\mathbf{a}_j - \mathbf{a}_{j+1} = \mathbf{e}_j$ (because $\mathbf{a}_j$ and $\mathbf{a}_{j+1}$ have the same entries except for the jth row), $\mathbf{b}_j = \mathbf{e}_j - \mathbf{e}_{j+1}$ and $\mathbf{a}_n = \mathbf{b}_n = \mathbf{e}_n$.

To show that $AB = I$, it suffices to show that $A\mathbf{b}_j = \mathbf{e}_j$ for each j. For $j = 1, \ldots, n-1$,

$$A\mathbf{b}_j = A(\mathbf{e}_j - \mathbf{e}_{j+1}) = A\mathbf{e}_j - A\mathbf{e}_{j+1} = \mathbf{a}_j - \mathbf{a}_{j+1} = \mathbf{e}_j$$

and $A\mathbf{b}_n = A\mathbf{e}_n = \mathbf{a}_n = \mathbf{e}_n$. Next, observe that $\mathbf{a}_j = \mathbf{e}_j + \cdots + \mathbf{e}_n$ for each j. Thus,

$$B\mathbf{a}_j = B(\mathbf{e}_j + \cdots + \mathbf{e}_n) = \mathbf{b}_j + \cdots + \mathbf{b}_n$$
$$= (\mathbf{e}_j - \mathbf{e}_{j+1}) + (\mathbf{e}_{j+1} - \mathbf{e}_{j+2}) + \cdots + (\mathbf{e}_{n-1} - \mathbf{e}_n) + \mathbf{e}_n = \mathbf{e}_j$$

This proves that $BA = I$. Combined with the first part, this proves that $B = A^{-1}$.

37. There are many possibilities for C, but $C = \begin{bmatrix} 1 & 1 & -1 \\ -1 & 1 & 0 \end{bmatrix}$ is the only one whose entries are

1, −1, and 0. With only three possibilities for each entry, the construction of C can be done by trial and error. This is probably faster than setting up a system of 4 equations in 6 unknowns. The fact that A cannot be invertible follows from Exercise 25 in Section 2.1, because A is not square.

MATLAB   Constructing A^{-1}

If A is a 5×5 matrix, then the command **M = [A eye(5)]** creates the augmented matrix $[A\ \ I]$. Use **gauss, swap, bgauss,** and **scale** to reduce $[A\ \ I]$. See page 1-17.

   MATLAB has other commands that row reduce matrices, invert matrices, and solve equations $A\mathbf{x} = \mathbf{b}$. They will be introduced later, after you have studied the concepts and algorithms in this section.

2.3   CHARACTERIZATION OF INVERTIBLE MATRICES

In many linear algebra texts, the equivalent of Chapter 4 is extremely difficult for students. But you won't have problems if you prepare well now, because you are already learning basic ideas that will be presented again in Chapter 4. Review the major concepts from the previous sections, and plan for more study time here than you ordinarily spend on one section.

KEY IDEAS

The Invertible Matrix Theorem (IMT) only applies to square matrices. However, some groups of these statements in the IMT are also equivalent for rectangular matrices. The following table will help you remember other important theorems as well as the IMT. (See Theorem 4 in Section 1.4, Theorems 11 and 12 in Section 1.9, Theorem 5 in Section 2.2, and Theorem 9 in Section 2.3.) All of the statements in the table are equivalent when A is square ($m = n = p$).

STATEMENTS FROM THE INVERTIBLE MATRIX THEOREM

Equivalent statements for an $m \times n$ matrix A.	Equivalent statements for an $n \times n$ square matrix A.	Equivalent statements for an $n \times p$ matrix A.
k. There is a matrix D such that $AD = I$.	a. A is an invertible matrix.	j. There is a matrix C such that $CA = I$.
*. A has a pivot position in every row.	c. A has n pivot positions.	*. A has a pivot position in every column.
h. The columns of A span $\mathbb{R}^m$.	b. A is row equivalent to the identity matrix.	e. The columns of A are linearly independent.
g. The equation $Ax = b$ has at least one solution for each b in $\mathbb{R}^m$.	*. The equation $Ax = b$ has a unique solution for each b in $\mathbb{R}^n$.	d. The equation $Ax = 0$ has only the trivial solution.
i. The transformation $x \mapsto Ax$ maps $\mathbb{R}^n$ onto $\mathbb{R}^m$.	*. The transformation $x \mapsto Ax$ is invertible. l. A^T is invertible.	f. The transformation $x \mapsto Ax$ is one-to-one.

The four statements denoted by (*) were not listed in the text as part of the IMT, mainly to avoid intimidating you with so many statements in one theorem. Note: the text did not actually prove that for a *rectangular* matrix, statements (j) and (k) are each equivalent to the other statements in their respective columns. (Exercises 23, 24, and 26 in Section 2.1 contain most of the facts needed to prove this.) A matrix C such that $CA = I$ is called a **left-inverse** of A, and a matrix D such that $AD = I$ is called a **right-inverse** of A.

Checkpoint: What can you say about the statements in the first column when A has more rows than columns? (Why?) What about the statements in the third column when A has more columns than rows? (Why?)

A question such as the one in the box below is one way I test whether my students know the IMT. Test yourself. Cover up the IMT, write your answers, and then check your work. The answers are given at the end of this section.

Test Question:

Let A be an $n \times n$ matrix. Write 6 statements from the Invertible Matrix Theorem, each equivalent to the statement that A is invertible. Use the following concepts, one in each statement: (*i*) row equivalent, (*ii*) the equation $AD = I$, (*iii*) columns, (*iv*) the equation $Ax = 0$, and (*v*) linear transformation.

SOLUTIONS TO EXERCISES

1. The columns of the matrix $\begin{bmatrix} 5 & 7 \\ -3 & -6 \end{bmatrix}$ are not multiples, so they are linearly independent. By (e) in the IMT, the matrix is invertible. Also, the matrix is invertible by Theorem 4 in Section 2.2 because the determinant is nonzero.

7. $\begin{bmatrix} -1 & -3 & 0 & 1 \\ 3 & 5 & 8 & -3 \\ -2 & -6 & 3 & 2 \\ 0 & -1 & 2 & 1 \end{bmatrix} \sim \begin{bmatrix} -1 & -3 & 0 & 1 \\ 0 & -4 & 8 & 0 \\ 0 & 0 & 3 & 0 \\ 0 & -1 & 2 & 1 \end{bmatrix} \sim \begin{bmatrix} -1 & -3 & 0 & 1 \\ 0 & -4 & 8 & 0 \\ 0 & 0 & 3 & 0 \\ 0 & 0 & 0 & 1 \end{bmatrix}$

The 4×4 matrix has four pivot positions and so is invertible by (c) of the IMT.

11. Study the Invertible Matrix Theorem. The statements there are true *only for an invertible matrix*. Also, if one of the statements is true about a *square* matrix A, then all statements in the theorem are true; if one of the statements is false, then all are false.

a. See statements (d) and (b) of the IMT.

b. See statements (h) and (e).   **c.** See statement (g).

d. See statements (d) and (c).   **e.** See statement (1).

Study Tip: Learn how to recognize when a square matrix is *not* invertible. If A is an $n \times n$ matrix, then each of the following statements is true if and only if A is **not** invertible.

- The matrix A has *fewer* than n pivot positions.
- The equation $Ax = 0$ has a *nontrivial* (nonzero) solution.
- The columns of A are linearly *dependent*.
- The linear transformation $x \mapsto Ax$ is *not* one-to-one.
- The equation $Ax = b$ has *no* solution (is *inconsistent*) for *some* b in $\mathbb{R}^n$.
- The equation $Ax = b$ has *more than one* solution for some b in $\mathbb{R}^n$.
- The columns of A *do not* span $\mathbb{R}^n$.
- The linear transformation $x \mapsto Ax$ *does not* map $\mathbb{R}^n$ onto $\mathbb{R}^n$.

13. If a square upper triangular $n \times n$ matrix has nonzero diagonal entries, then because it is already in echelon form, the matrix is row equivalent to I_n and hence is invertible, by the IMT. Conversely, if the matrix is invertible, it has n pivots on the diagonal and hence the diagonal entries are nonzero.

Study Tip: If you check your answer for odd exercises between 13 and 33, be careful not to read any other answers or hints. You *must* try to write your own solutions first.

19. By (e) of the IMT, D is invertible. Thus the equation $D\mathbf{x} = \mathbf{b}$ has a solution for each $\mathbf{b}$ in $\mathbb{R}^7$, by (g) of the IMT. Even better, the equation $D\mathbf{x} = \mathbf{b}$ has a *unique* solution for each $\mathbf{b}$ in $\mathbb{R}^7$, by Theorem 5 in Section 2.2. (See the paragraph following the proof of the IMT.)

25. Suppose that A is square and $AB = I$. Then A is invertible, by the (k) of the IMT. Left-multiplying each side of the equation $AB = I$ by A^{-1}, one has

$$A^{-1}AB = A^{-1}I, \quad IB = A^{-1}, \quad \text{and} \quad B = A^{-1}.$$

By Theorem 6 in Section 2.2, the matrix B (which is A^{-1}) is invertible, and its inverse is $(A^{-1})^{-1} = A$. Note: Exercise 25 makes a good test question.

27. Let W be the inverse of AB. Then $ABW = I$ and $A(BW) = I$. This equation, *by itself*, does not prove that A is invertible. However, since A is *square*, the IMT does apply and by statement (k), A is invertible.

   Of course, in this exercise set there is an overall assumption that matrices in this section are square unless otherwise stated. So, with that given, you do not really have to mention here that A is square. However, I put that question "Why not?" in the answer to make you think about this. Look back at Exercise 38 in Section 2.2. There, $AD = I$, which certainly makes AD invertible, yet A is not invertible.

31. Since the equation $A\mathbf{x} = \mathbf{b}$ has a solution for each $\mathbf{b}$, the matrix A has a pivot in each row (Theorem 4 in Section 1.4). Since A is square, A has a pivot in each column, and so there are no free variables in the equation $A\mathbf{x} = \mathbf{b}$, which shows that the solution is unique.

   The preceding argument shows that the (square) shape of A plays a crucial role. A less revealing proof is to use the "pivot in each row" and the IMT to conclude that A is invertible. Then Theorem 5 in Section 2.2 shows that the solution of $A\mathbf{x} = \mathbf{b}$ is unique.

33. The standard matrix of T is $A = \begin{bmatrix} -5 & 9 \\ 4 & -7 \end{bmatrix}$, which is invertible because $\det A \neq 0$. By Theorem 9, the transformation T is invertible and the standard matrix of T^{-1} is A^{-1}. From the formula for a 2×2 inverse, $A^{-1} = \begin{bmatrix} 7 & 9 \\ 4 & 5 \end{bmatrix}$. So

$$T^{-1}(x_1, x_2) = \begin{bmatrix} 7 & 9 \\ 4 & 5 \end{bmatrix}\begin{bmatrix} x_1 \\ x_2 \end{bmatrix} = (7x_1 + 9x_2, 4x_1 + 5x_2)$$

35. To show that T is one-to-one, suppose that $T(\mathbf{u}) = T(\mathbf{v})$ for some vectors $\mathbf{u}$ and $\mathbf{v}$ in $\mathbb{R}^n$. Then $S(T(\mathbf{u})) = S(T(\mathbf{v}))$, where S is the inverse of T. By Equation (1), $\mathbf{u} = S(T(\mathbf{u}))$ and $S(T(\mathbf{v})) = \mathbf{v}$, so $\mathbf{u} = \mathbf{v}$. Thus T is one-to-one. To show that T is onto, suppose $\mathbf{y}$ represents an arbitrary vector in $\mathbb{R}^n$ and define $\mathbf{x} = S(\mathbf{y})$. Then, using Equation (2), $T(\mathbf{x}) = T(S(\mathbf{y})) = \mathbf{y}$, which shows that T maps $\mathbb{R}^n$ onto $\mathbb{R}^n$.

   Second proof: By Theorem 9, the standard matrix A of T is invertible. By the IMT, the columns of A are linearly independent and span $\mathbb{R}^n$. By Theorem 12 in Section 1.9, T is one-to-one and maps $\mathbb{R}^n$ onto $\mathbb{R}^n$.

37. Let A and B be the standard matrices of T and U, respectively. Then AB is the standard matrix of the mapping $\mathbf{x} \mapsto T(U(\mathbf{x}))$, because of the way matrix multiplication is defined (in Section 2.1). By hypothesis, this mapping is the identity mapping, so $AB = I$. Since A and B are square, they are invertible, by the IMT, and $B = A^{-1}$. Thus, $BA = I$. This means that the mapping $\mathbf{x} \mapsto U(T(\mathbf{x}))$ is the identity mapping, i.e., $U(T(\mathbf{x})) = \mathbf{x}$ for all $\mathbf{x}$ in $\mathbb{R}^n$.

Answers to Checkpoint: If A has more rows than columns, then all statements in the first column of the table must be false, because they are equivalent and the statement about a pivot position in each row cannot be true. If A has more columns than rows, then all statements in the third column of the table must be false, because A cannot have a pivot in each of its columns.

Answers to Test Question: (*i*) A is row equivalent to I_n. (*ii*) There exists an $n \times n$ matrix D such that $AD = I$. (*iii*) The columns of A span $\mathbb{R}^n$. (*iv*) The equation $A\mathbf{x} = \mathbf{0}$ has only the trivial solution. (*v*) The linear transformation $\mathbf{x} \mapsto A\mathbf{x}$ maps $\mathbb{R}^n$ onto $\mathbb{R}^n$.

Another answer for (*iii*) is: The columns of A are linearly independent. Similarly, (*v*) has another answer. But the following statement is unacceptable as one of the answers to the test question:

A is invertible if and only if the columns of A span $\mathbb{R}^n$.

This statement is itself a (true) theorem (assuming A is square), not a statement that is true precisely when A is invertible.

Mastering Linear Algebra: Reviewing and Reflecting

Two important steps to mastery of linear algebra are periodic review of earlier material and reflection on its relation to new material. When you reread the basic conceptual material from Chapter 1, you may be surprised to discover new insights that you missed earlier. Your broader experience now should give you a better framework within which to understand concepts such as spanning and linear independence.

Compare the review you conducted in Section 1.9 (see the *Study Guide* appendix to that section) with the three-part table at the beginning of this *Study Guide* section. (You did carry out that review, didn't you?) The left and right columns of the table should match some of your "existence" and "uniqueness" statements, respectively.

If your review in Section 1.9 was thorough, you probably anticipated some of the content of the Invertible Matrix Theorem. Existence and uniqueness threads run through the fabric of linear algebra, and they intertwine when related to square matrices (the middle column of the table). A good review procedure now is to expand the table to include references to theorems, examples, and counterexamples. This will occupy several pages. The process of constructing this table is what will help you most.

MATLAB `inv, cond, and hilb`

Determining whether a matrix is invertible is not always a simple matter. A fast and fairly reliable method is to use the command `inv(A)`, which computes the inverse of A. A warning is given if the matrix is singular (noninvertible) or nearly singular.

For Exercises 41–44, the command `cond(A)` computes the condition number of a matrix A, using what are called the singular values of A (discussed in Section 7.4.) To perform the experiment described in Exercise 42, you can use the following MATLAB instructions

$$\texttt{x=rand(4,1);   b=A*x;   x1=inv(A)*b;   x-x1}$$

Use `format long`. Displaying the value of **x–x1** is the best way to compare **x** and **x1**. Press the up-arrow key ($\uparrow$) to repeat this instruction line.

For Exercise 45, the commands `format rat; hilb(n)` produce the $n \times n$ Hilbert matrix, with its entries displayed as rational numbers. Enter `format short` to return to the standard display of numbers as decimals.

2.4   PARTITIONED MATRICES

The ideas in this section are fairly simple. However, mark them for future reference, because you are likely to use this notation after you leave school. Partitioned matrices arise in theoretical discussions in essentially every field that makes use of matrices. Here are two examples.

1. The modern *state space* approach to control systems engineering depends on matrix calculations.[1] The problem of determining whether a system is *controllable* amounts to calculating the number of pivot positions in a *controllability matrix*

$$[B \quad AB \quad A^2B \quad \cdots \quad A^{n-1}B]$$

where A is $n \times n$, B has n rows, and the matrices come from an equation of the form (8) in the discussion preceding Exercise 19.

---

[1] An understanding of control systems is important in the design of filtering circuitry, robots, process control systems, and spacecraft. Thus a control systems course is often part of the undergraduate curriculum for electrical, mechanical, chemical, and aerospace engineering. See *Control Systems Engineering*, 3rd ed., by Norman S. Nise, John Wiley & Sons, New York, 2000.

2. Discussions of modern algorithms and computer software design for scientific computing naturally use the "language" of partitioned matrices. For instance, common techniques for parallel processing of large matrix calculations, such as *slicing* and *crinkling*, are described with partitioned matrices.[2] Also, the standard computer science reference on matrix calculations relies heavily on partitioned matrices.[3]

KEY IDEAS

The column-row evaluation of AB is the last of five different "views" of matrix multiplication. All five are special cases of the *block matrix* version of the *row-column* rule for matrix multiplication. Here they are:

(1) The definition of $A\mathbf{x}$ amounts to block multiplication of AB where B has only one column:

$$A\mathbf{x} = [\mathbf{a}_1 | \cdots | \mathbf{a}_n] \begin{bmatrix} x_1 \\ \vdots \\ x_n \end{bmatrix} = [x_1\mathbf{a}_1 + \cdots + x_n\mathbf{a}_n]$$

(2) Partition A as *one* row and *one* column. Then the definition of the usual product AB is a row-column block product:

$$AB = A[\mathbf{b}_1 | \mathbf{b}_2 | \cdots | \mathbf{b}_p] = [A\mathbf{b}_1 | A\mathbf{b}_2 | \cdots | A\mathbf{b}_p]$$

(3) Likewise, we observed in Section 2.1 that if B is partitioned as one row and one column, then

$$AB = \begin{bmatrix} \text{row}_1(A) \\ \text{row}_2(A) \\ \vdots \\ \text{row}_m(A) \end{bmatrix} B = \begin{bmatrix} \text{row}_1(A)B \\ \text{row}_2(A)B \\ \vdots \\ \text{row}_m(A)B \end{bmatrix}$$

(4) The next display can be viewed either as just the standard row-column rule in which each entry of AB is computed as the product of a row of A and a column of B, or as a multiplication of block matrices (with A having only one column of blocks and B having only one row of blocks):

---

[2] *Parallel Algorithms and Matrix Computations*, by Jagdish J. Modi, Oxford Applied Mathematics and Computing Science Series, Clarendon Press, Oxford, 1988, pp. 73–75.

[3] *Matrix Computations*, 3rd ed., by Gene H. Golub and Charles F. Van Loan, The Johns Hopkins Press, Baltimore, 1996.

$$AB = \begin{bmatrix} \text{row}_1(A) \\ \text{row}_2(A) \\ \vdots \\ \text{row}_m(A) \end{bmatrix} [\text{col}_1(B) \quad \text{col}_2(B) \quad \cdots \quad \text{col}_p(B)]$$

$$= \begin{bmatrix} \text{row}_1(A)\text{col}_1(B) & \cdots & \text{row}_1(A)\text{col}_j(B) & \cdots & \text{row}_1(A)\text{col}_p(B) \\ \vdots & & \vdots & & \vdots \\ \text{row}_i(A)\text{col}_1(B) & \cdots & \text{row}_i(A)\text{col}_j(B) & \cdots & \text{row}_i(A)\text{col}_p(B) \\ \vdots & & \vdots & & \vdots \\ \text{row}_m(A)\text{col}_1(B) & \cdots & \text{row}_m(A)\text{col}_j(B) & \cdots & \text{row}_m(A)\text{col}_p(B) \end{bmatrix}$$

(5) The final display is the column-row expansion of AB (Theorem 10 in this section). In this view, AB is expressed as a sum of *outer products* of the form $\mathbf{uv}^T$, with $\mathbf{u}$ a column of A and $\mathbf{v}^T$ a row of B. But the display can also be viewed as the block version of the row-column product in which A has one row (of blocks) and B has one column (of blocks):

$$AB = [\text{col}_1(A) \quad \text{col}_2(A) \quad \cdots \quad \text{col}_n(A)] \begin{bmatrix} \text{row}_1(B) \\ \text{row}_2(B) \\ \vdots \\ \text{row}_n(B) \end{bmatrix}$$

$$= \text{col}_1(A) \cdot \text{row}_1(B) + \cdots + \text{col}_n(A) \cdot \text{row}_n(B)$$

You might say that the row-column rule computes AB as an array of inner products (view 4 above), while the column-row expansion displays AB as a sum of arrays (view 5).

SOLUTIONS TO EXERCISES

1. Apply the row-column rule as if the matrix entries were numbers, but for each product (such as EA below), always write the entry of the left block-matrix on the *left*.

$$\begin{bmatrix} I & 0 \\ E & I \end{bmatrix}\begin{bmatrix} A & B \\ C & D \end{bmatrix} = \begin{bmatrix} IA+0C & IB+0D \\ EA+IC & EB+ID \end{bmatrix} = \begin{bmatrix} A & B \\ EA+C & EB+D \end{bmatrix}$$

This must be EA, not AE.

Checkpoint: Notice in Exercises 1 and 3 that $\begin{bmatrix} I & 0 \\ E & I \end{bmatrix}$ and $\begin{bmatrix} 0 & I \\ I & 0 \end{bmatrix}$ act as block-matrix generalizations of elementary matrices. What sort of 2×2 block matrix is the appropriate generalization of an elementary matrix that acts as a scaling operation? (Answer this carefully.)

7. Compute the left side of the equation:

$$\begin{bmatrix} X & 0 & 0 \\ Y & 0 & I \end{bmatrix} \begin{bmatrix} A & Z \\ 0 & 0 \\ B & I \end{bmatrix} = \begin{bmatrix} XA+0+0B & XZ+0+0 \\ YA+0+IB & YZ+0+I \end{bmatrix}$$

Set this equal to the right side of the equation:

$$\begin{bmatrix} XA & XZ \\ YA+B & YZ+I \end{bmatrix} = \begin{bmatrix} I & 0 \\ 0 & I \end{bmatrix} \quad \text{so that} \quad \begin{array}{ll} XA = I & XZ = 0 \\ YA+B=0 & YZ+I=I \end{array}$$

Since the (1, 1)-blocks are equal, $XA = I$. Since X and A are square, the IMT implies that A and X are invertible, and hence $X = A^{-1}$. From the (1, 2)-entries, $XZ = 0$. Since X is invertible, Z must be 0. Therefore, the (2, 2)-entries give no new information. Finally, from the (2, 1)-entries, $YA + B = 0$ and $YA = -B$. Right-multiplication by A^{-1} shows that $Y = -BA^{-1}$. The order of the factors for Y is crucial.

Study Tip: Problems such as 5–10 make good exam questions. Remember to mention the IMT when appropriate, and remember that matrix multiplication is generally not commutative.

11. a. See the subsection Addition and Scalar Multiplication.

b. See the paragraph before Example 3.

13. You are asked to establish an "if and only if" statement. First, suppose that A is invertible, and let $A^{-1} = \begin{bmatrix} D & E \\ F & G \end{bmatrix}$. Then

$$\begin{bmatrix} B & 0 \\ 0 & C \end{bmatrix} \begin{bmatrix} D & E \\ F & G \end{bmatrix} = \begin{bmatrix} BD & BE \\ CF & CG \end{bmatrix} = \begin{bmatrix} I & 0 \\ 0 & I \end{bmatrix}$$

Since B is square, the equation $BD = I$ implies that B is invertible, by the IMT. Similarly, $CG = I$ implies that C is invertible. Also, the equation $BE = 0$ implies that $E = B^{-1}0 = 0$. Similarly $F = 0$. Thus

$$A^{-1} = \begin{bmatrix} B & 0 \\ 0 & C \end{bmatrix}^{-1} = \begin{bmatrix} D & E \\ E & G \end{bmatrix} = \begin{bmatrix} B^{-1} & 0 \\ 0 & C^{-1} \end{bmatrix} \qquad (*)$$

This proves that A is invertible *only if* B and C are invertible. For the "*if*" part of the statement, suppose that B and C are invertible. Then (*) provides a likely candidate for A^{-1} which can be used to show that A is invertible. Compute:

$$\begin{bmatrix} B & 0 \\ 0 & C \end{bmatrix}\begin{bmatrix} B^{-1} & 0 \\ 0 & C^{-1} \end{bmatrix} = \begin{bmatrix} BB^{-1} & 0 \\ 0 & CC^{-1} \end{bmatrix} = \begin{bmatrix} I & 0 \\ 0 & I \end{bmatrix}$$

Since A is square, this calculation and the IMT imply that A is invertible. (Don't forget this final sentence. Without it, the argument is incomplete.) Instead of that sentence, you could add the equation:

$$\begin{bmatrix} B^{-1} & 0 \\ 0 & C^{-1} \end{bmatrix}\begin{bmatrix} B & 0 \\ 0 & C \end{bmatrix} = \begin{bmatrix} B^{-1}B & 0 \\ 0 & C^{-1}C \end{bmatrix} = \begin{bmatrix} I & 0 \\ 0 & I \end{bmatrix}$$

19. The matrix equation (8) in the text is equivalent to

$$(A - sI_n)\mathbf{x} + B\mathbf{u} = 0 \quad \text{and} \quad C\mathbf{x} + \mathbf{u} = \mathbf{y}$$

Rewrite the first equation as $(A - sI_n)\mathbf{x} = -B\mathbf{u}$. When $A - sI_n$ is invertible,

$$\mathbf{x} = (A - sI_n)^{-1}(-B\mathbf{u}) = -(A - sI_n)^{-1}B\mathbf{u}$$

Substitute this formula for $\mathbf{x}$ into the second equation above:

$$C(-(A - sI_n)^{-1}B\mathbf{u}) + \mathbf{u} = \mathbf{y}, \quad \text{so that} \quad I_m\mathbf{u} - C(A - sI_n)^{-1}B\mathbf{u} = \mathbf{y}$$

Thus $\mathbf{y} = (I_m - C(A - sI_n)^{-1}B)\mathbf{u}$. If $W(s) = I_m - C(A - sI_n)^{-1}B$, then $\mathbf{y} = W(s)\mathbf{u}$. The matrix $W(s)$ is the Schur complement of the matrix $A - sI_n$ in the system matrix in equation (8).

23. To prove a statement by induction, a good first step is to write the statement that depends on n but exclude the phrase "for all n," and label the statement for reference:

> *The product of two $n \times n$ lower triangular matrices is lower triangular.* (*)

Second, verify that the statement is true for $n = 1$. In this particular case, (*) is obviously true, because every 1×1 matrix is lower triangular. The "induction step" is next.

Suppose that (*) is true when n is some positive integer k, and consider any $(k+1) \times (k+1)$ lower-triangular matrices A_1 and B_1. Partition these matrices as

$$A_1 = \begin{bmatrix} a & \mathbf{0}^T \\ \mathbf{v} & A \end{bmatrix}, \quad B_1 = \begin{bmatrix} b & \mathbf{0}^T \\ \mathbf{w} & B \end{bmatrix}$$

where A and B are $k \times k$ matrices, $\mathbf{v}$ and $\mathbf{w}$ are in $\mathbb{R}^k$, and a and b are scalars. Since A_1 and B_1 are lower triangular, so are A and B. Now

$$A_1B_1 = \begin{bmatrix} a & \mathbf{0}^T \\ \mathbf{v} & A \end{bmatrix}\begin{bmatrix} b & \mathbf{0}^T \\ \mathbf{w} & B \end{bmatrix} = \begin{bmatrix} ab + \mathbf{0}^T\mathbf{w} & a\mathbf{0}^T + \mathbf{0}^TB \\ \mathbf{v}b + A\mathbf{w} & \mathbf{v}\mathbf{0}^T + AB \end{bmatrix} = \begin{bmatrix} ab & \mathbf{0}^T \\ b\mathbf{v} + A\mathbf{w} & AB \end{bmatrix}$$

Assuming (*) is true for $n = k$, AB must be lower triangular. The form of A_1B_1 shows that it, too, is lower triangular. Thus the statement (*) about lower triangular matrices is true for $n = k + 1$ if it is true for $n = k$. By the principle of induction, (*) is true for all $n \geq 1$.

25. First, visualize a partition of A as a 2×2 block-diagonal matrix, as below, and then visualize the (2,2) block-entry A_{22} itself as a block-diagonal matrix. That is,

$$A = \begin{bmatrix} 1 & 2 & 0 & 0 & 0 \\ 3 & 5 & 0 & 0 & 0 \\ 0 & 0 & 2 & 0 & 0 \\ 0 & 0 & 0 & 7 & 8 \\ 0 & 0 & 0 & 5 & 6 \end{bmatrix} = \begin{bmatrix} A_{11} & 0 \\ 0 & A_{22} \end{bmatrix}, \quad \text{where} \quad A_{22} = \begin{bmatrix} 2 & 0 & 0 \\ 0 & 7 & 8 \\ 0 & 5 & 6 \end{bmatrix} = \begin{bmatrix} 2 & 0 \\ 0 & B \end{bmatrix}$$

Observe that B is invertible and $B^{-1} = \dfrac{1}{2}\begin{bmatrix} 6 & -8 \\ -5 & 7 \end{bmatrix} = \begin{bmatrix} 3 & -4 \\ -2.5 & 3.5 \end{bmatrix}$. By Exercise 13, the block

diagonal matrix A_{22} is invertible, and

$$A_{22}^{-1} = \begin{bmatrix} .5 & 0 \\ 0 & \begin{matrix} 3 & -4 \\ -2.5 & 3.5 \end{matrix} \end{bmatrix} = \begin{bmatrix} .5 & 0 & 0 \\ 0 & 3 & -4 \\ 0 & -2.5 & 3.5 \end{bmatrix}$$

Next, observe that A_{11} is also invertible, with inverse $\begin{bmatrix} -5 & 2 \\ 3 & -1 \end{bmatrix}$. By Exercise 13, A itself is

invertible, and its inverse is block diagonal:

$$A^{-1} = \begin{bmatrix} A_{11}^{-1} & 0 \\ 0 & A_{22}^{-1} \end{bmatrix} = \begin{bmatrix} \begin{matrix} -5 & 2 \\ 3 & -1 \end{matrix} & 0 \\ 0 & \begin{matrix} .5 & 0 & 0 \\ 0 & 3 & -4 \\ 0 & -2.5 & 3.5 \end{matrix} \end{bmatrix} = \begin{bmatrix} -5 & 2 & 0 & 0 & 0 \\ 3 & -1 & 0 & 0 & 0 \\ 0 & 0 & .5 & 0 & 0 \\ 0 & 0 & 0 & 3 & -4 \\ 0 & 0 & 0 & -2.5 & 3.5 \end{bmatrix}$$

A somewhat less detailed solution would be to write (without formal proof) that the result of Exercise 13 seems to generalize to any block-diagonal matrix. Such a matrix A is invertible if and only if each of the diagonal blocks is invertible, and the inverse of A is the block-diagonal matrix formed from the inverses of the diagonal blocks. View the 5×5 matrix in this exercise as a 3×3 block matrix:

$$A = \begin{bmatrix} 1 & 2 & 0 & 0 & 0 \\ 3 & 5 & 0 & 0 & 0 \\ 0 & 0 & 2 & 0 & 0 \\ 0 & 0 & 0 & 7 & 8 \\ 0 & 0 & 0 & 5 & 6 \end{bmatrix} = \begin{bmatrix} A_{11} & 0 & 0 \\ 0 & A_{22} & 0 \\ 0 & 0 & A_{33} \end{bmatrix}$$

Finish by inverting each of the diagonal blocks and use the results to assemble A^{-1}, as above.

Answer to Checkpoint: The block diagonal matrices $\begin{bmatrix} E & 0 \\ 0 & I \end{bmatrix}$ and $\begin{bmatrix} I & 0 \\ 0 & E \end{bmatrix}$ are obvious choices.

Less obvious is the requirement that E be invertible, in order to make these block matrices invertible. (Recall that the invertibility of elementary matrices was essential for the theory in Section 2.2.)

Appendix: The Principle of Induction

Consider a statement "(*)" that depends on a positive integer n, as in Exercise 23. To prove "by induction" that (*) is true for all positive integers, you must prove two things:

   (a) Statement (*) is true for $n = 1$.

   (b) (The induction step) If (*) is true for any positive integer $n = k$, then (*) is also true for the next integer $n = k + 1$.

A property or axiom of the real number system, called the *principle of mathematical induction*, says that if (a) and (b) are true, then (*) is true for all integers $n \geq 1$. This is reasonable, because if (*) is true for $n = 1$, then (b) shows that (*) is true for $n = 2$. Applying (b) again with $n = 2$, we see that (*) is true for $n = 3$. Applying (b) repeatedly, we see that (*) is true for $2, 3, 4, 5, \ldots$.

MATLAB   Partitioned Matrices

MATLAB uses partitioned matrix notation. For example, if A, B, C, D, E, and F are matrices of appropriate sizes, then the command

```
M = [A B C; D E F]
```

creates a larger matrix of the form $M = \begin{bmatrix} A & B & C \\ D & E & F \end{bmatrix}$. Once M is formed, there is no record of the partition that was used to create M. For instance, although B was the $(1, 2)$-block used to form M, the number $M(1, 2)$ is the same as the $(1, 2)$-entry of A.

2.5   MATRIX FACTORIZATIONS

In a sense, Section 2.5 is the most up-to-date section in the text, because matrix factorizations lie at the heart of modern uses of matrix algebra. For instance, they are indispensable for the analysis of computational algorithms and research in parallel processing. The text focuses here on triangular factorizations, but the exercises introduce you to other important factorizations that you may encounter later.

KEY IDEAS

When a matrix A is factored as $A = LU$, the data in A are preprocessed in a way that makes the equation $A\mathbf{x} = \mathbf{b}$ easier to solve. Write $LU\mathbf{x} = \mathbf{b}$, or $L(U\mathbf{x}) = \mathbf{b}$, and let $\mathbf{y} = U\mathbf{x}$. Solve $L\mathbf{y} = \mathbf{b}$ for $\mathbf{y}$ and then solve $U\mathbf{x} = \mathbf{y}$ for $\mathbf{x}$. The two-step process is fast when L and U are triangular.

Finding L and U requires the same number of multiplications and divisions as row reducing A to an echelon form U (about $n^3/3$ operations when A is $n \times n$). After that, L and U are available for solving other equations involving A. The key to finding L is to place entries in L in such a way that the sequence of row operations reducing A to U also reduces L to the identity. In this case, LU must equal A. (See the top of page 145.)

The text discusses how to build L when no row interchanges are needed to reduce A to U. In this case, L can be unit lower triangular. An appendix below describes how to build L in permuted unit triangular form when row interchanges are needed (or desired, for numerical reasons).

SOLUTIONS TO EXERCISES

1. $L = \begin{bmatrix} 1 & 0 & 0 \\ -1 & 1 & 0 \\ 2 & -5 & 1 \end{bmatrix}$, $U = \begin{bmatrix} 3 & -7 & -2 \\ 0 & -2 & -1 \\ 0 & 0 & -1 \end{bmatrix}$, $\mathbf{b} = \begin{bmatrix} -7 \\ 5 \\ 2 \end{bmatrix}$. First, solve $L\mathbf{y} = \mathbf{b}$.

$$[L \ \ \mathbf{b}] = \begin{bmatrix} 1 & 0 & 0 & -7 \\ -1 & 1 & 0 & 5 \\ 2 & -5 & 1 & 2 \end{bmatrix} \sim \begin{bmatrix} 1 & 0 & 0 & -7 \\ 0 & 1 & 0 & -2 \\ 0 & -5 & 1 & 16 \end{bmatrix}$$ The only arithmetic is in column 4

$$\sim \begin{bmatrix} 1 & 0 & 0 & -7 \\ 0 & 1 & 0 & -2 \\ 0 & 0 & 1 & 6 \end{bmatrix}, \text{ so } \mathbf{y} = \begin{bmatrix} -7 \\ -2 \\ 6 \end{bmatrix}.$$

Next, solve $U\mathbf{x} = \mathbf{y}$, using back-substitution (with matrix notation).

$$[U \ \ \mathbf{y}] = \begin{bmatrix} 3 & -7 & -2 & -7 \\ 0 & -2 & -1 & -2 \\ 0 & 0 & -1 & 6 \end{bmatrix} \sim \begin{bmatrix} 3 & -7 & -2 & -7 \\ 0 & -2 & -1 & -2 \\ 0 & 0 & 1 & -6 \end{bmatrix} \sim \begin{bmatrix} 3 & -7 & 0 & -19 \\ 0 & -2 & 0 & -8 \\ 0 & 0 & 1 & -6 \end{bmatrix}$$

$$\sim \begin{bmatrix} 3 & -7 & 0 & -19 \\ 0 & 1 & 0 & 4 \\ 0 & 0 & 1 & -6 \end{bmatrix} \sim \begin{bmatrix} 3 & 0 & 0 & 9 \\ 0 & 1 & 0 & 4 \\ 0 & 0 & 1 & -6 \end{bmatrix} \sim \begin{bmatrix} 1 & 0 & 0 & 3 \\ 0 & 1 & 0 & 4 \\ 0 & 0 & 1 & -6 \end{bmatrix}$$

So $\mathbf{x} = (3, 4, -6)$.

Checkpoint: Exercise 12 in Section 2.2 shows how to compute $A^{-1}B$ by row reduction. Describe how you could speed up this calculation if you have an LU factorization of A available (and A is invertible).

7. Place the first pivot column of $\begin{bmatrix} 2 & 5 \\ -3 & -4 \end{bmatrix}$ into L, after dividing the column by 2 (the pivot), then add 3/2 times row 1 to row 2, yielding U.

$$A = \begin{bmatrix} ② & 5 \\ -3 & -4 \end{bmatrix} \sim \begin{bmatrix} 2 & 5 \\ 0 & ⑦/② \end{bmatrix} = U$$

$$\begin{bmatrix} ② \\ -3 \end{bmatrix} \quad \boxed{7/2}$$

$$+2 \quad +7/2$$

$$\begin{bmatrix} 1 \\ -3/2 & 1 \end{bmatrix}, \quad L = \begin{bmatrix} 1 & 0 \\ -3/2 & 1 \end{bmatrix}$$

13. $\begin{bmatrix} ① & 3 & -5 & -3 \\ -1 & -5 & 8 & 4 \\ 4 & 2 & -5 & -7 \\ -2 & -4 & 7 & 5 \end{bmatrix} \sim \begin{bmatrix} 1 & 3 & -5 & -3 \\ 0 & ② & 3 & 1 \\ 0 & -10 & 15 & 5 \\ 0 & 2 & -3 & -1 \end{bmatrix} \sim \begin{bmatrix} 1 & 3 & -5 & -3 \\ 0 & -2 & 3 & 1 \\ 0 & 0 & 0 & 0 \\ 0 & 0 & 0 & 0 \end{bmatrix} = U$ No more pivots!

$$\begin{bmatrix} ① \\ -1 \\ 4 \\ -2 \end{bmatrix} \quad \begin{bmatrix} -② \\ -10 \\ 2 \end{bmatrix}$$

Use the last two columns of I_4 to make L unit lower triangular.

$$+1 \quad +-2$$

$$\begin{bmatrix} 1 & & & \\ -1 & 1 & & \\ 4 & 5 & 1 & \\ -2 & -1 & 0 & 1 \end{bmatrix}, \quad L = \begin{bmatrix} 1 & 0 & 0 & 0 \\ -1 & 1 & 0 & 0 \\ 4 & 5 & 1 & 0 \\ -2 & -1 & 0 & 1 \end{bmatrix}$$

19. A good answer will require a written paragraph or two. If you have not tried to *write* your answer, do so now, *without reading the solution below.* Explain how you would row reduce [A I], knowing that A is lower triangular. Your answer to this question should contain some of the ideas shown below, although your wording might be quite different.

Let A be a lower-triangular $n \times n$ matrix with nonzero entries on the diagonal, and consider the augmented matrix $[A \quad I]$.

a. The $(1, 1)$-entry can be scaled to 1 and the entries below it can be changed to 0 by adding multiples of row 1 to the rows below. This affects only the first column of A and the first column of I. So the $(2, 2)$-entry in the new matrix is still nonzero and now is the only nonzero entry of row 2 in the first n columns (because A was lower triangular).

   The $(2, 2)$-entry can be scaled to 1, and the entries below it can be changed to 0 by adding multiples of row 2 to the rows below. This affects only columns 2 and $n + 2$ of the augmented matrix. Now the $(3, 3)$ entry in A is the only nonzero entry of the third row in the first n columns, so it can be scaled to 1 and then used as a pivot to zero out entries below it. Continuing in this way, A is eventually reduced to I, by scaling each row with a pivot and then using only row operations that add multiples of the pivot row to rows below.

b. The row operations just described only add rows to rows below, so the I on the right in $[A \quad I]$ changes into a lower triangular matrix. By Theorem 7 in Section 2.2, that matrix is A^{-1}.

21. Suppose $A = BC$, with B invertible. Then there exist elementary matrices $E_1, \ldots, E_p$ corresponding to row operations that reduce B to I, in the sense that $E_p \cdots E_1 B = I$. Applying the same sequence of row operations to A amounts to left-multiplying A by the product $E_p \cdots E_1$. By associativity of matrix multiplication,

$$E_p \cdots E_1 A = E_p \cdots E_1 BC = IC = C$$

so the same sequence of row operations reduces A to C.

25. $A = UDV^T$. Since U and V^T are square, the equations $U^T U = I$ and $V^T V = I$ imply that U and V^T are invertible, by the IMT, and hence $U^{-1} = U^T$ and $(V^T)^{-1} = V$. Since the diagonal entries $\sigma_1, \ldots, \sigma_n$ in D are nonzero, D is invertible, with the inverse of D being the diagonal matrix with $\sigma_1^{-1}, \ldots, \sigma_n^{-1}$ on the diagonal. Thus A is a product of invertible matrices. By Theorem 6, A is invertible and $A^{-1} = (UDV^T)^{-1} = (V^T)^{-1} D^{-1} U^{-1} = VD^{-1}U^T$.

Answer to Checkpoint: If A is an invertible $n \times n$ matrix, with an LU factorization $A = LU$, and if B is $n \times p$, then $A^{-1}B$ can be computed by first row reducing $[L \quad B]$ to a matrix $[I \quad Y]$ for some Y and then reducing $[U \quad Y]$ to $[I \quad A^{-1}B]$. One way to see that this algorithm works is to view $A^{-1}B$ as $[A^{-1}\mathbf{b}_1 \quad \cdots \quad A^{-1}\mathbf{b}_p]$ and use the LU algorithm to solve simultaneously the set of equations $A\mathbf{x} = \mathbf{b}_1, \ldots, A\mathbf{x} = \mathbf{b}_p$. MATLAB uses this approach to compute $A^{-1}B$ (after first finding L and U).

Appendix: Permuted LU Factorizations

Any $m \times n$ matrix A admits a factorization $A = LU$, with U in echelon form and L a *permuted unit lower triangular* matrix. That is, L is a matrix such that a permutation (rearrangement) of its rows (using row interchanges) will produce a lower triangular matrix with 1's on the diagonal.

The construction of L and U, illustrated below, depends on first using row replacements to reduce A to a *permuted echelon form V* and then using row interchanges to reduce V to an echelon form U. By watching the reduction of A to V, we can easily construct a permuted unit lower triangular matrix L with the property that the sequence of operations changing A into U also changes L into I. This property will guarantee that $A = LU$. (See the paragraph before Example 2 in the text.)

The following algorithm reduces any matrix to a permuted echelon form. In the algorithm when a row is covered, we ignore it in later calculations.

1. *Begin with the leftmost nonzero column. Choose any nonzero entry as the pivot. Designate the corresponding row as a pivot row.*

2. *Use row replacements to create zeros above and below the pivot (in all uncovered rows). Then cover that pivot row.*

3. *Repeat steps 1 and 2 on the uncovered submatrix, if any, until all nonzero entries are covered.*

This algorithm forces each pivot to be to the right of the preceding pivots; when the rows are rearranged with the pivots in stair-step fashion, all entries below each pivot will be zero. Thus, the algorithm produces a permuted echelon matrix. Whenever a pivot is selected, the column containing the pivot will be used to construct a column of L, as we shall see.

As an example, choose any entry in the first column of the following matrix as the first pivot, and use the pivot to create zeros in the rest of column 1. We choose the $(3, 1)$-entry.

$$
A = \begin{bmatrix} 1 & -1 & 5 & -8 & -7 \\ -2 & -1 & -4 & 9 & 1 \\ \textcircled{4} & 8 & -4 & 0 & -8 \\ 2 & 3 & 0 & -5 & 3 \end{bmatrix} \sim \begin{bmatrix} 0 & -3 & 6 & -8 & -5 \\ 0 & 3 & -6 & 9 & -3 \\ 4 & 8 & -4 & 0 & -8 \\ 0 & -1 & 2 & -5 & 7 \end{bmatrix}
$$

call this column **a** → (points to first column of A)

1st pivot row ← (points to third row of second matrix)

Row 3 is the first pivot row. Choose the (2, 2)-entry as the second pivot, and create zeros in the rest of column 2, excluding the first pivot row.

call this column b

$$\begin{bmatrix} 0 & -3 & 6 & -8 & -5 \\ 0 & ③ & -6 & 9 & -3 \\ 4 & 8 & -4 & 0 & -8 \\ 0 & -1 & 2 & -5 & 7 \end{bmatrix} \sim \begin{bmatrix} 0 & 0 & 0 & 1 & -8 \\ 0 & 3 & -6 & 9 & -3 \\ 4 & 8 & -4 & 0 & -8 \\ 0 & 0 & 0 & -2 & 6 \end{bmatrix}$$

← 2nd pivot row
← 1st pivot row

Cover row 2 and choose the (4, 4)-entry as the pivot. (The row index of the pivot is relative to the original matrix.) Create zeros in the other rows (in the pivot column), excluding the first two pivot rows.

call this column c   column d

$$\begin{bmatrix} 0 & 0 & 0 & 1 & -8 \\ 0 & 3 & -6 & 9 & -3 \\ 4 & 8 & -4 & 0 & -8 \\ 0 & 0 & 0 & -2 & 6 \end{bmatrix} \sim \begin{bmatrix} 0 & 0 & 0 & 0 & ⑤ \\ 0 & ③ & -6 & 9 & -3 \\ ④ & 8 & -4 & 0 & -8 \\ 0 & 0 & 0 & ⑤ & 6 \end{bmatrix}$$

← 4th pivot row
← 2nd pivot row
← 1st pivot row
← 3rd pivot row

Let V denote this permuted echelon form, and permute the rows of V to create an echelon form. The first pivot row goes to the top, the second pivot row goes next, and so on. The resulting echelon matrix U is

$$\begin{bmatrix} 4 & 8 & -4 & 0 & -8 \\ 0 & 3 & -6 & 9 & -3 \\ 0 & 0 & 0 & -2 & 6 \\ 0 & 0 & 0 & 0 & 5 \end{bmatrix} = U$$

The last step is to create L. Go back and watch the reduction of A to V. As each pivot is selected, take the pivot column, and divide the pivot into each entry in the column that is not yet in a pivot row. Place the resulting column into L. At the end, fill the holes in L with zeros.

Column: a   b   c   d

$$
\begin{bmatrix} 1 \\ -2 \\ \boxed{4} \\ 2 \end{bmatrix}
\quad
\begin{bmatrix} -3 \\ \boxed{3} \\ \\ -1 \end{bmatrix}
\quad
\begin{bmatrix} 1 \\ \\ \\ \boxed{-2} \end{bmatrix}
\quad
\begin{bmatrix} \boxed{-5} \\ \\ \\ \\ \end{bmatrix}
$$

$$
\begin{array}{cccc}
\div 4 & \div 3 & \div -2 & \div -5 \\
\downarrow & \downarrow & \downarrow & \downarrow
\end{array}
$$

$$
\begin{bmatrix}
1/4 & -1 & -1/2 & 1 \\
-1/2 & 1 & & \\
1 & & & \\
1/2 & -1/3 & 1 &
\end{bmatrix},
\quad
L =
\begin{bmatrix}
1/4 & -1 & -1/2 & 1 \\
-1/2 & 1 & 0 & 0 \\
1 & 0 & 0 & 0 \\
1/2 & -1/3 & 1 & 0
\end{bmatrix}
$$

You can check that $LU = A$. To see why this is so, observe that L is constructed so the operations that reduce A to V also reduce L to a permuted identity matrix. Since the pivots in L are in exactly the same rows as in V, the sequence of row interchanges that reduces V to U also reduces the permuted identity matrix to I. Thus, the full sequence of operations that reduces A to U also reduces L to I, so that $A = LU$. (See argument before the box on page 134 of the text.)

The next example illustrates what to do when V has one or more rows of zeros. The matrix is from the Practice Problem for Section 2.5. For the reduction of A to V, pivots were chosen to have the largest possible magnitude (the choice used for "partial pivoting"). Of course, other pivots could have been selected.

$$
A =
\begin{bmatrix}
2 & -4 & -2 & 3 \\
\boxed{6} & -9 & -5 & 8 \\
2 & -7 & -3 & 9 \\
4 & -2 & -2 & -1 \\
-6 & 3 & 3 & 4
\end{bmatrix}
\sim
\begin{bmatrix}
0 & -1 & -1/3 & 1/3 \\
6 & -9 & -5 & 8 \\
0 & -4 & -4/3 & 19/3 \\
0 & 4 & 4/3 & -19/3 \\
0 & \boxed{-6} & -2 & 12
\end{bmatrix}
\sim
\begin{bmatrix}
0 & 0 & 0 & -5/3 \\
6 & -9 & -5 & 8 \\
0 & 0 & 0 & -5/3 \\
0 & 0 & 0 & \boxed{5/3} \\
0 & -6 & -2 & 12
\end{bmatrix}
$$

$$
\sim V =
\begin{bmatrix}
0 & 0 & 0 & 0 \\
\boxed{6} & -9 & -5 & 8 \\
0 & 0 & 0 & 0 \\
0 & 0 & 0 & \boxed{5/3} \\
0 & \boxed{-6} & -2 & 21
\end{bmatrix}
\begin{array}{l}
\\ \leftarrow \text{1st pivot row} \\ \\ \leftarrow \text{3rd pivot row} \\ \leftarrow \text{2nd pivot row}
\end{array}
\qquad
\sim U =
\begin{bmatrix}
6 & -9 & -5 & 8 \\
0 & -6 & -2 & 12 \\
0 & 0 & 0 & 5/3 \\
0 & 0 & 0 & 0 \\
0 & 0 & 0 & 0
\end{bmatrix}
$$

The first three columns of L come from the three pivot columns above.

$$\begin{bmatrix} 2 \\ ⑥ \\ 2 \\ 4 \\ -6 \end{bmatrix} \quad \begin{bmatrix} -1 \\ \\ -4 \\ 4 \\ ⊖⑥ \end{bmatrix} \quad \begin{bmatrix} -5/3 \\ \\ -5/3 \\ ⑤/③ \\ \\ \end{bmatrix}$$

$$\div 6 \qquad \div -6 \qquad \div 5/3$$

$$\downarrow \qquad \downarrow \qquad \downarrow$$

$$\begin{bmatrix} 1/3 & 1/6 & -1 \\ 1 & & \\ 1/3 & 2/3 & -1 \\ 2/3 & -2/3 & 1 \\ -1 & 1 & \end{bmatrix}$$

← 1st pivot row

← 3rd pivot row

← 2nd pivot row

The matrix L needs two more columns. Use columns 1 and 3 of the 5×5 identity matrix to place 1's in the "nonpivot" rows 1 and 3. Fill in the remaining holes with zeros.

$$\begin{bmatrix} 1/3 & 1/6 & -1 & 1 & 0 \\ ① & 0 & 0 & 0 & 0 \\ 1/3 & 2/3 & -1 & 0 & 1 \\ 2/3 & -2/3 & ① & 0 & 0 \\ -1 & ① & 0 & 0 & 0 \end{bmatrix} \sim L = \begin{bmatrix} 1 & 0 & 0 & 0 & 0 \\ -1 & 1 & 0 & 0 & 0 \\ 2/3 & -2/3 & 1 & 0 & 0 \\ 1/3 & 1/6 & -1 & 1 & 0 \\ 1/3 & 2/3 & -1 & 0 & 1 \end{bmatrix}$$

Row reduction of L using only row replacements produces a permuted identity matrix. Moving the 1's in the "pivot rows" 2, 5, and 4 into rows 1, 2, and 3 of the identity requires the same row swaps as reducing V to U. If a further row interchange on the permuted identity is required, it will involve the bottom two rows, which came from the "nonpivot" rows 1 and 3. A corresponding interchange of the bottom two rows of U has no effect on U (and the product LU is unaffected). As a result, L is reduced to I by the same operations that reduce A to V and then to U. Check that $A = LU$.

**MATLAB   LU Factorization and the Backslash Operator **

Row reduction of A using the command **gauss** will produce the intermediate matrices needed for an LU factorization of A. You can try this on the matrix in Example 2, stored as Exercise 33 in the Laydata Toolbox. The matrices in (5) on page 145 in the text are produced by the commands

    U=gauss(A,1)   U has 0's below the first pivot
    U=gauss(U,2)   Now U has 0's below pivots 1 and 2
    U=gauss(U,3)   The echelon form

You can copy the information from the screen onto your paper, and divide by the pivot entries to produce L as in the text. For most text exercises, the pivots are integers and so are displayed accurately.

To construct a permuted LU factorization, use **U=gauss(U,r,v)**, where r is the row index of the pivot and **v** is a row vector that lists the rows to be changed by replacement operations. For example, if A has 5 rows and the first pivot is in row 4, use **U=gauss(A,4,[1 2 3 5])**. If the next pivot is in row 2, use **U=gauss(U,2,[1 3 5])**. To build the permuted matrix L, use full columns from A or the partially reduced U, divided by the pivots. Then change entries to zero if they are in a row already selected as a "pivot row."

The MATLAB command **[L U]=lu(A)** produces a permuted LU factorization for any square matrix A, but it does not handle the general case.

When A is invertible, the best way to solve $A\mathbf{x} = \mathbf{b}$ with MATLAB is to use the backslash command **x=A\\b**. MATLAB proceeds to compute a permuted LU factorization of A and then use L and U to compute **x**. The alternative command **x=inv(A)*b** is less efficient and can be less accurate. The command **inv(A)** uses the LU factorization to compute A^{-1} in the form $U^{-1}L^{-1}$.

2.6   THE LEONTIEF INPUT-OUTPUT MODEL ⎯⎯⎯⎯⎯⎯

If you are in economics, you definitely will need the material in this section for later work. Although most of the discussion concerns economics, the formula for the inverse of $I - C$ is used in a variety of applications.

⎯⎯⎯⎯⎯⎯

STUDY NOTES

The power of Leontief's model of the economy is that it compresses hundreds of equations in hundreds of variables into the simple matrix equation $(I - C)\mathbf{x} = \mathbf{d}$. You should know how to construct the consumption matrix C and know the algebra that leads from the matrix equation $\mathbf{x} = C\mathbf{x} + \mathbf{d}$ to its solution $\mathbf{x} = (I - C)^{-1}\mathbf{d}$, under the assumption that the column sums of C are less than one.

You may need to know the formula (8) for $(I-C)^{-1}$ on page 155. (Check with your instructor.) The formula is analogous to the formula for the sum of a geometric series of positive numbers:

$$1 + r + r^2 + r^3 + \cdots = (1-r)^{-1} \text{ when } |r| < 1.$$

SOLUTIONS TO EXERCISES

1. Fill in C one column at a time, since each column is a unit consumption vector for one sector. Make sure that the order of the sectors is the same for the rows and columns of C. From the way the data are presented, we use the order: manufacturing, agriculture, and services. Read the sentences carefully, to get the data arranged correctly.

Purchased from:	Unit consumption vectors		
	Manuf.	**Agric.**	**Serv.**
Manufacturing	.10	.60	.60
Agriculture	.30	.20	.00
Services	.30	.10	.10

The intermediate demands created by a production vector $\mathbf{x}$ are given by $C\mathbf{x}$. If agriculture plans to produce 100 units (and the other sectors plan to produce nothing), then the intermediate demand is

$$C\mathbf{x} = \begin{bmatrix} .10 & .60 & .60 \\ .30 & .20 & .00 \\ .30 & .10 & .10 \end{bmatrix} \begin{bmatrix} 0 \\ 100 \\ 0 \end{bmatrix} = \begin{bmatrix} 60 \\ 20 \\ 10 \end{bmatrix}$$

7. $C = \begin{bmatrix} .0 & .5 \\ .6 & .2 \end{bmatrix}$, $\mathbf{d} = \begin{bmatrix} 50 \\ 30 \end{bmatrix}$. Let $\mathbf{d}_1 = \begin{bmatrix} 1 \\ 0 \end{bmatrix}$, the demand for 1 unit of output of sector 1.

a. The production required to satisfy the demand $\mathbf{d}_1$ is the vector $\mathbf{x}_1$ such that $(I-C)\mathbf{x}_1 = \mathbf{d}_1$, namely, $\mathbf{x}_1 = (I-C)^{-1}\mathbf{d}_1$. From Exercise 5,

$$I - C = \begin{bmatrix} 1 & -.5 \\ -.6 & .8 \end{bmatrix} \quad \text{and} \quad (I-C)^{-1} = \begin{bmatrix} 1.6 & 1 \\ 1.2 & 2 \end{bmatrix}$$

so

$$\mathbf{x}_1 = \begin{bmatrix} 1.6 & 1 \\ 1.2 & 2 \end{bmatrix} \begin{bmatrix} 1 \\ 0 \end{bmatrix} = \begin{bmatrix} 1.6 \\ 1.2 \end{bmatrix}$$

b. For the final demand $\mathbf{d}_2 = \begin{bmatrix} 51 \\ 30 \end{bmatrix}$, the corresponding production $\mathbf{x}_2$ is given by

$$\mathbf{x}_2 = (I-C)^{-1}\mathbf{d}_2 = \begin{bmatrix} 1.6 & 1 \\ 1.2 & 2 \end{bmatrix} \begin{bmatrix} 51 \\ 30 \end{bmatrix} = \begin{bmatrix} 111.6 \\ 121.2 \end{bmatrix}$$

c. From Exercise 5, the production **x** corresponding to the demand **d** is given by $\mathbf{x} = \begin{bmatrix} 110 \\ 120 \end{bmatrix}$.

Observe from (a) and (b) that $\mathbf{x}_2 = \mathbf{x} + \mathbf{x}_1$. Also, as pointed out in the text, $\mathbf{d}_2 = \mathbf{d} + \mathbf{d}_1$. The sum of the production vectors **x** and $\mathbf{x}_1$ gives the production needed to satisfy the sum of the demands **d** and $\mathbf{d}_1$. This is expressing the *linearity* between final demand and production. This relation is true in general, because

$$\mathbf{x}_2 = (I - C)^{-1}\mathbf{d}_2 = (I - C)^{-1}(\mathbf{d} + \mathbf{d}_1)$$

$$= (I - C)^{-1}\mathbf{d} + (I - C)^{-1}\mathbf{d}_1$$

$$= \mathbf{x} + \mathbf{x}_1$$

Warning: In Exercise 9, don't multiply the consumption matrix C by 10, to get rid of the decimals. That changes the equation $C\mathbf{x} = \mathbf{x} + \mathbf{d}$ into $10C\mathbf{x} = \mathbf{x} + \mathbf{d}$, whose solution is different. However, you *may* multiply the *augmented matrix* $[(I - C)\ \ \mathbf{0}]$ by 10, because the solution of an equation is not affected when both sides are multiplied by a nonzero number.

11. Following the hint in the text, you should obtain $\mathbf{p}^T\mathbf{x} = \mathbf{p}^T C\mathbf{x} + \mathbf{v}^T\mathbf{x}$ (from the price equation). Then, from the production equation, $\mathbf{p}^T\mathbf{x} = \mathbf{p}^T(C\mathbf{x} + \mathbf{d}) = \mathbf{p}^T C\mathbf{x} + \mathbf{p}^T\mathbf{d}$. Equate the two expressions for $\mathbf{p}^T\mathbf{x}$ to yield $\mathbf{p}^T\mathbf{d} = \mathbf{v}^T\mathbf{x}$.

Another solution: Take transposes in the price equation,

$$\mathbf{p}^T = (C^T\mathbf{p})^T + \mathbf{v}^T = \mathbf{p}^T C + \mathbf{v}^T, \text{ so } \mathbf{v}^T = \mathbf{p}^T - \mathbf{p}^T C$$

and right-multiply by **x** to obtain

$$\mathbf{v}^T\mathbf{x} = \mathbf{p}^T\mathbf{x} - \mathbf{p}^T C\mathbf{x} = \mathbf{p}^T(I - C)\mathbf{x} = \mathbf{p}^T\mathbf{d} \quad \text{From the production equation}$$

13. The data for this exercise are in the Laydata Toolbox. To solve the equation $(I - C)\mathbf{x} = \mathbf{d}$, row reduce the augmented matrix $[(I - C)\ \ \mathbf{d}]$ rather than compute $(I - C)^{-1}$. (Another reasonable solution method is suggested in Exercise 15.) The numerical solution is given in the text.

2.7   APPLICATIONS TO COMPUTER GRAPHICS _____

According to my students over the past few years, this section is one of the most interesting application sections in the text, because it shows how matrix calculations, performed millions of times per second, can create the illusion of 3D-motion on a computer screen or in a movie theater. Of course, one short section cannot begin to indicate the vast scope of computer graphics. I encourage you to look at the book by Foley et al., referenced in your text. Chapters 5, 6, and 11 are filled with matrices! The rest of the 1100 pages in the book contains lots of interesting mathematics, detailed discussions of computer algorithms, and scores of spectacular (in some cases, almost unbelievable) color plates.

STUDY NOTES

When a graphical object is represented by a set of polygons, each vertex can be stored as one column of a data matrix D. When a linear transformation T acts on the graphical object, the transformed object is determined by the images of the vertices, because each line segment between vertices is transformed into a line segment between the image vertices. If A is the matrix of the transformation T, then the image vertices are the columns of the matrix AD, by definition of the product AD.

Homogeneous coordinates are needed to make translation act as a linear transformation. Translation by a vector $\mathbf{p}$ is illustrated by the computation $\begin{bmatrix} I & \mathbf{p} \\ \mathbf{0}^T & 1 \end{bmatrix}\begin{bmatrix} \mathbf{x} \\ 1 \end{bmatrix} = \begin{bmatrix} \mathbf{x}+\mathbf{p} \\ 1 \end{bmatrix}$. When several transformations are composed, the order of matrix products is important. See Example 6.

For 3D-graphics, homogeneous coordinates are used to compute perspective projections. The text only considers a perspective projection whose center of projection is at $(0, 0, d)$. The matrix for this is displayed just before Example 8. Check with your instructor whether you should memorize this matrix. Examples of good test questions can be found in Exercises 1–8 and 13–16.

SOLUTIONS TO EXERCISES

1. From Example 5, the matrix $\begin{bmatrix} 1 & .25 & 0 \\ 0 & 1 & 0 \\ 0 & 0 & 1 \end{bmatrix}$ has the same effect on homogeneous coordinates

for $\mathbb{R}^2$ that the matrix $\begin{bmatrix} 1 & .25 \\ 0 & 1 \end{bmatrix}$ of Example 2 has on ordinary vectors in $\mathbb{R}^2$. Partitioned

matrix notation explains why this is true. Let A be a 2×2 matrix. The following diagram

shows that the action of $\begin{bmatrix} A & 0 \\ 0^T & 1 \end{bmatrix}$ on $\begin{bmatrix} \mathbf{x} \\ 1 \end{bmatrix}$ corresponds to the action of A on $\mathbf{x}$.

$$\begin{array}{ccc} \mathbf{x} & \longrightarrow & A\mathbf{x} \quad \text{Coordinates in } \mathbb{R}^2 \\ \big\downarrow & & \big\uparrow \\ \begin{bmatrix} \mathbf{x} \\ 1 \end{bmatrix} \longmapsto \begin{bmatrix} A & 0 \\ 0^T & 1 \end{bmatrix}\begin{bmatrix} \mathbf{x} \\ 1 \end{bmatrix} & = \begin{bmatrix} A\mathbf{x}+0\cdot1 \\ 0^T\mathbf{x}+1\cdot1 \end{bmatrix} = \begin{bmatrix} A\mathbf{x} \\ 1 \end{bmatrix} & \text{Homogenous coordinates} \end{array}$$

7. A 60% rotation about the origin in $\mathbb{R}^2$ is given by

$\begin{bmatrix} \cos 60° & -\sin 60° \\ \sin 60° & \cos 60° \end{bmatrix} = \begin{bmatrix} 1/2 & -\sqrt{3}/2 \\ \sqrt{3}/2 & 1/2 \end{bmatrix}$, so the 3×3 matrix for rotation about $\begin{bmatrix} 6 \\ 8 \end{bmatrix}$ is

$$\begin{bmatrix} 1 & 0 & 6 \\ 0 & 1 & 8 \\ 0 & 0 & 1 \end{bmatrix} \begin{bmatrix} 1/2 & -\sqrt{3}/2 & 0 \\ \sqrt{3}/2 & 1/2 & 0 \\ 0 & 0 & 1 \end{bmatrix} \begin{bmatrix} 1 & 0 & 6 \\ 0 & 1 & 8 \\ 0 & 0 & 1 \end{bmatrix}$$

Finally,   Then, rotate   First,
translate   about the   translate
back   origin   by $-\mathbf{p}$

$$= \begin{bmatrix} 1 & 0 & 6 \\ 0 & 1 & 8 \\ 0 & 0 & 1 \end{bmatrix} \begin{bmatrix} 1/2 & -\sqrt{3}/2 & -3+4\sqrt{3} \\ \sqrt{3}/2 & 1/2 & -4-3\sqrt{3} \\ 0 & 0 & 1 \end{bmatrix} = \begin{bmatrix} 1/2 & -\sqrt{3}/2 & 3+4\sqrt{3} \\ \sqrt{3}/2 & 1/2 & 4-3\sqrt{3} \\ 0 & 0 & 1 \end{bmatrix}$$

13. The answer is given in the text. Notice that the order of the transformations is important. If the translation is done first (that is, if the matrix for the translation is on the right), then

$$\begin{bmatrix} A & \mathbf{0} \\ \mathbf{0}^T & 1 \end{bmatrix} \begin{bmatrix} I & \mathbf{p} \\ \mathbf{0}^T & 1 \end{bmatrix} = \begin{bmatrix} AI + \mathbf{0}\mathbf{0}^T & A\mathbf{p} + \mathbf{0}\cdot 1 \\ \mathbf{0}^T I + 1\mathbf{0}^T & \mathbf{0}^T \mathbf{p} + 1\cdot 1 \end{bmatrix} = \begin{bmatrix} A & A\mathbf{p} \\ \mathbf{0}^T & 1 \end{bmatrix} \neq \begin{bmatrix} A & \mathbf{p} \\ \mathbf{0}^T & 1 \end{bmatrix}$$

Here, $\mathbf{0}^T$ is a zero row vector, and so the outer product $\mathbf{0}\mathbf{0}^T$ is a zero matrix.

19. The matrix P for the perspective transformation with center of projection at $(0, 0, 10)$ and the data matrix D using homogeneous coordinates are shown below. The data matrix for the image of the triangle is PD:

$$PD = \begin{bmatrix} 1 & 0 & 0 & 0 \\ 0 & 1 & 0 & 0 \\ 0 & 0 & 0 & 0 \\ 0 & 0 & -.1 & 1 \end{bmatrix} \begin{bmatrix} 4.2 & 6 & 2 \\ 1.2 & 4 & 2 \\ 4 & 2 & 6 \\ 1 & 1 & 1 \end{bmatrix} = \begin{bmatrix} 4.2 & 6 & 2 \\ 1.2 & 4 & 2 \\ 0 & 0 & 0 \\ .6 & .8 & .4 \end{bmatrix}$$

The $\mathbb{R}^3$ coordinates of the image points come from the top three entries in each column, divided by the corresponding entries in the fourth row.

$$\begin{bmatrix} 4.2/.6 & 6/.8 & 2/.4 \\ 1.2/.6 & 4/.8 & 2/.4 \\ 0 & 0 & 0 \end{bmatrix} = \begin{bmatrix} 7 & 7.5 & 5 \\ 2 & 5 & 5 \\ 0 & 0 & 0 \end{bmatrix}$$

2.8   SUBSPACES OF $\mathbb{R}^n$

This section presents the basic ideas from Sections 4.1–4.3 that are needed for Chapters 5 to 7. You should study this section and the next only if your course will omit most or all of Chapter 4. If you reviewed carefully after Sections 1.9 and 2.3, you should be well prepared for the new material here.

KEY IDEAS

There are four fundamental concepts in this section: subspace, column space, null space, and basis. The best mental image for a subspace is a plane in $\mathbb{R}^3$ through the origin. (See Fig. 1 on page 168.) The distinguishing feature of such a plane is that the sum of any two vectors in the plane is another vector in the same plane (by the parallelogram rule), and any scalar multiple of any vector in the plane is also in the plane. Other subspaces of $\mathbb{R}^3$ are lines through the origin, the zero subspace, and $\mathbb{R}^3$ itself.

The main examples of subspaces of $\mathbb{R}^n$ are column spaces (defined explicitly) and null spaces (defined implicitly). Example 6 is probably the most important example in the section. It illustrates a type of computation needed frequently in Chapters 5 and 7.

Actually, there really is not so much to learn here, because you have already been using these concepts for several weeks, without the terminology. (For instance, the notion of a basis is a combination of the ideas of linear independence and spanning.) But you do need to know the precise definitions of these four terms, and you must move beyond mechanical computations. See "Mastering Linear Algebra Concepts" at the end of this section for help.

SOLUTIONS TO EXERCISES

1. The set is closed under sums but not under multiplication by a negative scalar. A counterexample to the subspace condition is shown at the right. You may also give an algebraic example, such as $\mathbf{x} = (2, 1)$ and $c = -1$. Then $\mathbf{x}$ is in H and $c\mathbf{x} = (-2, -1)$ is not in H. Ask your instructor what type of counterexample would be acceptable if this question were on a test.

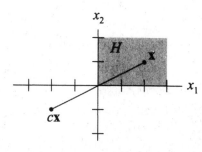

7. **a.** There are three vectors, $\mathbf{v}_1$, $\mathbf{v}_2$, and $\mathbf{v}_3$, in the set $\{\mathbf{v}_1, \mathbf{v}_2, \mathbf{v}_3\}$.

   b. There are infinitely many vectors in Span$\{\mathbf{v}_1, \mathbf{v}_2, \mathbf{v}_3\}$ = Col A.

   c. Deciding whether $\mathbf{p}$ is in Col A requires calculation:

$$[A \quad \mathbf{p}] \sim \begin{bmatrix} 2 & -3 & -4 & 6 \\ -8 & 8 & 6 & -10 \\ 6 & -7 & -7 & 11 \end{bmatrix} \sim \begin{bmatrix} 2 & -3 & -4 & 6 \\ 0 & -4 & -10 & 14 \\ 0 & 2 & 5 & -7 \end{bmatrix} \sim \begin{bmatrix} 2 & -3 & -4 & 6 \\ 0 & -4 & -10 & 14 \\ 0 & 0 & 0 & 0 \end{bmatrix}$$

The equation $A\mathbf{x} = \mathbf{p}$ has a solution, so $\mathbf{p}$ is in Col A.

13. To produce a vector in Col A, select any column of A. For Nul A, solve the equation $A\mathbf{x} = \mathbf{0}$. (Include an augmented column of zeros, to avoid errors.)

$$\begin{bmatrix} 3 & 2 & 1 & -5 & 0 \\ -9 & -4 & 1 & 7 & 0 \\ 9 & 2 & -5 & 1 & 0 \end{bmatrix} \sim \begin{bmatrix} 3 & 2 & 1 & -5 & 0 \\ 0 & 2 & 4 & -8 & 0 \\ 0 & -4 & -8 & 16 & 0 \end{bmatrix} \sim \begin{bmatrix} 3 & 2 & 1 & -5 & 0 \\ 0 & 2 & 4 & -8 & 0 \\ 0 & 0 & 0 & 0 & 0 \end{bmatrix}$$

$$\sim \begin{bmatrix} 3 & 2 & 1 & -5 & 0 \\ 0 & 1 & 2 & -4 & 0 \\ 0 & 0 & 0 & 0 & 0 \end{bmatrix} \sim \begin{bmatrix} 1 & 0 & -1 & 1 & 0 \\ 0 & 1 & 2 & -4 & 0 \\ 0 & 0 & 0 & 0 & 0 \end{bmatrix}, \qquad \begin{aligned} x_1 \quad - \quad x_3 + \quad x_4 &= 0 \\ x_2 + 2x_3 - 4x_4 &= 0 \\ 0 &= 0 \end{aligned}$$

The general solution is $x_1 = x_3 - x_4$, and $x_2 = -2x_3 + 4x_4$, with x_3 and x_4 free. The general solution in parametric vector form is not needed. All that is required here is one nonzero vector. So choose any values for x_3 and x_4 (not both zero). For instance, set $x_3 = 1$ and $x_4 = 0$ to obtain the vector $(1, -2, 1, 0)$ in Nul A. Another choice, setting $x_3 = 0$ and $x_4 = 1$, might be $(-1, 4, 0, 1)$.

19. No. The vectors cannot be a basis for $\mathbb{R}^3$ because they only span a plane in $\mathbb{R}^3$. Or, point out that the columns of the matrix $\begin{bmatrix} 1 & -5 \\ 1 & -1 \\ -2 & 2 \end{bmatrix}$ cannot possibly span $\mathbb{R}^3$ because the matrix cannot have a pivot in every row. So the columns are not a basis for $\mathbb{R}^3$. Be careful *not* to say that the vectors are a basis for $\mathbb{R}^2$. They are not *in* $\mathbb{R}^2$, because they each have three entries.

Warning: $\mathbb{R}^2$ is *not* a subspace of $\mathbb{R}^3$. The notation $\mathbb{R}^2$ refers explicitly to lists of numbers with exactly two entries. $\mathbb{R}^3$ is the set of all lists of three entries from $\mathbb{R}$.

21. a. Carefully read the definition at the beginning of the section. What is missing?

   b. See the paragraph before Example 4.

   c. See Theorem 12. The numbers m and n need not be equal.

   d. See Example 5.

   e. See the first part of the solution of Example 8.

23. $A = \begin{bmatrix} 4 & 5 & 9 & -2 \\ 6 & 5 & 1 & 12 \\ 3 & 4 & 8 & -3 \end{bmatrix} \sim \begin{bmatrix} 1 & 2 & 6 & -5 \\ 0 & 1 & 5 & -6 \\ 0 & 0 & 0 & 0 \end{bmatrix}$. The echelon form identifies columns 1 and 2 as

the pivot columns. A basis for Col A uses columns 1 and 2 of A: $\begin{bmatrix} 4 \\ 6 \\ 3 \end{bmatrix}, \begin{bmatrix} 5 \\ 5 \\ 4 \end{bmatrix}$. This is not the only choice, but it is the "standard" choice. A *wrong* choice is to select columns 1 and 2 of the echelon form. These columns have zero in the third entry and could not possibly generate the columns displayed in A.

For Nul A, obtain the *reduced* (and augmented) echelon form for $A\mathbf{x} = \mathbf{0}$:

$$\begin{bmatrix} 1 & 0 & -4 & 7 & 0 \\ 0 & 1 & 5 & -6 & 0 \\ 0 & 0 & 0 & 0 & 0 \end{bmatrix}.$$ This corresponds to: $\begin{aligned} x_1 \quad - 4x_3 + 7x_4 &= 0 \\ x_2 + 5x_3 - 6x_4 &= 0. \\ 0 &= 0 \end{aligned}$

Solve for the basic variables and write the solution of $A\mathbf{x} = \mathbf{0}$ in parametric vector form:

$$\begin{bmatrix} x_1 \\ x_2 \\ x_3 \\ x_4 \end{bmatrix} = \begin{bmatrix} 4x_3 - 7x_4 \\ -5x_3 + 6x_4 \\ x_3 \\ x_4 \end{bmatrix} = x_3 \begin{bmatrix} 4 \\ -5 \\ 1 \\ 0 \end{bmatrix} + x_4 \begin{bmatrix} -7 \\ 6 \\ 0 \\ 1 \end{bmatrix}. \quad \text{Basis for Nul } A: \begin{bmatrix} 4 \\ -5 \\ 1 \\ 0 \end{bmatrix}, \begin{bmatrix} -7 \\ 6 \\ 0 \\ 1 \end{bmatrix}$$

Note: A basis is a *set* of vectors. For simplicity, the answers here and in the text list the vectors without enclosing the list inside set brackets. Ask your instructor if this format is acceptable.

Warning: A common error is to confuse Col A with Nul A. This happens easily when the definitions of these spaces are not known precisely. Another error is to think that the nonpivot columns of an $m \times n$ matrix A form a basis for Nul A. This is not true in general, even when $m = n$.

25. $A = \begin{bmatrix} 1 & 4 & 8 & -3 & -7 \\ -1 & 2 & 7 & 3 & 4 \\ -2 & 2 & 9 & 5 & 5 \\ 3 & 6 & 9 & -5 & -2 \end{bmatrix} \sim \begin{bmatrix} 1 & 4 & 8 & 0 & 5 \\ 0 & 2 & 5 & 0 & -1 \\ 0 & 0 & 0 & 1 & 4 \\ 0 & 0 & 0 & 0 & 0 \end{bmatrix}.$

Basis for Col A: $\begin{bmatrix} 1 \\ -1 \\ -2 \\ 3 \end{bmatrix}, \begin{bmatrix} 4 \\ 2 \\ 2 \\ 6 \end{bmatrix}, \begin{bmatrix} -3 \\ 3 \\ 5 \\ -5 \end{bmatrix}.$

For Nul A, obtain the *reduced* (and augmented) echelon form for $A\mathbf{x} = \mathbf{0}$:

$$[A \quad \mathbf{0}] \sim \begin{bmatrix} 1 & 0 & -2 & 0 & 7 & 0 \\ 0 & 1 & 2.5 & 0 & -.5 & 0 \\ 0 & 0 & 0 & 1 & 4 & 0 \\ 0 & 0 & 0 & 0 & 0 & 0 \end{bmatrix}. \quad \begin{aligned} x_1 \quad - 2x_3 \quad + 7x_5 &= 0 \\ x_2 + 2.5x_3 \quad - .5x_5 &= 0 \\ x_4 + 4x_5 &= 0 \\ 0 &= 0 \end{aligned}$$

The solution in parametric vector form: $\begin{bmatrix} x_1 \\ x_2 \\ x_3 \\ x_4 \\ x_5 \end{bmatrix} = \begin{bmatrix} 2x_3 - 7x_5 \\ -2.5x_3 + .5x_5 \\ x_3 \\ -4x_5 \\ x_5 \end{bmatrix} = x_3 \begin{bmatrix} 2 \\ -2.5 \\ 1 \\ 0 \\ 0 \end{bmatrix} + x_5 \begin{bmatrix} -7 \\ .5 \\ 0 \\ -4 \\ 1 \end{bmatrix}.$
$$\qquad\qquad\qquad\qquad\qquad\qquad\qquad\qquad\uparrow\qquad\ \uparrow$$
$$\qquad\qquad\qquad\qquad\qquad\qquad\qquad\qquad\mathbf{u}\qquad \mathbf{v}$$

Basis for Nul A: $\{\mathbf{u}, \mathbf{v}\}$.

Note: This solution illustrates how you can save time on an exam and not copy the ten numbers in the basis vectors for your answer. Just label the basis vectors as **u** and **v**, and write something such as "**u** and **v** form a basis for Nul A." You might ask your instructor if this is acceptable.

Warning: Do not become too attached to the symbols commonly used for certain ideas. For instance, calling a vector "**b**" does not imply that it can only be the "right side" of an equation $A\mathbf{x} = \mathbf{b}$. In Exercise 29, you should be looking for a vector **b** such that $A\mathbf{b} = \mathbf{0}$.

29. A simple construction is to write any nonzero 3×3 matrix whose columns are obviously linearly dependent, and then make **b** a vector of weights that come from a linear dependence relation among the columns. For instance, if the first two columns of A are equal, then $\mathbf{a}_1 - \mathbf{a}_2 + 0\mathbf{a}_3 = \mathbf{0}$. So, **b** could be $(1, -1, 0)$.

31. The text has an answer. An answer such as "Nul F is a subset of $\mathbb{R}^5$" says something true, but not much. "Nul F is a subspace of $\mathbb{R}^5$" ought to be good for some partial credit, but this fact does not use the information given about the column space of F. Probably the best possible answer is that Nul F is a nonzero subspace of $\mathbb{R}^5$.

37. **[M]** Use the command that produces the reduced echelon form in one step (**ref** or **rref** depending on the program). By Theorem 13, the pivot columns of A form a basis for Col A.

$$A = \begin{bmatrix} 3 & -5 & 0 & -1 & 3 \\ -7 & 9 & -4 & 9 & -11 \\ -5 & 7 & -2 & 5 & -7 \\ 3 & -7 & -3 & 4 & 0 \end{bmatrix} \sim \begin{bmatrix} 1 & 0 & 2.5 & -4.5 & 3.5 \\ 0 & 1 & 1.5 & -2.5 & 1.5 \\ 0 & 0 & 0 & 0 & 0 \\ 0 & 0 & 0 & 0 & 0 \end{bmatrix}$$

Basis for Col A: $\begin{bmatrix} 3 \\ -7 \\ -5 \\ 3 \end{bmatrix}, \begin{bmatrix} -5 \\ 9 \\ 7 \\ -7 \end{bmatrix}$

For Nul A, obtain the solution of $A\mathbf{x} = \mathbf{0}$ in parametric vector form:

$$x_1 \quad + 2.5x_3 - 4.5x_4 + 3.5x_5 = 0$$
$$x_2 + 1.5x_3 - 2.5x_4 + 1.5x_5 = 0$$

Solution: $\begin{cases} x_1 = -2.5x_3 + 4.5x_4 - 3.5x_5 \\ x_2 = -1.5x_3 + 2.5x_4 - 1.5x_5 \\ x_3, x_4, \text{ and } x_5 \text{ are free} \end{cases}$

$$\mathbf{x} = \begin{bmatrix} x_1 \\ x_2 \\ x_3 \\ x_4 \\ x_5 \end{bmatrix} = \begin{bmatrix} -2.5x_3 + 4.5x_4 - 3.5x_5 \\ -1.5x_3 + 2.5x_4 - 1.5x_5 \\ x_3 \\ x_4 \\ x_5 \end{bmatrix} = x_3 \begin{bmatrix} -2.5 \\ -1.5 \\ 1 \\ 0 \\ 0 \end{bmatrix} + x_4 \begin{bmatrix} 4.5 \\ 2.5 \\ 0 \\ 1 \\ 0 \end{bmatrix} + x_5 \begin{bmatrix} -3.5 \\ -1.5 \\ 0 \\ 0 \\ 1 \end{bmatrix} = x_3\mathbf{u} + x_4\mathbf{v} + x_5\mathbf{w}$$

Basis for Nul A: $\{\mathbf{u}, \mathbf{v}, \mathbf{w}\}$.

Mastering Linear Algebra Concepts: Subspace, Column Space, Null Space, Basis

To form strong mental images of a subspace and the two main types of subspaces (Col A and Nul A), prepare a review sheet that covers the following categories:

• definitions	Pages 168 and 169
• equivalent descriptions	Sentence after Example 4, Theorem 12
• geometric interpretations	Fig. 1, Sentence after Example 1
• special cases	Margin figure on page 168, $\mathbb{R}^n$ itself, zero subspace
• examples and counterexamples	Examples 1, 2, 3, Exercises 1–4
• typical computations	Examples 4, Practice Problems 1 and 2, Exercises 5–14
• contrast between Col A and Nul A	Table on page 232 (in Chapter 4) and Examples on page 231

The concept of a basis deserves a separate review sheet. Use the categories below. Also, add notes to your review sheets for Span and Linear Independence that say when a spanning set is a basis for a subspace Span$\{v_1, \ldots, v_p\}$ and when a linearly independent set in $\mathbb{R}^n$ is a basis for $\mathbb{R}^n$.

*   definition   Page 170
*   geometric interpretations   Fig. 3
*   special cases   Example 5
*   examples and counterexamples   Warning, page 172, Exercises 15–20
*   typical computations   Examples 6 and 8, Exercises 23–26

MATLAB ref

The command **ref(A)** produces the r̲educed e̲chelon f̲orm of A. From that you can write a basis for Col A or write the homogeneous equations that describe Nul A. (Don't forget that A is a coefficient matrix, not an augmented matrix.) MATLAB has another command, **rref**, which works basically the same as **ref** but is often much slower, because it checks for rational entries in the matrix.

2.9   DIMENSION AND RANK

This section and Section 2.8 cover the ideas from Chapter 4 that you need for Chapters 5–7. There is no need to read this section if your course covers Chapter 4.

KEY IDEAS

The two fundamental concepts in this section are the dimension of a subspace and the rank of a matrix. Coordinate vectors are used to give an intuitive understanding of dimension and a geometric explanation of why a k-dimensional subspace of $\mathbb{R}^n$ behaves as if it were $\mathbb{R}^k$.

The Basis Theorem ties together the concepts of dimension, subspace, linear independence, span, and basis. So does the Invertible Matrix Theorem. Make sure you know the precise wording of the statements in these theorems. If you desire extra review material, you might look at Examples 2, 4, and 5 in Section 4.6 (pages 258 and 260).

The following table lists all statements that are in the Invertible Matrix Theorem at this point in the course. They are arranged in the scheme used in Section 2.3 of this *Study Guide*. As before, a few extra statements have been added to make the table more symmetrical.

STATEMENTS FROM THE INVERTIBLE MATRIX THEOREM

Equivalent statements for an $m \times n$ matrix A.	Equivalent statements for an $n \times n$ square matrix A.	Equivalent statements for an $n \times p$ matrix A.
k. There is a matrix D such that $AD = I$.	a. A is an invertible matrix.	j. There is a matrix C such that $CA = I$.
*. A has a pivot position in every row.	c. A has n pivot positions.	*. A has a pivot position in every column.
h. The columns of A span $\mathbb{R}^m$.	m. The columns of A form a basis for $\mathbb{R}^n$.	e. The columns of A are linearly independent.
g. The equation $Ax = b$ has at least one solution for each b in $\mathbb{R}^m$.	*. The equation $Ax = b$ has a unique solution for each b in $\mathbb{R}^n$.	d. The equation $Ax = 0$ has only the trivial solution.
i. The transformation $x \mapsto Ax$ maps $\mathbb{R}^n$ onto $\mathbb{R}^m$.	*. The transformation $x \mapsto Ax$ is invertible.	f. The transformation $x \mapsto Ax$ is one-to-one.
n. Col $A = \mathbb{R}^m$.	b. A is row equivalent to I.	q. Nul $A = \{0\}$.
o. dim Col $A = m$.	l. A^T is invertible.	r. dim Nul $A = 0$.
*. rank $A = m$.	p. rank $A = n$.	*. rank $A = p$.

With many concepts to learn in Sections 2.8 and 2.9, you need to be careful not to combine terms in ways that are undefined, even though they may sound reasonable to you. For example, after you finish your work on this section you should recognize that the following phrases (which have appeared on my students' papers) are meaningless: "the basis of a matrix," "the dimension of a basis," and "the rank of a basis."

SOLUTIONS TO EXERCISES

1. If $[x]_B = \begin{bmatrix} 3 \\ 2 \end{bmatrix}$, then x is formed from b_1 and b_2 using weights 3 and 2:

$$x = 3b_1 + 2b_2 = 3\begin{bmatrix} 1 \\ 1 \end{bmatrix} + 2\begin{bmatrix} 2 \\ -1 \end{bmatrix} = \begin{bmatrix} 7 \\ 1 \end{bmatrix}$$

7. The figure in the text suggests that $\mathbf{w} = 2\mathbf{b}_1 - \mathbf{b}_2$ and $\mathbf{x} = 1.5\mathbf{b}_1 + .5\mathbf{b}_2$, in which case,

$$[\mathbf{w}]_B = \begin{bmatrix} 2 \\ -1 \end{bmatrix} \text{ and } [\mathbf{x}]_B = \begin{bmatrix} 1.5 \\ .5 \end{bmatrix}.$$

To confirm $[\mathbf{x}]_B$, compute

$$1.5\mathbf{b}_1 + .5\mathbf{b}_2 = 1.5\begin{bmatrix} 3 \\ 0 \end{bmatrix} + .5\begin{bmatrix} -1 \\ 2 \end{bmatrix} = \begin{bmatrix} 4 \\ 1 \end{bmatrix} = \mathbf{x}$$

Note: The figures in the text for Exercises 7 and 8 display what Section 4.4 calls *B*-graph paper. See pages 247–248.

Study Tip: Exercises 9–16 make good test questions because they do not require much arithmetic. The problem of finding a basis for a null space is particularly important, because this skill is needed throughout Chapters 5 and 7.

13. The four vectors span the column space H of a matrix that can be reduced to echelon form:

$$\begin{bmatrix} 1 & -3 & 2 & -4 \\ -3 & 9 & -1 & 5 \\ 2 & -6 & 4 & -3 \\ -4 & 12 & 2 & 7 \end{bmatrix} \sim \begin{bmatrix} 1 & -3 & 2 & -4 \\ 0 & 0 & 5 & -7 \\ 0 & 0 & 0 & 5 \\ 0 & 0 & 10 & -9 \end{bmatrix} \sim \begin{bmatrix} 1 & -3 & 2 & -4 \\ 0 & 0 & 5 & -7 \\ 0 & 0 & 0 & 5 \\ 0 & 0 & 0 & 5 \end{bmatrix} \sim \begin{bmatrix} 1 & -3 & 2 & -4 \\ 0 & 0 & 5 & -7 \\ 0 & 0 & 0 & 5 \\ 0 & 0 & 0 & 0 \end{bmatrix}$$

Columns 1, 3, and 4 of the original matrix form a basis for H, so dim $H = 3$.

17. a. Check the definition of coordinates relative to a basis.

b. Dimension is defined only for a subspace.   **c.** See the sentence before Example 3.

d. See the Rank Theorem.   **e.** See the Basis Theorem.

19. The text answer uses the Rank Theorem, which is fine. However, you can also answer Exercises 19–22 without explicit reference to the Rank Theorem. For instance, in Exercise 19, if the null space of a matrix A is three-dimensional, then the equation $A\mathbf{x} = \mathbf{0}$ has three free variables, and three of the columns of A are nonpivot columns. Since a 5×7 matrix has seven columns, A must have four pivot columns (which form a basis of Col A). So rank $A = $ dim Col $A = 4$.

22. The wording of this problem is poor, because the phrase "it spans a four-dimensional sub-space" may be unclear. Here is a revision that I will put in later printings of the third edition:

Show that a set $\{\mathbf{v}_1, ..., \mathbf{v}_5\}$ in $\mathbb{R}^n$ is linearly dependent if dim Span$\{\mathbf{v}_1, ..., \mathbf{v}_5\} = 4$.

25. The text has the solution. This question provides a good way to test knowledge of the Basis Theorem.

27. a. Start with $B = [\mathbf{b}_1 \ \cdots \ \mathbf{b}_p]$ and $A = [\mathbf{a}_1 \ \cdots \ \mathbf{a}_q]$, where $q > p$. For $j = 1, ..., q$, the vector $\mathbf{a}_j$ is in W. Since the columns of B span W, the vector $\mathbf{a}_j$ is in the column space of B. That is, $\mathbf{a}_j = B\mathbf{c}_j$ for some vector $\mathbf{c}_j$ of weights. Note that $\mathbf{c}_j$ is in $\mathbb{R}^p$ because B has p columns.

b. Let $C = [\mathbf{c}_1 \ \cdots \ \mathbf{c}_q]$. Then C is a $p \times q$ matrix because each of the q columns is in $\mathbb{R}^p$. By hypothesis, q is larger than p, so C has more columns than rows. By a theorem, the columns of C are linearly dependent and there exists a nonzero vector $\mathbf{u}$ in $\mathbb{R}^q$ such that $C\mathbf{u} = \mathbf{0}$.

c. From part (a) and the definition of matrix multiplication

$$A = [\mathbf{a}_1 \ \cdots \ \mathbf{a}_q] = [B\mathbf{c}_1 \ \cdots \ B\mathbf{c}_q] = BC$$

From part (b), $A\mathbf{u} = (BC)\mathbf{u} = B(C\mathbf{u}) = B\mathbf{0} = \mathbf{0}$. Since $\mathbf{u}$ is nonzero, the columns of A are linearly dependent.

Mastering Linear Algebra Concepts: Dimension and Rank

The concepts of dimension and rank are relatively simple, but they are used so often later that they deserve a review sheet. Pay attention to how they are used in the sentences of the Invertible Matrix Theorem. "Dimension" is always attached to a subspace (not a matrix or vector or basis), and "rank" is attached to a matrix (not a subspace or other object).

• definitions	Pages 177 and 178
• geometric interpretations	Fig. 1, a coordinate system on a 2-dim subspace
• special cases	Paragraph before Example 2
• examples and counterexamples	Examples 2 and 3
• typical computations	Practice Problem 1, Exercises 9–16 and 19–21

MATLAB   rank

You can use **ref(A)** to check the rank of A, but roundoff error or an extremely small pivot entry can produce an incorrect echelon form. A more reliable command is **rank(A)**.

Chapter 2   SUPPLEMENTARY EXERCISES ————————————

7. Since $A^{-1}B$ is the solution of $AX = B$, row reduction of $[A \quad B]$ to $[I \quad X]$ produces $X = A^{-1}B$.

   See Exercise 12 in Section 2.2. In fact, $A^{-1}B = \begin{bmatrix} 10 & -1 \\ 9 & 10 \\ -5 & -3 \end{bmatrix}$.

11. **c.** When $x_1, \ldots, x_n$ are distinct, the columns of V are linearly independent, by (b). By the Invertible Matrix Theorem, V is invertible and its columns span $\mathbb{R}^n$. So, for every vector $\mathbf{y} = (y_1, \ldots, y_n)$ in $\mathbb{R}^n$, there is a vector $\mathbf{c}$ such that $V\mathbf{c} = \mathbf{y}$. Let p be the polynomial whose coefficients are listed in $\mathbf{c}$. Then, by (a), p is an interpolating polynomial for $(x_1, y_1), \ldots, (x_n, y_n)$.

17. The text has a solution. In addition, note that it is possible that BA is invertible. For example, let C be an invertible 4×4 matrix and construct $A = \begin{bmatrix} C \\ 0 \end{bmatrix}$ and $B = [C^{-1} \quad 0]$. Then $BA = I_4$, which is invertible.

Chapter 2   GLOSSARY CHECKLIST ————————————

Check your knowledge by attempting to write definitions of the terms below. Then compare your work with the definitions given in the text's Glossary. Ask your instructor which definitions, if any, might appear on a test.

associative law of multiplication: . . . .

basis (for a subspace H of $\mathbb{R}^n$, §2.8):   A set $\mathcal{B} = \{\mathbf{v}_1, \ldots, \mathbf{v}_p\}$ in $\mathbb{R}^n$ such that: . . . .

block matrix: *See* partitioned matrix.

block matrix multiplication: The . . . multiplication of . . . as if . . . .

column space (of an $m \times n$ matrix A, §2.8):   The set Col A of . . . .

column sum: The sum of . . . .

commuting matrices: Two matrices A and B such that . . . .

composition of linear transformations: A mapping produced by applying . . . .

conformable for block multiplication: Two partitioned matrices A and B such that . . . .

consumption matrix: A matrix in the . . . model whose columns are . . . .

coordinate vector of x relative to a basis $B = \{b_1, ..., b_p\}$ **(§2.9):**   The vector $[x]_B$ whose entries $c_1, ..., c_p$ satisfy . . . .

diagonal entries (in a matrix):   Entries having . . . .

diagonal matrix: A square matrix . . . whose entries are . . . .

dimension (of a subspace, §2.9):   The number . . . .

determinant $\left(\text{of } A = \begin{bmatrix} a & b \\ c & d \end{bmatrix} \right)$:   The number . . . , denoted by . . . .

distributive laws: (left) . . . (right) . . . .

elementary matrix: An invertible matrix that results by . . . .

final demand vector (or **bill of final demands**):   The vector **d** in the . . . model that lists . . . . The vector **d** can represent . . . .

flexibility matrix: A matrix whose jth column gives . . . of an elastic beam at specified points when . . . is applied at . . . .

Householder reflection: A transformation $x \mapsto Qx$, where $Q = $ . . . .

identity matrix: The $n \times n$ matrix I or I_n with . . . .

inner product: A matrix product . . . where **u** and **v** are . . . .

input-output matrix: *See* consumption matrix.

input-output model: *See* Leontief input-output model.

intermediate demands: Demands for goods or services that . . . .

inverse (of an $n \times n$ matrix A):   An $n \times n$ matrix A^{-1} such that . . . .

invertible linear transformation: A linear transformation $T: \mathbb{R}^n \rightarrow \mathbb{R}^n$ such that there exists . . . .

invertible matrix: A square matrix that . . . .

ladder network: An electrical network assembled by connecting . . . .

left inverse (of A):   Any rectangular matrix C such that . . . .

Leontief input-output model (or **Leontief production equation**):   The equation . . . , where . . . .

lower triangular matrix: A matrix with . . . .

lower triangular part (of A):   A . . . matrix whose entries on . . . .

LU factorization: The representation of a matrix A in the form $A = LU$, where L is . . . and U is . . . .

main diagonal (of a matrix):   The location of the . . . .

null space (of an $m \times n$ matrix A, §2.8):   The set Nul A of all . . . .

outer product: A matrix product . . . where **u** and **v** are . . . .

partitioned matrix: A matrix whose entries are . . . . Sometimes called . . . .

permuted lower triangular matrix: A matrix such that . . . .

permuted LU factorization: The representation of a matrix A in the form $A = LU$, where L is . . . and U is . . . .

production vector: The vector in the . . . model that lists . . . .

rank (of a matrix A, §2.9):   . . . .

right inverse (of A):   Any rectangular matrix C such that . . . .

row-column rule: The rule for computing a product AB in which . . . .

Schur complement: A certain matrix formed from the blocks of a 2×2 partitioned matrix $A = [A_{ij}]$. If A_{11} is invertible, its Schur complement is given by . . . . If A_{22} is invertible, its Schur complement is given by . . . .

stiffness matrix: The inverse of a . . . matrix. The jth column of a stiffness matrix gives . . . at specified points on an elastic beam in order to produce . . . .

subspace (of $\mathbb{R}^n$, §2.8):   A subset H of $\mathbb{R}^n$ with the properties: . . . .

transfer matrix: A matrix A associated with an electrical circuit having input and output terminals, such that . . . .

transpose (of A):   An $n \times m$ matrix A^T whose . . . are the corresponding . . . of . . . .

unit consumption vector: A column vector in the . . . model that lists . . . .

unit lower triangular matrix: A matrix with . . . .

upper triangular matrix: A matrix U with . . . .

Vandermonde matrix: An $n \times n$ matrix V or its transpose, of the form . . . .

3 Determinants

3.1   INTRODUCTION TO DETERMINANTS

This section is relatively short and easy.   Some exercises provide computational practice and others allow you to discover properties of determinants to be studied in the next section.   You will enjoy Section 3.2 more if you finish your work on this section first.

KEY IDEAS

The second paragraph of the section sets the stage for what follows.   Read it quickly, without worrying about the details of the row operations.   The main idea is that a multiple of the determinant of A is a number that appears along the diagonal of an echelon form of A, and this number is nonzero if and only if the matrix is invertible.   Later (in Section 3.3) you will see why this idea is important.   For now, this 3×3 case is only used to motivate the definition of det A.

Determinants are defined here via a cofactor expansion along the first row.   Since the cofactors involve determinants of smaller matrices, the definition is said to be *recursive*.   For each $n \geq 2$, the determinant of an $n \times n$ matrix is based on the definition of the determinant of an $(n-1) \times (n-1)$ matrix. There are other equivalent definitions of the determinant, but we shall not digress to discuss them.

Study Tip: Watch how parentheses are used in Example 2 to avoid a common mistake.   The cofactor expansion puts a minus sign in front of a_{32} because $(-1)^{3+2} = -1$.   Since a_{32} happens to be negative, the correct term in the expansion is $-(-2)$ det A_{32}, *not* -2 det A_{32}.

SOLUTIONS TO EXERCISES

1. By definition, det A is computed via a cofactor expansion along the first row:

$$\begin{vmatrix} 3 & 0 & 4 \\ 2 & 3 & 2 \\ 0 & 5 & -1 \end{vmatrix} = 3\begin{vmatrix} 3 & 2 \\ 5 & -1 \end{vmatrix} - 0\begin{vmatrix} 2 & 2 \\ 0 & -1 \end{vmatrix} + 4\begin{vmatrix} 2 & 3 \\ 0 & 5 \end{vmatrix}$$

$$= 3(-3-10) - 0 + 4(10-0) = -39 + 40 = 1$$

For comparison, a cofactor expansion down the second column yields

$$\begin{vmatrix} 3 & 0 & 4 \\ 2 & 3 & 2 \\ 0 & 5 & -1 \end{vmatrix} = (-1)^{1+2} \cdot 0 \begin{vmatrix} 2 & 2 \\ 0 & -1 \end{vmatrix} + (-1)^{2+2} \cdot 3 \begin{vmatrix} 3 & 4 \\ 0 & -1 \end{vmatrix} + (-1)^{3+2} \cdot 5 \begin{vmatrix} 3 & 4 \\ 2 & 2 \end{vmatrix}$$

$$= 0 + 3(-3-0) - 5(6-8) = -9 + 10 = 1$$

Study Tip: To save time, omit the zero terms in a cofactor expansion, but be careful to use the proper plus or minus signs with the nonzero terms.

7. By definition,

$$\begin{vmatrix} 4 & 3 & 0 \\ 6 & 5 & 2 \\ 9 & 7 & 3 \end{vmatrix} = 4\begin{vmatrix} 5 & 2 \\ 7 & 3 \end{vmatrix} - 3\begin{vmatrix} 6 & 2 \\ 9 & 3 \end{vmatrix} = 4(15-14) - 3(18-18) = 4$$

Using the second column of A instead,

$$\begin{vmatrix} 4 & 3 & 0 \\ 6 & 5 & 2 \\ 9 & 7 & 3 \end{vmatrix} = -3\begin{vmatrix} 6 & 2 \\ 9 & 3 \end{vmatrix} + 5\begin{vmatrix} 4 & 0 \\ 9 & 3 \end{vmatrix} - 7\begin{vmatrix} 4 & 0 \\ 6 & 2 \end{vmatrix}$$

$$= -3(18-18) + 5(12-0) - 7(8-0) = 0 + 60 - 56 = 4$$

13. Row 2 or column 2 are the best choices because they contain the most zeros. We'll use row 2. Since the only nonzero entry in that row is 2, the determinant is $(-1)^{2+3} \cdot 2 \cdot 2A_{23}$.

$$\det A = (-1)^{2+3} \cdot 2 \begin{vmatrix} 4 & 0 & -7 & 3 & -5 \\ 0 & 0 & 2 & 0 & 0 \\ 7 & 3 & -6 & 4 & -8 \\ 5 & 0 & 5 & 2 & -3 \\ 0 & 0 & 9 & -1 & 2 \end{vmatrix} = (-2) \cdot \begin{vmatrix} 4 & 0 & 3 & -5 \\ 7 & 3 & 4 & -8 \\ 5 & 0 & 2 & -3 \\ 0 & 0 & -1 & 2 \end{vmatrix}$$

The best choice for this 4×4 determinant is to expand down the second column. Notice that the cofactor associated with the 3 in the (2, 2) position is the (2, 2)-cofactor of the 4×4 matrix. The original location of the "3" in the 5×5 matrix is irrelevant.

$$\det A = (-2) \cdot (-1)^{2+2}(3) \cdot \begin{vmatrix} 4 & 3 & -5 \\ 5 & 2 & -3 \\ 0 & -1 & 2 \end{vmatrix} = -6 \begin{vmatrix} 4 & 3 & -5 \\ 5 & 2 & -3 \\ 0 & -1 & 2 \end{vmatrix}$$

Finally, use column 1 (although row 3 would work as well).

$$\det A = (-6) \cdot \left(4 \begin{vmatrix} 2 & -3 \\ -1 & 2 \end{vmatrix} - 5 \begin{vmatrix} 3 & -5 \\ -1 & 2 \end{vmatrix} + 0 \right)$$

$$= -6[4(4-3) - 5(6-5)] = -6(4-5) = 6$$

Checkpoint: Try to complete the following statement: "If the kth column of the $n \times n$ identity matrix is replaced by a column vector $\mathbf{x}$ whose entries are $x_1, ..., x_n$, then the determinant of the resulting matrix is _____." To discover the answer, compute the determinants of the following matrices:

a. $\begin{bmatrix} 1 & 3 & 0 & 0 \\ 0 & 4 & 0 & 0 \\ 0 & 5 & 1 & 0 \\ 0 & 6 & 0 & 1 \end{bmatrix}$
b. $\begin{bmatrix} 1 & 0 & 3 & 0 \\ 0 & 1 & 4 & 0 \\ 0 & 0 & 5 & 0 \\ 0 & 0 & 6 & 1 \end{bmatrix}$
c. $\begin{bmatrix} 1 & 0 & 3 & 0 & 0 \\ 0 & 1 & 4 & 0 & 0 \\ 0 & 0 & 5 & 0 & 0 \\ 0 & 0 & 6 & 1 & 0 \\ 0 & 0 & 7 & 0 & 1 \end{bmatrix}$

19. $\det \begin{bmatrix} a & b \\ c & d \end{bmatrix} = ad - bc$, and $\det \begin{bmatrix} c & d \\ a & b \end{bmatrix} = cb - da = -\det \begin{bmatrix} a & b \\ c & d \end{bmatrix}$.

Interchanging two rows reverses the sign of the determinant, at least for the 2×2 case. Perhaps this is true for larger matrices.

25. The matrix is triangular, so use Theorem 2.

$$\det \begin{bmatrix} 1 & 0 & 0 \\ 0 & 1 & 0 \\ 0 & k & 1 \end{bmatrix} = 1 \cdot 1 \cdot 1 = 1 \qquad \text{Product of the diagonal entries}$$

31. A 3×3 row replacement matrix has one of the following forms:

$$\begin{bmatrix} 1 & 0 & 0 \\ k & 1 & 0 \\ 0 & 0 & 1 \end{bmatrix}, \quad \begin{bmatrix} 1 & 0 & 0 \\ 0 & 1 & 0 \\ k & 0 & 1 \end{bmatrix}, \quad \begin{bmatrix} 1 & 0 & 0 \\ 0 & 1 & 0 \\ 0 & k & 1 \end{bmatrix},$$

$$\begin{bmatrix} 1 & k & 0 \\ 0 & 1 & 0 \\ 0 & 0 & 1 \end{bmatrix}, \quad \begin{bmatrix} 1 & 0 & k \\ 0 & 1 & 0 \\ 0 & 0 & 1 \end{bmatrix}, \quad \begin{bmatrix} 1 & 0 & 0 \\ 0 & 1 & k \\ 0 & 0 & 1 \end{bmatrix}$$

In each case the matrix is triangular with l's on the diagonal, so its determinant equals 1. The determinant of a row replacement matrix is l, at least for the 3×3 case.   Perhaps this is true for larger matrices.

37. $\det A = \det \begin{bmatrix} 3 & 1 \\ 4 & 2 \end{bmatrix} = 3(2) - 1(4) = 6 - 4 = 2.$ Since $5A = \begin{bmatrix} 5 \cdot 3 & 5 \cdot 1 \\ 5 \cdot 4 & 5 \cdot 2 \end{bmatrix}$,

$$\det 5A = (5 \cdot 3)(5 \cdot 2) - (5 \cdot 1)(5 \cdot 4) = 150 - 100 = 50.$$

So, $\det 5A \neq 5 \cdot \det A$.   Can you see what the true relation between $\det 5A$ and $\det A$ really is, at least for this example? What about $\det 5A$ for any 2×2 matrix? Try to guess (and perhaps verify) a formula for $\det rA$, where r is any scalar and A is any $n \times n$ matrix.

39. a. See the paragraph preceding the definition of det A.

   b. See the definition of cofactor, preceding Theorem 1.

41. $\det[\mathbf{u} \quad \mathbf{v}] = \det \begin{bmatrix} 3 & 1 \\ 0 & 2 \end{bmatrix} = 6$, $\det[\mathbf{u} \quad \mathbf{x}] = \det \begin{bmatrix} 3 & x \\ 0 & 2 \end{bmatrix} = 6$, and the areas of the parallelograms

determined by $[\mathbf{u} \quad \mathbf{v}]$ and $[\mathbf{u} \quad \mathbf{x}]$ both equal 6.   To see why the areas are equal, consider the parallelograms determined by $\mathbf{u} = (3, 0)$ and $\mathbf{v} = (1, 2)$ and by $\mathbf{u}$ and $\mathbf{x} = (x, 2)$:

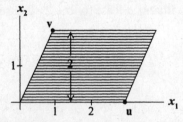 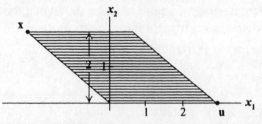

The parallelogram on the left is determined by $\mathbf{u}$ and $\mathbf{v}$ (and the vertices $\mathbf{u} + \mathbf{v}$ and $\mathbf{0}$). Its base is 3 and its altitude is 2, so the area is (base)(altitude) = 6. The parallelogram on the right, determined by $\mathbf{u}$ and $x = (x, 2)$, has the same base. Also, the altitude is 2 for any value of x, so the area again equals $3 \cdot 2 = 6$.

Answer to Checkpoint: **a.** 4 **b.** 5 **c.** 5 "If the kth column of the $n \times n$ identity matrix is replaced by a column vector $\mathbf{x}$ whose entries are x_1, ..., x_n, then the determinant of the resulting matrix is x_k." Can you explain why this is true? You'll learn the answer when you begin Section 3.3.

3.2 PROPERTIES OF DETERMINANTS

This section presents the main properties of determinants, gives an efficient method of computation, and proves that a matrix is invertible if and only if its determinant is nonzero.

KEY IDEAS

It is not surprising that row operations relate nicely to determinants. After all, we found the definition of a 3×3 determinant by row reducing a 3×3 matrix. Theorem 3 can be rephrased informally as follows:

 a. *Adding a multiple of one row (or column) of A to another does not change the determinant.*

 b. *Interchanging two rows (or columns) of A reverses the sign of the determinant.*

 c. *A constant may be factored out of one row (or column) of the determinant of A.*

The other properties to learn are stated in Theorems 4, 5, and 6, together with the boxed formula for det *A* on page 194. Theorem 4 is sometimes stated as: *A square matrix is nonsingular if and only if its determinant is nonzero.* Theorems 4 and 6 will be used extensively in Chapter 5.

 Your instructor may or may not want you to know the (multi-) linearity property on page 197. This property is important in more advanced courses but is not used later in the text. *Warning*: in general, $\det(A + B)$ is *unequal* to det *A* + det *B*.

SOLUTIONS TO EXERCISES

1. Rows 1 and 2 are interchanged, so the determinant changes sign.

7.
$$\begin{vmatrix} 1 & 3 & 0 & 2 \\ -2 & -5 & 7 & 4 \\ 3 & 5 & 2 & 1 \\ 1 & -1 & 2 & -3 \end{vmatrix} = \begin{vmatrix} 1 & 3 & 0 & 2 \\ 0 & 1 & 7 & 8 \\ 0 & -4 & 2 & -5 \\ 0 & -4 & 2 & -5 \end{vmatrix} = \begin{vmatrix} 1 & 3 & 0 & 2 \\ 0 & 1 & 7 & 8 \\ 0 & 0 & 30 & 27 \\ 0 & 0 & 30 & 27 \end{vmatrix} = \begin{vmatrix} 1 & 3 & 0 & 2 \\ 0 & 1 & 7 & 8 \\ 0 & 0 & 30 & 27 \\ 0 & 0 & 0 & 0 \end{vmatrix} = 0$$

Note, the second array already shows that the determinant is zero, because two rows are equal, as in Example 3.

Study Tip: In general, computation of a 3×3 determinant by row reduction takes 10 multiplications (and divisions), but cofactor expansion only takes 9 multiplications. At $n = 4$, the advantage switches to row reduction, which requires 23 multiplications, cofactor expansion 40 (9 for four 3×3 determinants, plus 4 multiplications of a_{ij} times det A_{ij}). Often, the best strategy is to combine the two techniques, as in Exercises 11–14.

13. Use row or column operations whenever convenient to create a row or column that has only one nonzero entry. (I recommend using only row operations, because you already have experience with them.) Then use a cofactor expansion to reduce the size of the matrix.

$$
\begin{vmatrix} 2 & 5 & 4 & 1 \\ 4 & 7 & 6 & 2 \\ 6 & -2 & -4 & 0 \\ -6 & 7 & 7 & 0 \end{vmatrix} = \begin{vmatrix} 2 & 5 & 4 & 1 \\ 0 & -3 & -2 & 0 \\ 6 & -2 & -4 & 0 \\ -6 & 7 & 7 & 0 \end{vmatrix}
$$

Zero created
in column 4

$$
= - \begin{vmatrix} 0 & -3 & -2 \\ 6 & -2 & -4 \\ -6 & 7 & 7 \end{vmatrix}
$$

Result of cofactor
expansion down column 4

$$
= - \begin{vmatrix} 0 & -3 & -2 \\ 6 & -2 & -4 \\ 0 & 5 & 3 \end{vmatrix}
$$

Zero created
in column 1

$$
= -(-6) \begin{vmatrix} -3 & -2 \\ 5 & 3 \end{vmatrix}
$$

Result of cofactor
expansion down column 1

$$
= 6 \cdot (-9 + 10) = 6
$$

19.
$$
\begin{vmatrix} a & b & c \\ 2d+a & 2e+b & 2f+c \\ g & h & i \end{vmatrix} = \begin{vmatrix} a & b & c \\ 2d & 2e & 2f \\ g & h & i \end{vmatrix}
$$

$(-1) \cdot$ row 1 added
to row 2

$$
= 2 \cdot \begin{vmatrix} a & b & c \\ d & e & f \\ g & h & i \end{vmatrix}
$$

2 factored
out of row 2

$$
= 2 \cdot 7 = 14
$$

25. By Theorem 4 and the IMT, the set $\{v_1, v_2, v_3\}$ is linearly independent if and only if $\det[v_1 \ \ v_2 \ \ v_3] \neq 0$. Rather than use row operations on $[v_1 \ \ v_2 \ \ v_3]$, you might choose to expand the determinant by cofactors of the third column:

$$
\begin{vmatrix} 7 & -8 & 7 \\ -4 & 5 & 0 \\ -6 & 7 & -5 \end{vmatrix} = 7 \begin{vmatrix} -4 & 5 \\ -6 & 7 \end{vmatrix} + (-5) \begin{vmatrix} 7 & -8 \\ -4 & 5 \end{vmatrix} = 7(-28+30) - 5(35-32)
$$

$$
= 7(2) - 5(3) = -1
$$

The determinant is nonzero, so the vectors are linearly independent.

Study Tip: For 3×3 matrices, some students tend to prefer the special trick suggested for Exercises 15–18 in Section 3.1, even though in general there are 12 multiplications instead of the 9 multiplications needed for cofactor expansion. Note, however, that numbers in the special method can sometimes be large. For comparison, here are those computations for the matrix studied above in Exercise 25:

$$\begin{vmatrix} 7 & -8 & 7 \\ -4 & 5 & 0 \\ -6 & 7 & -5 \end{vmatrix} \begin{matrix} 7 & -8 \\ -4 & 5 \\ -6 & 7 \end{matrix}$$

$$\det\begin{bmatrix} \mathbf{v}_1 & \mathbf{v}_2 & \mathbf{v}_3 \end{bmatrix} = 7(5)(-5) + (-8)(0)(-6) + 7(-4)(7)$$
$$- (-6)(5)(7) - 7(0)(7) - (-5)(-4)(-8)$$

$$= -175 + 0 + (-196) - (-210) - 0 - (-160) = -1$$

27. a. See Theorem 3.

    b. See the paragraph following Example 2.

    c. See the remark following Theorem 4.

    d. See the warning after Example 5.

31. Since the determinant is multiplicative (Theorem 6),

$$(\det A)(\det A^{-1}) = \det(AA^{-1}) = \det I = 1. \quad \text{So } \det A^{-1} = 1/\det A.$$

Study Tip: The result of Exercise 31 might be useful on a test.

33. By Theorem 6 (twice), $\det AB = (\det A)(\det B) = (\det B)(\det A) = \det BA$.

35. By Theorem 5, $\det U^T = \det U$. So, by Theorem 6,

$$\det U^T U = (\det U^T)(\det U) = (\det U)^2$$

If $U^T U = I$, then $(\det U)^2 = \det I = 1$, which implies that $\det U = \pm 1$.

37. The solution is in the text. (The determinant of a triangular matrix is the product of the entries on the main diagonal.)

Study Tip: Exercises 15–26, 39, and 40 make good test questions because they check your knowledge of determinant properties without requiring much computation. Exercise 39(b) is the one most likely to be answered incorrectly. What would be the answer to 39(b) if A were 4×4?

43. Compute $\det A$ by a cofactor expansion down column 3:

$$\det A = (u_1 + v_1) \cdot \det A_{13} - (u_2 + v_2) \cdot \det A_{23} + (u_3 + v_3) \cdot \det A_{33}$$

$$= u_1 \cdot \det A_{13} - u_2 \cdot \det A_{23} + u_3 \cdot \det A_{33} + v_1 \cdot \det A_{13} - v_2 \cdot \det A_{23} + v_3 \cdot \det A_{33}$$

$$= u_1 \cdot \det B_{13} - u_2 \cdot \det B_{23} + u_3 \cdot \det B_{33} + v_1 \cdot \det C_{13} - v_2 \cdot \det C_{23} + v_3 \cdot \det C_{33}$$

$$= \det B + \det C$$

45. Suppose A is $m \times n$ with more columns than rows. Then $A^T A$ is $n \times n$ and must be singular. If A is generated with random entries, then AA^T will be nonsingular (invertible) practically all the time. Try to explain why these statements should be true. (Use the IMT.)

MATLAB   Computing Determinants

To compute det A, set $\mathbf{U = A}$ and then repeatedly use the commands $\mathbf{U\ =\ gauss(U,r)}$ and $\mathbf{U\ =\ swap(U,r,s)}$ as needed to reduce A to an echelon form U. Then, except for a + or − sign (depending on how many times you swap rows), the determinant of A is given by the command

```
prod(diag(U))
```

The command $\mathbf{diag(U)}$ extracts the diagonal entries of U and places them in a column vector, and $\mathbf{prod}$ computes the product of those entries. You can also use $\mathbf{det(A)}$ to check your work, but the longer sequence of commands helps you think about the *process* of computing det A.

3.3   CRAMER'S RULE, VOLUME, AND LINEAR TRANSFORMATIONS

This section will be a valuable reference for students who plan to take a course in multivariable calculus. Mathematics and statistics majors probably will encounter the material here several times. Also, economics students and engineers (particularly electrical engineers) are likely to need Cramer's rule and some of the supplementary exercises in later courses.

KEY IDEAS

The main results of the section are stated in Theorems 7, 8, 9, and 10. The proof of Theorem 7 is simple and yet involves three important ideas: the definition of a matrix product, the multiplicative property of the determinant, and the evaluation of a determinant by cofactors. Check with your instructor about whether you should be able to reproduce the proof of Theorem 7.

A heuristic proof of Theorem 9 for 2×2 matrices is given in an appendix at the end of this section. Theorem 10 provides a key idea in calculus and physics needed for the study of double and triple integrals. The determinant used there in calculus is called a *Jacobian*.

STUDY NOTE

In Exercise 25, you are asked to use Theorem 9 to explain why the determinant of a 3×3 matrix A is zero if and only if A is not invertible. (A similar explanation holds for the 2×2 case.) The answer is in the text, so be sure to work on this before looking at the answer section. *Work on Exercise 25, even if it is not assigned.*

Remember, learning *does* take place when you think hard about an exercise, even when you are unsuccessful, if you try to look at the problem from different angles, browse back through the text, and perhaps look at earlier exercises. *Write* your solution, don't just talk to yourself about what you would write if you had to.

SOLUTIONS TO EXERCISES

1. The system is equivalent to $A\mathbf{x} = \mathbf{b}$, where $A = \begin{bmatrix} 5 & 7 \\ 2 & 4 \end{bmatrix}$ and $\mathbf{b} = \begin{bmatrix} 3 \\ 1 \end{bmatrix}$. Write

$$A_1(\mathbf{b}) = \begin{bmatrix} 3 & 7 \\ 1 & 4 \end{bmatrix}, \quad A_2(\mathbf{b}) = \begin{bmatrix} 5 & 3 \\ 2 & 1 \end{bmatrix}$$
$$\qquad\quad \uparrow \qquad\qquad\qquad \uparrow$$
$$\qquad\quad \mathbf{b} \qquad\qquad\qquad \mathbf{b}$$

and compute

$$\det A = 20 - 14 = 6, \quad \det A_1(\mathbf{b}) = 12 - 7 = 5, \ \det A_2(\mathbf{b}) = 5 - 6 = -1$$

$$x_1 = \frac{\det A_1(\mathbf{b})}{\det A} = \frac{5}{6}, \quad x_2 = \frac{\det A_2(\mathbf{b})}{\det A} = \frac{-1}{6} = \frac{-1}{6}$$

7. The system is equivalent to $A\mathbf{x} = \mathbf{b}$, where $A = \begin{bmatrix} 6s & 4 \\ 9 & 2s \end{bmatrix}$ and $\mathbf{b} = \begin{bmatrix} 5 \\ -2 \end{bmatrix}$. Write

$$A_1(\mathbf{b}) = \begin{bmatrix} 5 & 4 \\ -2 & 2s \end{bmatrix}, \quad A_2(\mathbf{b}) = \begin{bmatrix} 6s & 5 \\ 9 & -2 \end{bmatrix}$$

and compute

$$\det A = 12s^2 - 36 = 12(s^2 - 3) = 12(s - \sqrt{3})(s + \sqrt{3})$$

$$\det A_1(\mathbf{b}) = 10s + 8, \quad \det A_2(\mathbf{b}) = -12s - 45$$

The system has a unique solution when $\det A \neq 0$, that is, when $s \neq \pm\sqrt{3}$. For such a system, the solution is $\mathbf{x} = (x_1, x_2)$, where

$$x_1 = \frac{\det A_1(\mathbf{b})}{\det A} = \frac{10s + 8}{12(s^2 - 3)} = \frac{5s + 4}{6(s^2 - 3)}$$

$$x_2 = \frac{\det A_2(\mathbf{b})}{\det A} = \frac{-12s - 45}{12(s^2 - 3)} = \frac{-4s - 15}{4(s^2 - 3)}$$

13. First, find the cofactors of $A = \begin{bmatrix} 3 & 5 & 4 \\ 1 & 0 & 1 \\ 2 & 1 & 1 \end{bmatrix}$.

$$C_{11} = + \begin{vmatrix} 0 & 1 \\ 1 & 1 \end{vmatrix} = -1, \qquad C_{12} = - \begin{vmatrix} 1 & 1 \\ 2 & 1 \end{vmatrix} = 1, \qquad C_{13} = + \begin{vmatrix} 1 & 0 \\ 2 & 1 \end{vmatrix} = 1$$

$$C_{21} = - \begin{vmatrix} 5 & 4 \\ 1 & 1 \end{vmatrix} = -1, \qquad C_{22} = + \begin{vmatrix} 3 & 4 \\ 2 & 1 \end{vmatrix} = -5, \qquad C_{23} = - \begin{vmatrix} 3 & 5 \\ 2 & 1 \end{vmatrix} = 7$$

$$C_{31} = + \begin{vmatrix} 5 & 4 \\ 0 & 1 \end{vmatrix} = 5, \qquad C_{32} = - \begin{vmatrix} 3 & 4 \\ 1 & 1 \end{vmatrix} = 1, \qquad C_{33} = + \begin{vmatrix} 3 & 5 \\ 1 & 0 \end{vmatrix} = -5$$

Then, arrange the *transpose* of the array of cofactors into the adjugate of A.

$$\text{adj}\, A = \begin{bmatrix} -1 & -1 & 5 \\ 1 & -5 & 1 \\ 1 & 7 & -5 \end{bmatrix}$$

Were you to compute det A now, you could write A^{-1}, but you would still need to check whether your calculations are correct. To build in this check, compute

$$A \cdot \text{adj}\, A = \begin{bmatrix} 3 & 5 & 4 \\ 1 & 0 & 1 \\ 2 & 1 & 1 \end{bmatrix} \begin{bmatrix} -1 & -1 & 5 \\ 1 & -5 & 1 \\ 1 & 7 & -5 \end{bmatrix} = \begin{bmatrix} 6 & 0 & 0 \\ 0 & 6 & 0 \\ 0 & 0 & 6 \end{bmatrix}$$

If any off-diagonal entries in the product are nonzero, or if the diagonal entries are not all the same, then some errors have been made, and you can recheck your cofactor calculations. (One possible mistake is to forget the ± signs in front of the 2 × 2 determinants. Another error is to *not* transpose the array of cofactors.) In this case, the calculations above *are* correct and det A must be 6. So

$$A^{-1} = \frac{1}{\det A} \text{adj}\, A = \frac{1}{6} \begin{bmatrix} -1 & -1 & 5 \\ 1 & -5 & 1 \\ 1 & 7 & -5 \end{bmatrix}$$

19. The parallelogram with vertices (0, 0), (5, 2), (6, 4), (11, 6) is shown below. If no vertex were zero, we would have to translate the parallelogram to the origin by subtracting one vertex from all four vertices. Since one vertex already is zero, use the two vertices adjacent to the origin to construct the columns of A, and compute $|\det A|$.

$$A = \begin{bmatrix} 5 & 6 \\ 2 & 4 \end{bmatrix}, \quad \left(\begin{array}{c}\text{area of the}\\\text{parallelogram}\end{array}\right) = |\det A| = |20 - 12| = 8$$

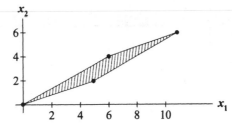

25. The answer is in the text. I hope you took the advice at the beginning of this *Study Guide* section and worked the problem (or at least tried hard to work the problem) before checking the answer section. If you were successful, you should be proud of yourself; you are mastering the material—not only determinants but also linear dependence!

31. Let $\mathbf{x} = \begin{bmatrix} x_1 \\ x_2 \\ x_3 \end{bmatrix}$, $\mathbf{u} = \begin{bmatrix} u_1 \\ u_2 \\ u_3 \end{bmatrix}$, and $A = \begin{bmatrix} a & 0 & 0 \\ 0 & b & 0 \\ 0 & 0 & c \end{bmatrix}$. Also, let S be the unit ball in $\mathbb{R}^3$, whose bounding surface consists of all vectors $\mathbf{u}$ such that $u_1^2 + u_2^2 + u_3^2 = 1$, and let S' be the image of S under the mapping $\mathbf{u} \mapsto A\mathbf{u}$.

a. If $\mathbf{x}$ is in S', then $\mathbf{x} = A\mathbf{u}$ for some $\mathbf{u}$ in S, and $\mathbf{u} = A^{-1}\mathbf{x} = \begin{bmatrix} x_1/a \\ x_2/b \\ x_3/c \end{bmatrix}$.

The condition on u_1, u_2, u_3 shows that $\left(\dfrac{x_1}{a}\right)^2 + \left(\dfrac{x_2}{b}\right)^2 + \left(\dfrac{x_3}{c}\right)^2 = 1$.

b. Since the volume of the unit ball bounded by S is $4\pi/3$ and the determinant of A is abc, Theorem 10 shows that the volume of the region bounded by S' is $4\pi abc/3$.

Appendix: A Geometric Proof of a Determinant Property

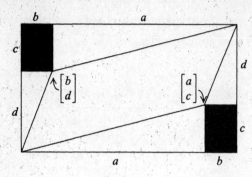

$$\det \begin{bmatrix} a & b \\ c & d \end{bmatrix} = ad - bc$$

$$(a + b)(c + d) = ac + ad + bc + bd$$

$$= \quad -2bc - bd$$

$$= -ac$$

$$\text{(Area of the Parallelogram)} \qquad = \quad ad - bc$$

Chapter 3   GLOSSARY CHECKLIST _____

Check your knowledge by attempting to write definitions of the terms below. Then compare your work with the definitions given in the text's Glossary. Ask your instructor which definitions, if any, might appear on a test.

adjugate (or **classical adjoint**):   The matrix adj A formed from a square matrix A by replacing the (i, j)-entry of A by . . . .

cofactor: A number $C_{ij} = $ . . . , called the (i, j)-*cofactor of A*, where A_{ij} is the submatrix formed by deleting . . . .

cofactor expansion: A formula for det A using cofactors associated with one row or one column, such as for row 1: . . . .

Cramer's Rule: A formula for each entry in . . . .

interchange (matrix):   An elementary matrix obtained by interchanging . . . .

row replacement (matrix):   An elementary matrix obtained from the identity matrix by . . . .

scale by r (matrix):   An elementary matrix obtained by multiplying . . . .

4 Vector Spaces

4.1 VECTOR SPACES AND SUBSPACES

The main focus of the chapter is on $\mathbb{R}^n$ and its subspaces. However, Section 4.1 builds a framework within which the theory for $\mathbb{R}^n$ rests. Most of the exercises in the chapter concern subspaces of $\mathbb{R}^n$, but some are designed to help you learn gradually about other important vector spaces.

KEY IDEAS

A *vector space* is any collection of objects that behave as vectors do in $\mathbb{R}^n$. (The precise meaning of "behave" is described by the axioms on page 217.) A *vector* is simply any object that belongs to a vector space. Arrows, polynomials, and infinite sequences of numbers are all examples of vectors, in different vector spaces.

The most important vector spaces in this text are subspaces of $\mathbb{R}^n$. Visualize subspaces as lines or planes *through the origin*, the origin by itself, or the entire space $\mathbb{R}^n$. In this section, Theorem 1 is a useful tool to show that a set is a vector space. To show that a set is *not* a subspace, show that one of the properties in the subspace definition is violated. (See Exercises 1–4.)

STUDY NOTES

Parts of this chapter are somewhat more theoretical than the earlier chapters, but that is necessary in order to give a solid foundation for the rest of the course. Learn the key definitions and theorems as they appear in the text (rather than waiting until just before an exam). You need this knowledge to get through the conceptual exercises and to be prepared for later sections.

In Example 5, the concept of a function as a single "vector" in a vector space is difficult to absorb on a first reading, and you should not expect to master it in a few days.

The word subspace should usually be accompanied by a phrase such as "of V", or "of $\mathbb{R}^n$". Without such a phrase, the nature of the elements in the subspace is unknown. For instance, the statement "H is a subspace" does not say whether the elements in H are pairs of numbers, or polynomials, or something else. But the phrase "H is a subspace of $\mathbb{P}_3$" includes the information that each vector in H is a polynomial of degree three or less.

Set Notation: The notation introduced in Example 8 is sometimes used in this chapter as an efficient way to describe a set. The symbols and phrases inside the set brackets describe the set. The part to the left of the colon displays the basic nature of the elements in the set (such as vectors in $\mathbb{R}^3$) while the part to the right adds any qualifying conditions that must be met in order for an element to belong to the set (such as all vector entries must be positive). For instance, the set in Example 8 can also be written as

$$H = \left\{ \begin{bmatrix} s \\ t \\ u \end{bmatrix} : s \text{ and } t \text{ are real, and } u = 0 \right\}$$

As another example, the set in Exercise ll can be written as

$$W = \left\{ \begin{bmatrix} a \\ b \\ c \end{bmatrix} : a = 5b + 2c, \text{ where } b \text{ and } c \text{ are arbitrary} \right\}$$

Here, and elsewhere, reference to the scalars b and c as "arbitrary" means that the scalars can be any real numbers.

Study Tip: Review Sections 1.3–1.5 and 1.7 before you reach Section 4.3.

SOLUTIONS TO EXERCISES

1. **a.** V is a subset of $\mathbb{R}^2$, The defining property of V is that the entries of every vector in V are nonnegative. So, if $\mathbf{u}$ and $\mathbf{v}$ are in V, their entries are nonnegative. Since a sum of nonnegative numbers is nonnegative, the vector $\mathbf{u} + \mathbf{v}$ satisfies the condition that defines V. That is, $\mathbf{u} + \mathbf{v}$ is in V.

   b. The text's solution gives a specific $\mathbf{u}$ and c. One specific "counterexample" suffices to show that V is not a vector space. However, you could also simply say, "If any nonzero vector $\mathbf{v}$ in V is multiplied by a negative scalar c, then $c\mathbf{v}$ is not in V because at least one of its entries is negative."

7. Most examples in the text involve integers, to make calculations simple, and one can easily overlook the fact that a subspace must be closed under multiplication by *all* real numbers.

Checkpoint: Let H be the set of all points (x, y, z) in $\mathbb{R}^3$ that satisfy the condition $x^2 + y^2 - z^2 = 0$. Is H a subspace of $\mathbb{R}^3$?

13. **a.** **w** is certainly not one of the three vectors in $\{\mathbf{v}_1, \mathbf{v}_2, \mathbf{v}_3\}$.

   b. Span$\{\mathbf{v}_1, \mathbf{v}_2, \mathbf{v}_3\}$ contains infinitely many vectors.

   c. **w** is in the subspace (of $\mathbb{R}^3$) spanned by $\mathbf{v}_1, \mathbf{v}_2, \mathbf{v}_3$ if and only if the equation $x_1\mathbf{v}_1 + x_2\mathbf{v}_2 + x_3\mathbf{v}_3 = \mathbf{w}$ is consistent (has a solution). Row reduce the augmented matrix:

$$
\begin{bmatrix} 1 & 2 & 4 & 3 \\ 0 & 1 & 2 & 1 \\ -1 & 3 & 6 & 2 \end{bmatrix} \sim \begin{bmatrix} 1 & 2 & 4 & 3 \\ 0 & 1 & 2 & 1 \\ 0 & 5 & 10 & 5 \end{bmatrix} \sim \begin{bmatrix} 1 & 2 & 4 & 3 \\ 0 & 1 & 2 & 1 \\ 0 & 0 & 0 & 0 \end{bmatrix}
$$

   There is no pivot in the augmented column, so the vector equation is consistent, and **w** is in Span$\{\mathbf{v}_1, \mathbf{v}_2, \mathbf{v}_3\}$.

19. Let H be the set of all functions described in (5). Then H is a subset of the vector space V of all real-valued functions, and H consists of all linear combinations of the functions cos wt and sin wt. By Theorem 1, H is a subspace of V, and hence is a vector space.

23. **a.** See Example 5.

   b. See the definition of a vector.

   c. See Exercises 1, 2, or 3.

   d. See the paragraph before Example 6.

   e. See Example 3.

25. Axiom 4 (plus Axiom 2) shows that $\mathbf{0} + \mathbf{w} = \mathbf{w}$. Exercises 25–30 show how facts that we take for granted in $\mathbb{R}^n$ depend only on a few basic properties of $\mathbb{R}^n$, properties that are now axioms for a general vector space.

31. Let H be a subspace of V that contains the vectors **u** and **v**. Since H is closed under multiplication by scalars, H must contain all scalar multiples of **u** and **v**. Since H is also closed under vector addition, it contains all sums of scalar multiples of **u** and **v**. That is, H contains all vectors in Span$\{\mathbf{u}, \mathbf{v}\}$.

Answer to Checkpoint: *H* is not a subspace. Counterexample: $(1, 0, 1)$ and $(0, 1, 1)$ are in *H* and yet their sum, $(1, 1, 2)$, is not in *H* (because $1^2 + 1^2 - 2^2 \neq 0$). Many counterexamples are possible. Only one is needed.

MATLAB   Graphing Functions

The following commands will graph the function *f* in Exercise 37.

```
t = linspace(0,2*pi);
f = t.^0 - 8*cos(t).^2 + 8*cos(t).^4;
grid on
plot(t,f)
```

Here, `t` is a vector with 101 entries, the endpoints of 100 equal subintervals of $[0, 2\pi]$; `cos(t)` is a vector whose entries are the cosines of the corresponding entries in `t`, and `t.^0` is a vector of 1's. In general, `t.^k` is a vector whose entries are the *k*th powers of the corresponding entries in `t`. The command `grid on` (or `off`) turns on (or off) gridlines the next time a display is created. Use `hold on` if you want the next graph to appear on the current display. The command `hold off` makes each graph appear in a separate display. See the topic "Basic Plotting" in MATLAB's `Help` menu for examples of plotting options.

4.2   NULL SPACES, COLUMN SPACES, AND LINEAR TRANSFORMATIONS

Many problems in linear algebra involve a subspace in one way or another. This section provides an opportunity to become comfortable with the concept. The foundation for this section was laid in Sections 1.3 and 1.5. Have you reviewed those sections yet?

KEY IDEAS

Theorems 2 and 3 describe the main types of subspaces. (The proof of Theorem 2 makes a good exam question, because it tests both the definition of Nul *A* and the definition of a subspace.)

Theorem 3 actually has two conclusions: Col *A* is a subspace of $\mathbb{R}^m$.

The first phrase tells us that linear combinations of vectors in Col *A* remain in Col *A*. The phrase " of $\mathbb{R}^m$ " reminds us that each vector has *m* entries (because *A* has *m* rows). A similar remark applies to the statement from Theorem 2 that Nul *A* is a subspace of $\mathbb{R}^n$.

The box after Example 4 shows that the statement " Col $A = \mathbb{R}^m$ " can be added to the list of equivalent statements in Theorem 4 of Section 1.4.

STUDY NOTES

In Example 3, the statement "Nul A = Span{$\mathbf{u}$, $\mathbf{v}$, $\mathbf{w}$}" would be an *explicit* description of Nul A (provided you specify what $\mathbf{u}$, $\mathbf{v}$, and $\mathbf{w}$ are).

In some applications, it is important to know that for a given $m \times n$ matrix A, every equation $A\mathbf{x} = \mathbf{b}$ has a solution (assuming $\mathbf{b}$ is in $\mathbb{R}^m$). Yet it may require some effort to determine whether this is the case. If not every equation $A\mathbf{x} = \mathbf{b}$ has a solution, then not every $\mathbf{b}$ belongs to Col A, and hence Col A is a proper subspace of $\mathbb{R}^m$. One of the goals of the next few sections is to obtain a method for determining when Col $A = \mathbb{R}^m$.

Checkpoint 1:   How many pivot positions does an $m \times n$ matrix A have if Col $A = \mathbb{R}^m$?

Study Tip: Theorems 1, 2, and 3 are the main tools for showing that a set is a vector space (that is, a subspace of some known vector space). Review these theorems now, before starting the exercises.

SOLUTIONS TO EXERCISES

1. Now is the time to learn the *definition* of Nul A. A vector $\mathbf{x}$ is in Nul A precisely when the product $A\mathbf{x}$ is defined and $A\mathbf{x} = \mathbf{0}$. Given $\mathbf{x}$, simply compute $A\mathbf{x}$ to determine whether $A\mathbf{x}$ is zero.

$$A\mathbf{x} = \begin{bmatrix} 3 & -5 & -3 \\ 6 & -2 & 0 \\ -8 & 4 & 1 \end{bmatrix} \begin{bmatrix} 1 \\ 3 \\ -4 \end{bmatrix} = \begin{bmatrix} 0 \\ 0 \\ 0 \end{bmatrix}, \quad \text{so} \quad \begin{bmatrix} 1 \\ 3 \\ -4 \end{bmatrix} \text{ is in Nul } A.$$

Warning: In Exercises 3–6, writing an equation $\mathbf{x} = c\mathbf{u} + d\mathbf{v}$ is not the same as listing, say, the vectors $\mathbf{u}$ and $\mathbf{v}$ that span the null space. The appropriate answer for these exercises is a list of a small finite number of vectors that span Nul A, not a description of *all* the vectors in Nul A.

Study Tip: Try Practice Problem 1 before you work on Exercises 7–14. If you can't find *two* ways to work the practice problem, reread the first paragraph of Section 4.2 (but don't look at it until you have attempted the practice problem).

7. The set W is a subset of $\mathbb{R}^3$. If W were a vector space (under the standard operations in $\mathbb{R}^3$), it would be a subspace of $\mathbb{R}^3$. But W fails *every* property of a subspace, so it is not a vector space. For instance, the vector $(0, 0, 0)$ does not satisfy the condition $a + b + c = 2$, and so the zero vector is not in W.

13. A typical element of W can be written as follows:

$$\begin{bmatrix} c-6d \\ d \\ c \end{bmatrix} = \begin{bmatrix} c \\ 0 \\ c \end{bmatrix} + \begin{bmatrix} -6d \\ d \\ 0 \end{bmatrix} = c \begin{bmatrix} 1 \\ 0 \\ 1 \end{bmatrix} + d \begin{bmatrix} -6 \\ 1 \\ 0 \end{bmatrix} = \begin{bmatrix} 1 & -6 \\ 0 & 1 \\ 1 & 0 \end{bmatrix} \begin{bmatrix} c \\ d \end{bmatrix}$$

$$\quad\quad\quad\quad\quad\quad\quad\quad\quad\quad\quad\quad\quad\quad\quad\quad \uparrow \quad\quad \uparrow$$

$$\quad\quad\quad\quad\quad\quad\quad\quad\quad\quad\quad\quad\quad\quad\quad\quad \mathbf{u} \quad\quad \mathbf{v}$$

Since c and d are any real numbers, this calculation shows that W is a subspace of $\mathbb{R}^3$ (and hence is a vector space) by Theorem 3. Alternatively, this calculation shows that W is the same as Span$\{\mathbf{u}, \mathbf{v}\}$, so W is a subspace of $\mathbb{R}^3$, by Theorem 1.

Checkpoint 2:   Why is W a subspace "of $\mathbb{R}^3$"?

19. The matrix A is 2×5, so vectors in Nul A must have 5 entries and vectors in Col A must have 2 entries. Thus Nul A is a subspace of $\mathbb{R}^5$ and Col A is a subspace of $\mathbb{R}^2$.

Study Tip: Exercises 17–20 may seem simple, but they will help you in Section 4.5.

25. a. Check the definition before Example 1.
  b. See Theorem 2.
  c. See the remark just before Example 4.
  d. See the table that contrasts Nul A and Col A.
  e. See Fig. 2.
  f. See the remark after Theorem 3.

31. The solution in the text shows that T is a linear transformation. If $T(\mathbf{p})$ is the zero vector, then $\mathbf{p}(0) = 0$ and $\mathbf{p}(1) = 0$, by definition of T. One such polynomial is $\mathbf{p}(t) = t(t-1)$. Any other *quadratic* polynomial that vanishes at 0 and 1 must be a multiple of $\mathbf{p}$, so $\mathbf{p}$ spans the kernel of T.

For the range of T, observe that the image of the constant 1 function is $\begin{bmatrix} 1 \\ 1 \end{bmatrix}$, and the image of the polynomial t is $\begin{bmatrix} 0 \\ 1 \end{bmatrix}$. Denote these two images by $\mathbf{u}$ and $\mathbf{v}$, respectively. Since the range of T is a subspace of $\mathbb{R}^2$ that contains $\mathbf{u}$ and $\mathbf{v}$, the range must contain all linear combinations of $\mathbf{u}$ and $\mathbf{v}$. By inspection, $\mathbf{u}$ and $\mathbf{v}$ are linearly independent, so they span $\mathbb{R}^2$. Thus the range of T must contain all of $\mathbb{R}^2$.

37. The vector $\mathbf{w}$ is in Col A because row reduction of the augmented matrix $[A \quad \mathbf{w}]$ shows that the equation $A\mathbf{x} = \mathbf{w}$ is consistent. The vector $\mathbf{w}$ is not in Nul A because $A\mathbf{x}$ is not the zero vector.

Answers to Checkpoints:

1. An $m \times n$ matrix A must have m pivot positions in order for Col A to be all of $\mathbb{R}^m$.
2. W is a subspace " of $\mathbb{R}^3$ " because each vector in W has three entries.

Mastering Linear Algebra Concepts: Vector Space, Subspace, and Column and Null Spaces

A vector space is a nonempty set of objects on which are defined two operations, called addition and multiplication by scalars, that satisfy ten axioms. Make one review sheet for vector space and subspace. Include as much of the definition of a vector space as you are required to know. (Ask your instructor.) Include the full definition of a subspace.

Organize what you have learned in Sections 4.1 and 4.2 (together with Sections 1.3–1.5), using the categories listed below. Your examples of vector spaces will be the same as your examples of subspaces. (See the paragraph just before Example 6 in Section 4.1.) The column space and null space of a matrix should be among your examples of subspaces.

- definitions of vector space, subspace   Pages 217 and 220

- equivalent descriptions   Paragraph before Example 6 in Sec. 4.1

- geometric interpretations   Figs. 6, 7, 9, in Sec. 4.1

- subspaces defined explicitly   Theorems 1, 3; Exercises 15–18 in Sec. 4.1

- subspaces defined implicitly   Theorem 2; Exercises 20(b), 22 in Sec. 4.1; Example 2 in Sec. 4.2

- examples and counterexamples   Examples and exercises in Sec. 4.1, 4.2

- algorithms and computations   Exercises 7–14 in Sec. 4.2

- connections with other concepts   Fig. 2 and Example 8 in Sec. 4.2

The references above are not exhaustive. You can find more facts and examples in the exercises for Sections 4.1 and 4.2. In addition to the review sheet for vector spaces, you should make another sheet for column space and null space. List the two basic definitions, equivalent formulations of each definition, and examples of computations. Also, you could attach a copy of the table on page 232 that contrasts the two types of subspaces. My students say that the table is quite helpful.

4.3   LINEARLY INDEPENDENT SETS; BASES _____

The definition of linear independence carries over from $\mathbb{R}^n$ to any vector space. The geometric interpretations in Chapter 1 of linearly independent and dependent sets should help you visualize these concepts here.

KEY IDEAS

In general, you cannot use an ordinary matrix equation $A\mathbf{x} = \mathbf{0}$ to study linear dependence of $\{\mathbf{v}_1, \ldots, \mathbf{v}_p\}$. You have to work with the vector equation $c_1\mathbf{v}_1 + \cdots + c_p\mathbf{v}_p = \mathbf{0}$, or with Theorem 4, unless the vectors happen to be n-tuples of numbers.

A set $\{\mathbf{v}\}$ with $\mathbf{v} \neq \mathbf{0}$ is linearly independent because the vector equation $x_1\mathbf{v} = \mathbf{0}$ has only the trivial solution. (See Exercise 30 in Section 4.1.) The set $\{\mathbf{0}\}$ is linearly dependent, because the equation $x_1\mathbf{0} = \mathbf{0}$ has many nontrivial solutions.

Theorem 6 is important for later work, but its proof is rather subtle. Study Examples 8 and 9 carefully, as well as the proof of the theorem.

A basis for a vector space V is a set in V that is large enough to span V and small enough to be linearly independent. See also the subsection *Two Views of a Basis*. The plural of *basis* is *bases*.

Warning: If a set in V does not span V, the set may or may not be linearly dependent.

SOLUTIONS TO EXERCISES

1. The complete solution is in the text. For a general set of n vectors in $\mathbb{R}^n$, row operations on a matrix will usually be needed to determine if the matrix has n pivot positions.

7. Again, the solution is in the text. Any set in $\mathbb{R}^n$ with fewer than n vectors cannot span $\mathbb{R}^n$ and therefore cannot be a basis for $\mathbb{R}^n$. Such a set may or may not be linearly independent. What similar statement can you make about a set in $\mathbb{R}^n$ with more than n vectors? See Exercise 8 for ideas.

Study Tip: Theorem 4 in Section 1.4 may help you decide whether a set of vectors spans $\mathbb{R}^n$.

13. The matrix B is in echelon form and displays the pivot columns. A basis for Col A consists

of columns 1 and 2 of A: $\begin{bmatrix} -2 \\ 2 \\ -3 \end{bmatrix}, \begin{bmatrix} 4 \\ -6 \\ 8 \end{bmatrix}$. This is not the only correct choice, but it is the

"standard" choice. A *wrong* choice would be columns 1 and 2 of B. See the Warning after Theorem 6.

For the null space, solve $A\mathbf{x} = \mathbf{0}$:

$$[A \quad \mathbf{0}] \sim [B \quad \mathbf{0}] = \begin{bmatrix} 1 & 0 & 6 & 5 & 0 \\ 0 & 2 & 5 & 3 & 0 \\ 0 & 0 & 0 & 0 & 0 \end{bmatrix} \sim \begin{bmatrix} 1 & 0 & 6 & 5 & 0 \\ 0 & 1 & 5/2 & 3/2 & 0 \\ 0 & 0 & 0 & 0 & 0 \end{bmatrix}$$

Then $x_1 = -6x_3 - 5x_4$, $x_2 = -(5/2)x_3 - (3/2)x_4$, with x_3 and x_4 free. The general solution is

$$\mathbf{x} = \begin{bmatrix} x_1 \\ x_2 \\ x_3 \\ x_4 \end{bmatrix} = \begin{bmatrix} -6x_3 - 5x_4 \\ -(5/2)x_3 - (3/2)x_4 \\ x_3 \\ x_3 \end{bmatrix} = x_3 \begin{bmatrix} -6 \\ -5/2 \\ 1 \\ 0 \end{bmatrix} + x_4 \begin{bmatrix} -5 \\ -3/2 \\ 0 \\ 1 \end{bmatrix}$$

This equation describes *all* vectors in Nul A, not just a basis for Nul A. For a basis, the "standard" choice is $(-6, -5/2, 1, 0)$ and $(-5, -3/2, 0, 1)$. Another choice is $(-12, -5, 2, 0)$ and $(-10, -3, 0, 2)$, which avoids fractions.

Warning: You really need to know the definition of Nul A and the definition of Col A, not just the procedures for finding bases for these spaces. The definitions will help you avoid the fairly common mistake of attempting to use the null space procedure to find a basis for a column space, or vice-versa.

19. You can solve the equation $4\mathbf{v}_1 + 5\mathbf{v}_2 - 3\mathbf{v}_3 = \mathbf{0}$ for any one of the three vectors in terms of the others. By the Spanning Set Theorem, the set spanned by all three vectors is the same as the set spanned by any two of the vectors—any one of the three vectors can be discarded. If you discard $\mathbf{v}_3$, then $\mathbf{v}_1$ and $\mathbf{v}_2$ span H and are obviously linearly independent. Hence $\{\mathbf{v}_1, \mathbf{v}_2\}$ is a basis for H. The same reasoning applies to $\{\mathbf{v}_1, \mathbf{v}_3\}$ and $\{\mathbf{v}_2, \mathbf{v}_3\}$. These are the most likely answers. But, once you realize that $\{\mathbf{v}_1, \mathbf{v}_2\}$ is a basis for H, you can make others, such as $\{\mathbf{v}_1, \mathbf{v}_2 + r\mathbf{v}_1\}$ for any scalar r. Showing that this set *is* a basis for H does take some work, however. You might try to do this. In some courses that spend time on theoretical questions, this problem could appear on a test.

21. a. See the paragraph preceding Theorem 4.

 b. See the definition of a basis.

 c. See Example 3.

 d. See the subsection *Two Views of a Basis*.

 e. See the box before Example 9.

23. Let $A = [\mathbf{v}_1 \quad \mathbf{v}_2 \quad \mathbf{v}_3 \quad \mathbf{v}_4]$. Since A is square and its columns span $\mathbb{R}^4$, the columns of A must be linearly independent, by the Invertible Matrix Theorem. So $\{\mathbf{v}_1, \mathbf{v}_2, \mathbf{v}_3, \mathbf{v}_4\}$ is a basis for $\mathbb{R}^4$.

Checkpoint: Suppose $\{\mathbf{v}_1, ..., \mathbf{v}_n\}$ is a basis for $\mathbb{R}^n$ and A is an invertible $n \times n$ matrix. Explain why $\{A\mathbf{v}_1, ..., A\mathbf{v}_n\}$ is a basis for $\mathbb{R}^n$.

25. The displayed equation shows only that Span $\{\mathbf{v}_1, \mathbf{v}_2, \mathbf{v}_3\}$ *contains* H. In fact, the vectors $\mathbf{v}_1, \mathbf{v}_2, \mathbf{v}_3$ are not all in H, so Span$\{\mathbf{v}_1, \mathbf{v}_2, \mathbf{v}_3\}$ cannot be H. Therefore $\{\mathbf{v}_1, \mathbf{v}_2, \mathbf{v}_3\}$ cannot be a basis for H. (It is easy to check that $\{\mathbf{v}_1, \mathbf{v}_2, \mathbf{v}_3\}$ is a basis for $\mathbb{R}^3$.)

31. (This generalizes Exercise 31 in Section 1.8.) Suppose $\{v_1, \ldots, v_p\}$ is linearly dependent. Then there exist $c_1, \ldots, c_p$, not all zero, such that

$$c_1 v_1 + \cdots + c_p v_p = 0$$

Then, since T is linear,

$$T(c_1 v_1 + \cdots + c_p v_p) = T(0) = 0 \qquad \text{See the boxed statement on page 77.}$$

and

$$c_1 T(v_1) + \cdots + c_p T(v_p) = 0$$

Since not all the c_i are zero, $\{T(v_1), \ldots, T(v_p)\}$ is linearly dependent.

Study Tip: The solution of Exercise 31 illustrates how to use linear dependence: If $\{v_1, \ldots, v_p\}$ is known to be linearly dependent, then you can write $c_1 v_1 + \cdots + c_p v_p = 0$, assume that not all the c_k are zero, and use this equation in some way.

Answer to Checkpoint: Let $B = [v_1 \; \cdots \; v_n]$. Then B is invertible because $\{v_1, \ldots, v_n\}$ is a basis for $\mathbb{R}^n$. Since A is also an invertible $n \times n$ matrix, so is AB; hence the columns of AB, namely $Av_1, \ldots, Av_n$, form a basis for $\mathbb{R}^n$.

Mastering Linear Algebra Concepts: Basis

To the review sheet(s) you have on linear independence, add Examples 1, 2, and 6 from this section. The definition and geometric interpretations are unchanged. Add a note about not using the matrix equation $Ax = 0$ in the general case.

Start a separate review sheet for "basis", even though it involves two other concepts (span and linear independence) already being reviewed.

- definition   Page 238

- geometric interpretation   Fig. 1

- special cases   Standard bases for $\mathbb{R}^n$ and $\mathbb{R}_n$

- examples and counterexamples   Example 10, Exercise 25

- algorithms and computations   Examples 7 and 9; Example 3 in Sec. 4.2

- connections with other concepts   Invertible Matrix Theorem

                                                       Unique Representation Theorem (in Sec. 4.4)

MATLAB `ref` and `cos`

The command `ref(A)` produces the reduced echelon form of A. From that you can write a basis for Col A or write the homogeneous equations that describe Nul A. (Don't forget that A is a coefficient matrix, not an augmented matrix.) MATLAB has another command, `rref`, that works basically the same as `ref` but often is much slower because it checks for rational entries in the matrix.

In some cases, roundoff error or an extremely small pivot entry can cause `ref` to produce an incorrect echelon form. The more reliable singular value decomposition (see Section 7.4) can produce bases for Col A and Nul A, but `ref` is satisfactory for our purposes.

For Exercise 38, see the MATLAB box for Section 4.1.

4.4   COORDINATE SYSTEMS ⸻⸻⸻⸻⸻⸻⸻⸻⸻⸻⸻⸻

This section contains a variety of geometric and algebraic explanations of the idea of a coordinate system for a vector space.

KEY IDEAS

The coordinate mapping from a vector space V (with a basis of n elements) onto $\mathbb{R}^n$ is a rule for giving "$\mathbb{R}^n$-names" to vectors in V in such a way that the vector space structure of V is still visible in $\mathbb{R}^n$. Every vector space calculation in V is precisely mirrored by the same calculation in $\mathbb{R}^n$.

An important special case is when V is itself $\mathbb{R}^n$, and each vector $\mathbf{x}$ and its coordinate vector $[\mathbf{x}]_B$ are related by a matrix equation $\mathbf{x} = P_B[\mathbf{x}]_B$.

Everything in the section depends on the Unique Representation Theorem. The proof of that theorem could appear on an exam because it shows precisely why the two properties of a basis B are important, and it illustrates how linear independence can be used in an argument:

Any vector in V has "coordinates" because B spans V, and the coordinates are uniquely determined because B is linearly independent.

If you are asked to prove the theorem, make sure your proof shows exactly where each property of a basis is needed in the proof. Also, be careful *not* to use a matrix in the proof. The vectors $\mathbf{v}_1, \ldots, \mathbf{v}_n$ cannot be arranged as the columns of an ordinary matrix when the vectors are in some abstract vector space.

Checkpoint: Let $B = \{\mathbf{b}_1, \ldots, \mathbf{b}_n\}$ be a basis for $\mathbb{R}^n$. Apply the Invertible Matrix Theorem to the matrix $A = [\mathbf{b}_1 \ \cdots \ \mathbf{b}_n]$ and deduce the Unique Representation Theorem for the case when $V = \mathbb{R}^n$.

STUDY NOTES

Be careful to distinguish between $\mathbf{x}$ and $[\mathbf{x}]_B$. They are *not equal* in general, even if $\mathbf{x}$ itself is in $\mathbb{R}^n$ (unless B is the standard basis for $\mathbb{R}^n$).

Theorem 8 and Exercises 25 and 26 show that the coordinate mapping translates vector space statements or calculations in V into equivalent (and familiar) calculations in $\mathbb{R}^n$. The table below lists some examples of typical linear algebra statements.

CORRESPONDING STATEMENTS IN ISOMORPHIC VECTOR SPACES

Linear Algebra in V	Matrix Algebra in $\mathbb{R}^n$
a. $\mathbf{u}$, $\mathbf{v}$, and $\mathbf{w}$ are in V	$[\mathbf{u}]_B$, $[\mathbf{v}]_B$, and $[\mathbf{w}]_B$ are in $\mathbb{R}^n$
b. $\mathbf{w}$ is in Span$\{\mathbf{u}, \mathbf{v}\}$, or $\mathbf{w}$ is in the subspace of V spanned by $\mathbf{u}$ and $\mathbf{v}$	$[\mathbf{w}]_B$ is in Span$\{[\mathbf{u}]_B, [\mathbf{v}]_B\}$, or $[\mathbf{w}]_B$ is in the subspace of $\mathbb{R}^n$ spanned by $[\mathbf{u}]_B$ and $[\mathbf{v}]_B$
c. $\mathbf{w} = c\mathbf{u} + d\mathbf{v}$	$[\mathbf{w}]_B = c[\mathbf{u}]_B + d[\mathbf{v}]_B$
d. $\{\mathbf{v}_1, \ldots, \mathbf{v}_p\}$ is lin. indep.	$\{[\mathbf{v}_1]_B, \ldots, [\mathbf{v}_p]_B\}$ is lin. indep.
e. $\{\mathbf{v}_1, \ldots, \mathbf{v}_p\}$ spans V	$\{[\mathbf{v}_1]_B, \ldots, [\mathbf{v}_p]_B\}$ spans $\mathbb{R}^n$
f. $\{\mathbf{v}_1, \ldots, \mathbf{v}_n\}$ is a basis for V	$\{[\mathbf{v}_1]_B, \ldots, [\mathbf{v}_n]_B\}$ is a basis for $\mathbb{R}^n$

SOLUTIONS TO EXERCISES

1. Since $[\mathbf{x}]_B = \begin{bmatrix} 5 \\ 3 \end{bmatrix}$, we have $\mathbf{x} = 5\mathbf{b}_1 + 3\mathbf{b}_2 = 5\begin{bmatrix} 3 \\ -5 \end{bmatrix} + 3\begin{bmatrix} -4 \\ 6 \end{bmatrix} = \begin{bmatrix} 3 \\ -7 \end{bmatrix}$.

7. The B-coordinates of $\mathbf{x}$ are scalars c_1, c_2, c_3 that satisfy $c_1\mathbf{b}_1 + c_2\mathbf{b}_2 + c_3\mathbf{b}_3 = \mathbf{x}$. To solve this vector equation, row reduce the augmented matrix:

$$\begin{bmatrix} 1 & -3 & 2 & 8 \\ -1 & 4 & -2 & -9 \\ -3 & 9 & 4 & 6 \\ \uparrow & \uparrow & \uparrow & \uparrow \\ \mathbf{b}_1 & \mathbf{b}_2 & \mathbf{b}_3 & \mathbf{x} \end{bmatrix} \sim \begin{bmatrix} 1 & -3 & 2 & 8 \\ 0 & 1 & 0 & -1 \\ 0 & 0 & 10 & 30 \end{bmatrix} \sim \begin{bmatrix} 1 & -3 & 0 & 2 \\ 0 & 1 & 0 & -1 \\ 0 & 0 & 1 & 3 \end{bmatrix} \sim \begin{bmatrix} 1 & 0 & 0 & -1 \\ 0 & 1 & 0 & -1 \\ 0 & 0 & 1 & 3 \end{bmatrix}$$

So $[\mathbf{x}]_B = \begin{bmatrix} -1 \\ -1 \\ 3 \end{bmatrix}$.

13. The $\mathcal{B}$–coordinates of **p** are scalars c_1, c_2, c_3 that satisfy

$$c_1(1 + t^2) + c_2(t + t^2) + c_3(1 + 2t + t^2) = \mathbf{p}(t) = 1 + 4t + 7t^2 \tag{1}$$

Multiply out terms on the left:

$$c_1 + c_1 t^2 \quad + \quad c_2 t + c_2 t^2 \quad + \quad c_3 + 2c_3 t + c_3 t^2 = 1 + 4t + 7t^2$$

On the left, group the constant terms, the terms involving t, and the terms involving t^2:

$$(c_1 + c_3) + (c_2 + 2c_3)t + (c_1 + c_2 + c_3)t^2 = 1 + 4t + 7t^2$$

Equate coefficients of like powers of t to obtain the system of equations:

$$
\begin{aligned}
c_1 \quad\quad\ + \ \ c_3 &= 1 \\
c_2 + 2c_3 &= 4 \\
c_1 + c_2 + \ \ c_3 &= 7
\end{aligned}
$$

Row reduce the augmented matrix to obtain

$$
\begin{bmatrix} 1 & 0 & 1 & 1 \\ 0 & 1 & 2 & 4 \\ 1 & 1 & 1 & 7 \end{bmatrix}
\sim
\begin{bmatrix} 1 & 0 & 1 & 1 \\ 0 & 1 & 2 & 4 \\ 0 & 0 & -2 & 2 \end{bmatrix}
\sim
\begin{bmatrix} 1 & 0 & 0 & 2 \\ 0 & 1 & 0 & 6 \\ 0 & 0 & 1 & -1 \end{bmatrix}.
\text{ Thus, } [\mathbf{p}]_{\mathcal{B}} = \begin{bmatrix} 2 \\ 6 \\ -1 \end{bmatrix} \tag{2}
$$

Perhaps you can skip writing the second and third displayed equations and mentally go from (1) directly to the systems of equations. That will save writing time, but mistakes can occur.

A shorter solution uses Theorem 8 and the fact that a calculation in $\mathbb{P}_2$ can be done instead with coordinate vectors relative to the standard basis $\{1, t, t^2\}$. Using this idea, you go directly from equation (1) to the equivalent equation using coordinate vectors:

$$
c_1 \begin{bmatrix} 1 \\ 0 \\ 1 \end{bmatrix} + c_2 \begin{bmatrix} 0 \\ 1 \\ 1 \end{bmatrix} + c_3 \begin{bmatrix} 1 \\ 2 \\ 1 \end{bmatrix} = \begin{bmatrix} 1 \\ 4 \\ 7 \end{bmatrix} \tag{3}
$$

This vector equation is, of course, equivalent to the system of equations above, and you solve it by row reducing the augmented matrix as in (2).

Warning: The second solution for Exercise 13 is faster, but students can easily forget what their calculations mean. I expect my students to *write* about what they are doing, to show that they understand what they are calculating. For an acceptable solution to Exercise 13, write that the $\mathcal{B}$–coordinates of **p** are scalars c_1, c_2, c_3 that satisfy equation (1), write something about using coordinate vectors to express (1) in the equivalent form (3), and then solve (3) by the calculations in (2).

15. a. See the definition of a $\mathcal{B}$-coordinate vector.

    b. See equation (4).

    c. See Example 5.

4-14CHAPTER 4 • Vector Spaces

19. The set S spans V because every $\mathbf{x}$ in V has a representation as a (unique) linear combination of elements of S. To show linear independence, suppose that $S = \{\mathbf{v}_1, \ldots, \mathbf{v}_n\}$ and $c_1\mathbf{v}_1 + \cdots + c_n\mathbf{v}_n = \mathbf{0}$ for some scalars $c_1, \ldots, c_n$. The case when $c_1 = \cdots = c_n = 0$ is one possibility. By hypothesis, this is the *only* possible representation of the zero vector as a linear combination of the elements of S. So S is linearly independent and hence is a basis for V.

23. Suppose that $[\mathbf{u}]_B = [\mathbf{w}]_B$ for some $\mathbf{u}$ and $\mathbf{w}$ in V, and denote the entries in this coordinate vector by $c_1, \ldots, c_n$. By definition of the coordinate vectors,

$$\mathbf{u} = c_1\mathbf{b}_1 + \cdots + c_n\mathbf{b}_n \text{ and } \mathbf{w} = c_1\mathbf{b}_1 + \cdots + c_n\mathbf{b}_n$$

which shows that $\mathbf{u} = \mathbf{w}$. Since $\mathbf{u}$ and $\mathbf{w}$ were arbitrary elements of V, this shows that the coordinate mapping is one-to-one.

25. Since the coordinate mapping is one-to-one, the following equations have the same solutions, $c_1, \ldots, c_p$:

$$c_1\mathbf{u}_1 + \cdots + c_p\mathbf{u}_p = \mathbf{0} \qquad \text{(the zero vector in } V) \qquad (1)$$

$$[c_1\mathbf{u}_1 + \cdots + c_p\mathbf{u}_p]_B = [\mathbf{0}]_B \qquad \text{(the zero vector in } \mathbb{R}^n) \qquad (2)$$

Since the coordinate mapping is linear, (2) is equivalent to

$$c_1[\mathbf{u}_1]_B + \cdots + c_p[\mathbf{u}_p]_B = \begin{bmatrix} 0 \\ \vdots \\ 0 \end{bmatrix} \qquad (3)$$

Hence $c_1, \ldots, c_p$ satisfy (1) if and only if they satisfy (3). So (1) has only the trivial solution if and only if (3) has only the trivial solution. It follows that $\{\mathbf{u}_1, \ldots, \mathbf{u}_p\}$ is linearly independent if and only if $\{[\mathbf{u}_1]_B, \ldots, [\mathbf{u}_p]_B\}$ is linearly independent. (This fact is also an immediate consequence of Exercises 31 and 32 in Section 4.3.)

Study Tip: Exercises 27–34 tell you to explain your work and justify your conclusions. This requirement is to help you think about your calculations. Check with your instructor. A similar requirement is likely to appear on exam questions.

Warning: The standard mistake in Exercises 27–34 is to write the coordinate vectors as the *rows* of a matrix and then to check the linear independence of the *columns*. This is completely wrong! Since we mainly work with column vectors, it is wise to write the coordinate vectors first (as columns) and afterwards write a matrix that can be row reduced to check for linear independence. See the second solution of Exercise 13 and the sample solution for Exercise 27, below.

27. The coordinate vectors of the polynomials are $\mathbf{v}_1 = \begin{bmatrix} 1 \\ 0 \\ 0 \\ 1 \end{bmatrix}$, $\mathbf{v}_2 = \begin{bmatrix} 3 \\ 1 \\ -2 \\ 0 \end{bmatrix}$, $\mathbf{v}_3 = \begin{bmatrix} 0 \\ -1 \\ 3 \\ -1 \end{bmatrix}$ (relative to

the standard basis). To check linear independence of these vectors in $\mathbb{R}^4$, compute

$$\begin{bmatrix} 1 & 3 & 0 & 0 \\ 0 & 1 & -1 & 0 \\ 0 & -2 & 3 & 0 \\ 1 & 0 & -1 & 0 \end{bmatrix} \sim \begin{bmatrix} 1 & 0 & -1 & 0 \\ 0 & 1 & -1 & 0 \\ 0 & -2 & 3 & 0 \\ 2 & 3 & -1 & 0 \end{bmatrix} \sim \cdots \sim \begin{bmatrix} 1 & 0 & -1 & 0 \\ 0 & 1 & -1 & 0 \\ 0 & 0 & 1 & 0 \\ 0 & 0 & 0 & 0 \end{bmatrix}$$

The three coordinate vectors are linearly independent in $\mathbb{R}^4$, because the equation $x_1\mathbf{v}_1 + x_2\mathbf{v}_2 + x_3\mathbf{v}_3 = \mathbf{0}$ has only the trivial solution. By the isomorphism with $\mathbb{P}_3$, the three polynomials are linearly independent in $\mathbb{P}_3$.

Note: When you write that an equation has only the trivial solution, you must indicate in some way what equation you have in mind.

Study Tip: Exercises 27–30 could be expanded to ask whether the given polynomials form a basis for $\mathbb{P}_3$. You might see such a question on an exam.

31. In each part, place the coordinate vectors of the polynomials into the columns of a matrix and reduce the matrix to echelon form:

a. $\begin{bmatrix} 1 & -3 & -4 & 1 \\ -3 & 5 & 5 & 0 \\ 5 & -7 & -6 & -1 \end{bmatrix} \sim \begin{bmatrix} 1 & -3 & -4 & 1 \\ 0 & -4 & -7 & 3 \\ 0 & 8 & 14 & -6 \end{bmatrix} \sim \begin{bmatrix} 1 & -3 & -4 & 1 \\ 0 & -4 & -7 & 3 \\ 0 & 0 & 0 & 0 \end{bmatrix}$. The four coordinate

vectors do not span $\mathbb{R}^3$ because there is no pivot in row 3. Because of the isomorphism between $\mathbb{R}^3$ and $\mathbb{P}_2$, the corresponding polynomials do not span $\mathbb{P}_2$.

b. $\begin{bmatrix} 0 & 1 & -3 & 2 \\ 5 & -8 & 4 & -3 \\ 1 & -2 & 2 & 0 \end{bmatrix} \sim \begin{bmatrix} 1 & -2 & 2 & 0 \\ 5 & -8 & 4 & -3 \\ 0 & 1 & -3 & 2 \end{bmatrix} \sim \begin{bmatrix} 1 & -2 & 2 & 0 \\ 0 & 2 & -6 & -3 \\ 0 & 1 & -3 & 2 \end{bmatrix} \sim \begin{bmatrix} 1 & -2 & 2 & 0 \\ 0 & 2 & -6 & -3 \\ 0 & 0 & 0 & 3.5 \end{bmatrix}$.

The four coordinate vectors span $\mathbb{R}^3$ because there is a pivot in each row. Because of the isomorphism between $\mathbb{R}^3$ and $\mathbb{P}_2$, the corresponding polynomials span $\mathbb{P}_2$.

Study Tip: Carefully study the solutions in Exercise 31, because some student papers have a discussion that is far removed from a correct answer. After creating the matrix from the coordinate vectors and row reducing, a student might write something such as

"They do not span because there is not a pivot in each row."

Error #1: "They" is a pronoun with no mention of what "they" is. Does the student mean the polynomials, the coordinate vectors, or the columns of the matrix?

Error #2: Span what? The verb "span" requires an object, such as $\mathbb{P}_2$ or $\mathbb{R}^3$.

Error #3. There is no mention of "isomorphism." The phrase "not a pivot in each row" has relevance only to the fact that the coordinate vectors do not span $\mathbb{R}^3$. The only way to get from this fact to the polynomials in the exercise is to use the isomorphism between $\mathbb{P}_2$ and $\mathbb{R}^3$. (Some instructors may permit the term "correspondence" instead of the more precise "isomorphism.")

Answer to Checkpoint: The columns of the matrix $A = [\mathbf{b}_1 \; \cdots \; \mathbf{b}_n]$ form a basis for $\mathbb{R}^n$, so A is invertible, by the Invertible Matrix Theorem. By Theorem 5 in Section 2.2, for each $\mathbf{x}$ in $\mathbb{R}^n$ there exists a unique vector $\mathbf{c} = (c_1, \ldots, c_n)$ such that $\mathbf{x} = A\mathbf{c}$, that is, $\mathbf{x} = c_1\mathbf{b}_1 + \cdots + c_n\mathbf{b}_n$.

**MATLAB   The Backslash Operator **

If an equation $A\mathbf{x} = \mathbf{b}$ has a unique solution, MATLAB will automatically produce $\mathbf{x}$ if you use the command

```
x = A\b
```

In this section, the equation probably will have the form $P\mathbf{u} = \mathbf{x}$, with $\mathbf{u}$ the $\mathcal{B}$-coordinate vector of $\mathbf{x}$, and the command will be `u = P\x`.

The "backslash" command works in two different ways. When A is square, the command `A\b` causes MATLAB to create an LU factorization of A (see Section 2.5); if A is invertible, the factorization is used to produce the unique solution to $A\mathbf{x} = \mathbf{b}$; and if A is not invertible, MATLAB gives the error message *"matrix is singular"* (even if the system $A\mathbf{x} = \mathbf{b}$ has a solution). When A is not square, `A\b` creates a least-squares solution (see Section 6.5).

4.5   THE DIMENSION OF A VECTOR SPACE

This short section provides a convenient way to compare the "sizes" of various subspaces of a vector space. The notion of dimension will be used frequently throughout the rest of the text.

KEY IDEAS

Theorem 10 shows that the dimension of a finite-dimensional vector space does not depend on the particular basis for the space. Example 4 shows how to visualize subspaces of various dimensions. Theorems 9 and 12 are important for later theory and applications. You might remember Theorem 9 more easily in this form:

> In an *n*-dimensional vector space, any set of more than *n* vectors must be linearly dependent.

Theorem 12, the Basis Theorem, may be restated as follows:

> If *V* is a *p*-dimensional vector space, with $p \geq 1$, and if *S* is a subset of *V* that contains exactly *p* elements, then *S* is linearly independent if and only if *S* spans *V*.

Warning: Theorem 11 shows that any basis of a subspace *H* of a finite-dimensional space *V* can be extended to a basis of *V*. But it is *not* true that any basis of *V* can be cut down to a basis for *H*. That is, if *S* is a basis for *V*, it is not likely that a subset of *S* is a basis for *H*. For instance, suppose *S* is the standard basis for $\mathbb{R}^3$ and *H* is a plane in $\mathbb{R}^3$ that contains the origin but none of the coordinate axes. Then no subset of *S* can be a basis for *H*.

SOLUTIONS TO EXERCISES

1. Since $\begin{bmatrix} s-2t \\ s+t \\ 3t \end{bmatrix} = s\begin{bmatrix} 1 \\ 1 \\ 0 \end{bmatrix} + t\begin{bmatrix} -2 \\ 1 \\ 3 \end{bmatrix}$ for all *s*, *t*, the set $\left\{ \begin{bmatrix} 1 \\ 1 \\ 0 \end{bmatrix}, \begin{bmatrix} -2 \\ 1 \\ 3 \end{bmatrix} \right\}$ certainly spans the subspace,

call it *H*. Also, the set is obviously linearly independent (because the vectors are not multiples), so the set is a basis for *H*. Hence, dim *H* = 2.

3. The given subspace, call it *H*, is the set of all linear combinations of the vectors

$$\mathbf{v}_1 = \begin{bmatrix} 0 \\ 1 \\ 0 \\ 1 \end{bmatrix}, \mathbf{v}_2 = \begin{bmatrix} 0 \\ -1 \\ 1 \\ 2 \end{bmatrix}, \mathbf{v}_3 = \begin{bmatrix} 2 \\ 0 \\ -3 \\ 0 \end{bmatrix}$$

First determine if $\{\mathbf{v}_1, \mathbf{v}_2, \mathbf{v}_3\}$ is linearly independent. One way to do this is to row reduce the augmented matrix $[\mathbf{v}_1 \quad \mathbf{v}_2 \quad \mathbf{v}_3 \quad \mathbf{0}]$. A faster way is to use Theorem 4 in Section 4.3. Clearly, $\mathbf{v}_1 \neq \mathbf{0}$, $\mathbf{v}_2$ is not a multiple of $\mathbf{v}_1$, and $\mathbf{v}_3$ is not a linear combination of the vectors $\mathbf{v}_1$, $\mathbf{v}_2$ that precede it, because the first entry in $\mathbf{v}_3$ is not zero. Hence $\{\mathbf{v}_1, \mathbf{v}_2, \mathbf{v}_3\}$ is linearly independent and thus is a basis for the space *H* it spans. Thus dim *H* = 3.

7. Standard calculations show that the set of solutions of the homogeneous system consists of only the trivial solution. So the subspace is {**0**}, and it has no basis. (The vector **0** spans the space, but {**0**} is a linearly dependent set.) By definition, the dimension is zero. [*Note*: Instructors who want every subspace to have a basis often define the empty set to be a basis for {**0**}. The number of vectors in this basis is zero, so the dimension of {**0**} is still zero.]

13. *A* has three pivot columns, so dim Col *A* = 3. There are two columns without pivot positions, so the equation *A***x** = **0** has two free variables, and dim Nul *A* = 2.

19. **a.** See the box before Example 5.   **b.** Read Example 4 carefully.

    c. See Example 1.   **d.** See Theorem 10.

    e. See Practice Problem 2. (You should be working the practice problems before you start the exercises.)

21. Form the matrix whose columns are the coordinate vectors of the Hermite polynomials, relative to the standard basis $\{1, t, t^2, t^3\}$:

$$A = \begin{bmatrix} 1 & 0 & -2 & 0 \\ 0 & 2 & 0 & -12 \\ 0 & 0 & 4 & 0 \\ 0 & 0 & 0 & 8 \end{bmatrix}$$

The matrix has four pivots and hence is invertible. So its columns, the coordinate vectors, are linearly independent. Hence the Hermite polynomials themselves are linearly independent in $\mathbb{P}_3$. Since there are four Hermite polynomials, and dim $\mathbb{P}_3 = 4$, we conclude from The Basis Theorem that the Hermite polynomials form a *basis* for $\mathbb{P}_3$.

Note: You could, of course, say that the columns of the matrix *A* span $\mathbb{R}^4$. But you cannot stop with that assertion, because you need the polynomials to span $\mathbb{P}_3$. You have to go on and point out that because of the isomorphism between $\mathbb{P}_3$ and $\mathbb{R}^4$, a set of vectors spans $\mathbb{P}_3$ if and only if the set of coordinate vectors (the columns of *A*) spans $\mathbb{R}^4$. So the solution is shorter if you appeal to the Basis Theorem.

25. Note that $n \geq 1$, because *S* cannot have fewer than 0 vectors. If dim $V = n \geq 1$, then $V \neq \{0\}$. If *S* spans *V*, then a subset S' of *S* is a basis for *V*, by the Spanning Set Theorem. But if *S* has fewer than *n* vectors, then S' also has fewer than *n* vectors. This is impossible, by Theorem 10, because dim $V = n$. So *S* cannot span *V*.

27. If $\mathbb{P}$ were finite-dimensional, then Theorem 11 would imply that $n + 1 = \dim \mathbb{P}_n \leq \dim \mathbb{P}$ for each *n*, because each $\mathbb{P}_n$ is a subspace of $\mathbb{P}$. This is impossible, so $\mathbb{P}$ must be infinite-dimensional.

29. **a.** True. Apply the Spanning Set Theorem to the set $\{v_1, \ldots, v_p\}$ and produce a basis for *V*. This basis will have no more than *p* elements in it, so dim *V* must be no more than *p*.

b. True. By Theorem 11, $\{\mathbf{v}_1, \ldots, \mathbf{v}_p\}$ can be expanded to a basis for V. The basis will have at least p elements in it, so dim V must be at least p.

c. True. Take any basis (of p vectors) for V and adjoin the zero vector. Spanning sets can be arbitrarily large. The dimension of V being p only keeps spanning sets from having *fewer* than p elements.

31. Since H is a nonzero subspace of a finite-dimensional space, H is finite-dimensional and has a basis, say, $\mathbf{v}_1, \ldots, \mathbf{v}_p$. Any vector in $T(H)$ has the form $T(\mathbf{y})$ for some $\mathbf{y}$ in H. Since $\{\mathbf{v}_1, \ldots, \mathbf{v}_p\}$ spans H, there exist scalars $c_1, \ldots, c_p$ such that $\mathbf{y} = c_1\mathbf{v}_1 + \cdots + c_p\mathbf{v}_p$. Since T is linear, $T(\mathbf{y}) = c_1 T(\mathbf{v}_1) + \cdots + c_p T(\mathbf{v}_p)$. This shows that $\{T(\mathbf{v}_1), \ldots, T(\mathbf{v}_p)\}$ spans $T(H)$. By Exercise 29(a), dim $T(H) \le p =$ dim H.

　　Second proof: Let $k =$ dim $T(H)$. If $k = 0$, then $k <$ dim H. Otherwise, $T(H)$ has a basis, which can be written in the form $T(\mathbf{v}_1), \ldots, T(\mathbf{v}_k)$ for some vectors $\mathbf{v}_1, \ldots, \mathbf{v}_k$ in H. Since $\{T(\mathbf{v}_1), \ldots, T(\mathbf{v}_k)\}$ is linearly independent, so is $\{\mathbf{v}_1, \ldots, \mathbf{v}_k\}$, by Exercise 31 in Section 4.3. Since $\mathbf{v}_1, \ldots, \mathbf{v}_k$ are in H, the dimension of H must be at least k.

Hint for Exercise 32:　Use an exercise in Section 4.3.

Study Tip:　The next section is quite important.　Do your best to get caught up now. Otherwise, you may have difficulty relating the various concepts and facts about matrices that will be reviewed in Section 4.6.

4.6　RANK

This section gives you a chance to put together most of the ideas of the chapter in the same way that Section 2.3 collected the main ideas of the sections that preceded it.

KEY IDEAS

The Rank Theorem is the main result. By definition, rank $A =$ dim Col A. But because rank A is also the dimension of Row A, the displayed equation in the theorem leads to the equation: dim Row A + dim Nul $A = n$.

Equivalent Descriptions of Rank

The rank of an $m \times n$ matrix A may be described in several ways:

- the dimension of the column space of A, (our definition)
- the number of pivot positions in A, (from Theorem 6)
- the maximum number of linearly independent columns in A,
- the dimension of the row space of A, (from the Rank Theorem)
- the maximum number of linearly independent rows in A,
- the number of nonzero rows in an echelon form of A,
- the maximum number of columns in an invertible submatrix of A. (Supplementary Exercise 17 at the end of the chapter)

Pay attention to how Theorem 13 differs from the results in Section 4.3 about Col A: If you are interested in *rows* of A, use the nonzero rows of an echelon form B as a basis for Row A; if you are interested in the *columns* of A, only use B to obtain *information* about A (namely, to identify the pivot columns), and use the pivot columns of A as a basis for Col A. For Nul A, it is important to use the *reduced* echelon form of A.

When a matrix A is changed into a matrix B by one or more elementary row operations, the row space, null space, and column space of A may or may not be the same as the corresponding subspaces for B. The following table summarizes what can happen in this situation.

Effects of Elementary Row Operations

- Row operations do not affect the linear dependence relations among the columns. (That is, the columns of A have exactly the same linear dependence relations as the columns of any matrix that is row-equivalent to A.)
- Row operations usually change the column space.
- Row operations never change the row space.
- Row operations never change the null space.

The four subspaces shown in Figure 1 in the text are called the *fundamental subspaces* determined by A. (See Exercises 27–29.) The main difficulty here is to avoid confusion between Row A, Nul A, and Col A. The fourth subspace will appear again in Sections 6.1 and 7.4.

The following table lists all statements that are in the Invertible Matrix Theorem at this point in the course, arranged in the scheme used in Section 2.3 of this *Study Guide*. The statements in all three columns are equivalent when A is square ($m = n = p$). As before, a few extra statements have been added to make the table more symmetrical.

STATEMENTS FROM THE INVERTIBLE MATRIX THEOREM

Equivalent statements for an $m \times n$ matrix A.	Equivalent statements for an $n \times n$ square matrix A.	Equivalent statements for any $n \times p$ matrix A.
k. There is a matrix D such that $AD = I$.	a. A is an invertible matrix.	j. There is a matrix C such that $CA = I$.
*. A has a pivot position in every row.	c. A has n pivot positions.	*. A has a pivot position in every column.
h. The columns of A span $\mathbb{R}^m$.	m. The columns of A form a basis for $\mathbb{R}^n$.	e. The columns of A are linearly independent.
g. The equation $A\mathbf{x} = \mathbf{b}$ has at least one solution for each $\mathbf{b}$ in $\mathbb{R}^m$.	*. The equation $A\mathbf{x} = \mathbf{b}$ has a unique solution for each $\mathbf{b}$ in $\mathbb{R}^n$.	d. The equation $A\mathbf{x} = \mathbf{0}$ has only the trivial solution.
i. The transformation $\mathbf{x} \mapsto A\mathbf{x}$ maps $\mathbb{R}^n$ onto $\mathbb{R}^m$.	*. The transformation $\mathbf{x} \mapsto A\mathbf{x}$ is invertible.	f. The transformation $\mathbf{x} \mapsto A\mathbf{x}$ is one-to-one.
n. Col $A = \mathbb{R}^m$.	b. A is row equivalent to I.	q. Nul $A = \{\mathbf{0}\}$.
o. dim Col $A = m$.	l. A^T is invertible.	r. dim Nul $A = 0$.
*. rank $A = m$.	p. rank $A = n$.	*. rank $A = p$.

With so many concepts in your linear algebra vocabulary, you need to be careful not to combine terms in ways that are undefined, even though they may sound reasonable to you. For example, after you finish your work on this section, you should recognize that the following phrases (which have appeared on my students' papers) are meaningless: "the basis of a matrix," "the dimension of a basis," and "the rank of a basis."

SOLUTIONS TO EXERCISES

1. $A = \begin{bmatrix} 1 & -4 & 9 & -7 \\ -1 & 2 & -4 & 1 \\ 5 & -6 & 10 & 7 \end{bmatrix} \sim B = \begin{bmatrix} 1 & 0 & -1 & 5 \\ 0 & -2 & 5 & -6 \\ 0 & 0 & 0 & 0 \end{bmatrix}$

Look at B, and conclude that A has two pivot columns and the equation $A\mathbf{x} = \mathbf{0}$ has two free variables. So rank $A = 2$ and dim Nul $A = 2$. In fact, the first two columns of A are pivot columns, so

Basis for Col A: $\left\{ \begin{bmatrix} 1 \\ -1 \\ 5 \end{bmatrix}, \begin{bmatrix} -4 \\ 2 \\ -6 \end{bmatrix} \right\}$

For the row space, use the rows in the echelon form B. That is,

   Basis for Row A: $\{(1, 0, -1, 5), (0, -2, 5, -6)\}$

For the null space, use the *reduced* echelon form of A to solve $A\mathbf{x} = \mathbf{0}$:

$$A \sim B \sim \begin{bmatrix} 1 & 0 & -1 & 5 \\ 0 & 1 & -5/2 & 3 \\ 0 & 0 & 0 & 0 \end{bmatrix}; \quad \begin{aligned} x_1 \quad - \quad x_3 + 5x_4 &= 0 \\ x_2 - (5/2)x_3 + 3x_4 &= 0 \\ 0 &= 0 \end{aligned}$$

Thus $x_1 = x_3 - 5x_4$, $x_2 = (5/2)x_3 - 3x_4$, with x_3, x_4 free. The general solution of $A\mathbf{x} = \mathbf{0}$ is

$$\begin{bmatrix} x_1 \\ x_2 \\ x_3 \\ x_4 \end{bmatrix} = \begin{bmatrix} x_3 - 5x_4 \\ (5/2)x_3 - 3x_4 \\ x_3 \\ x_4 \end{bmatrix} = x_3 \underset{\mathbf{u}}{\begin{bmatrix} 1 \\ 5/2 \\ 1 \\ 0 \end{bmatrix}} + x_4 \underset{\mathbf{v}}{\begin{bmatrix} -5 \\ -3 \\ 0 \\ 1 \end{bmatrix}}$$

Thus $\{\mathbf{u}, \mathbf{v}\}$ is a basis for Nul A.

Study Tip: Because rank $A = 2$ in Exercise 1, *any* two linearly independent columns of A form a basis for Col A, and any two linearly independent rows of A form a basis for Row A. When the rank of a matrix exceeds 2, selecting bases in this way is not so easy. (That is why you examine an echelon form of A.) On an exam, you should always choose the pivot columns of A as the basis for Col A and the nonzero rows of an echelon form of A as the basis for Row A. This will show that you can handle matrices with any rank.

7. Yes, Col $A = \mathbb{R}^4$, because Col A is a 4-dimensional subspace of $\mathbb{R}^4$ and hence coincides with $\mathbb{R}^4$. No, Nul A cannot be $\mathbb{R}^3$, because the vectors in Nul A have 7 entries. Nul A is a 3-dimensional subspace of $\mathbb{R}^7$, by the Rank Theorem.

13. If A is either a 7×5 matrix or a 5×7 matrix, then A has at most 5 pivot positions. So 5 is the largest possible value for rank A.

17. **a.** See the paragraph before Example 1.

   b. See the warning after Example 2.

   c. See the Rank Theorem.

   d. See the Rank Theorem.

   e. See the Numerical Note before the Practice Problem.

19. Visualize the system as $A\mathbf{x} = \mathbf{0}$ where A is a 5×6 matrix. The information in the problem implies that the solution space is one-dimensional. By the Rank Theorem, rank $A = 6 - 1 = 5$. So dim Col $A = 5$. But Col A is a subspace of $\mathbb{R}^5$. Hence Col $A = \mathbb{R}^5$. Thus $A\mathbf{x} = \mathbf{b}$ has a solution for all $\mathbf{b}$.

Study Tip: Exercises 19–25 make good exam questions.

21. Visualize the system as $Ax = b$, where A is 9×10 matrix. You are told that the system has a solution for all b, so A must have a pivot position in each row. (That is, rank $A = 9$.) Since A has 10 columns, the Rank Theorem implies that dim Nul $A = 1$. So it is not possible to find two linearly independent vectors in Nul A.

23. The set of interest is the null space of a 12×8 matrix A. The description of this set implies that dim Nul $A = 2$. By the Rank Theorem, rank $A = 8 - 2 = 6$. So the equation $Ax = 0$ is equivalent to $Bx = 0$, where B is an echelon form of A with 6 nonzero rows. The answer to the question is that six homogeneous equations are sufficient.

25. Let A be the 10×12 coefficient matrix. By hypothesis, there are three free variables in the system $Ax = b$, so dim Nul $A = 3$. By the Rank Theorem, dim Col $A = 12 - 3 = 9$. Since Col A is a subspace of $\mathbb{R}^{10}$ (because A has 10 rows), Col A cannot be all of $\mathbb{R}^{10}$, so some nonhomogeneous equations $Ax = b$ will *not* have solutions.

31. The solution is in the text.

33. Let $A = [\mathbf{u} \quad \mathbf{u}_2 \quad \mathbf{u}_3]$. If $\mathbf{u} \neq \mathbf{0}$, then $\mathbf{u}$ must be a basis for Col A, since Col A is one-dimensional. Hence there exist scalars r and s such that $\mathbf{u}_2 = r\mathbf{u}$ and $\mathbf{u}_3 = s\mathbf{u}$, so that

$$A = [\mathbf{u} \quad r\mathbf{u} \quad s\mathbf{u}] = \mathbf{u}[1 \quad r \quad s] = \mathbf{u}\mathbf{v}^T, \text{ where } \mathbf{v} = \begin{bmatrix} 1 \\ r \\ s \end{bmatrix}$$

If the first column of A is zero and the second column, call it $\mathbf{u}$, is nonzero, then $A = [\mathbf{0} \quad \mathbf{u} \quad r\mathbf{u}]$ for some r. In this case, take $\mathbf{v} = (0, 1, r)$. If $A = [\mathbf{0} \quad \mathbf{0} \quad \mathbf{u}]$, take $\mathbf{v} = (0, 0, 1)$.

35. **a.** Let C and N be the matrices you construct whose columns are bases for Col A and Nul A, respectively. For the specific 5×7 matrix A in this problem, rank $A = 4$ and dim Nul $A = 3$. So C should be 5×4 (because Col A is a four-dimensional subspace of $\mathbb{R}^5$), and N should be 7×3 (because Nul A is a three-dimensional subspace of $\mathbb{R}^7$). Also, if the *rows* of R form a basis for Row A, then R should be 4×7, because dim Row A = rank $A = 4$ and Row A is a subspace of $\mathbb{R}^7$. (Make sure you understand these statements.)

   b. The matrix $S = [R^T \quad N]$ is 7×7, because R^T is 7×4 and N is 7×3. If M is a matrix whose columns form a basis for Nul A^T, then M should be 5×1, because Nul A^T is a one-dimensional subspace of $\mathbb{R}^5$, by Exercise 28(b). Since C is 5×4, the matrix $T[C \quad M]$ should be 5×5.

   In general, the matrix S is $n \times n$ because the dimensions of Row A (spanned by the columns of R^T) and Nul A add up to n, the number of columns of A, and both Row A and Nul A are subspaces of $\mathbb{R}^n$. The matrix T is $m \times m$ because the dimensions of the column space and Nul A^T add up to m, the number of rows of A, and both Col A and Nul A are subspaces of $\mathbb{R}^m$.

Mastering Linear Algebra Concepts: Eight Basic Ideas

Sometime between now and when you finish the chapter, you should do a major review of the eight key concepts introduced in this chapter: vector space, subspace, column space, null space, basis, coordinate vector, dimension, and rank. (The row space of A is not really a separate concept; it is just the column space of A^T.) Study your old review sheets, and prepare new summary sheets for coordinate vector, dimension, and rank. Use as many of the standard categories (special cases, examples, algorithms, etc.) as possible. The tables in this section will be helpful. Also, add cross-references about dimension and rank to other sheets (subspace, column space, etc.) and update your summary sheet for the Invertible Matrix Theorem.

MATLAB   ref, rank, and randomint

In this course, you can use either **ref(A)** or **rank(A)** to check the rank of A. In practical work, you should use the more reliable command **rank(A)**, based on the singular value decomposition (Section 7.4).

The Laydata command **randomint(m,n)** produces an $m \times n$ matrix with integer entries between -9 and 9. (The former name for this command was **randint**, but MATLAB now uses that for a slightly different command in its Communications Toolbox.)

4.7   CHANGE OF BASIS

This section will help you better understand coordinate systems. A review of Section 4.4 now is strongly recommended.

KEY IDEAS

Figure 1 and the accompanying discussion will help you visualize the main idea of the section. Imagine superimposing the C-graph paper (Figure 1-b) on the B-graph paper (Figure 1-a). Can you see where $\mathbf{b}_1$ will lie on the C-coordinate system? Four units in the $\mathbf{c}_1$-direction and one unit in the $\mathbf{c}_2$-direction. That is the geometric interpretation of the equation $[\mathbf{b}_2]_C = \begin{bmatrix} 4 \\ 1 \end{bmatrix}$ in Example 1.

Similarly, since $[\mathbf{b}_2]_C = \begin{bmatrix} -6 \\ 1 \end{bmatrix}$, $\mathbf{b}_2$ lies six units in the negative $\mathbf{c}_1$-direction and one unit in the $\mathbf{c}_2$-direction.

In general, the locations of $\mathbf{b}_1$ and $\mathbf{b}_2$ on the C-graph paper are precisely what you must find in order to build the columns of the change-of-coordinates matrix:

$$\underset{C \leftarrow B}{P} = \begin{bmatrix} [\mathbf{b}_1]_C & [\mathbf{b}_2]_C \end{bmatrix}$$

The notation for this matrix should help you remember the basic equation for changing $\mathcal{B}$-coordinates into $\mathcal{C}$-coordinates:

$$[\mathbf{x}]_\mathcal{C} = \underset{\mathcal{C}\leftarrow\mathcal{B}}{P} = [\mathbf{x}]_\mathcal{B}$$

The calculations are simple when $\mathcal{B}$ and $\mathcal{C}$ are bases for $\mathbb{R}^n$. The box after Example 2 illustrates the algorithm for computing the change-of-coordinates matrix. In general,

$$[\mathbf{c}_1 \quad \cdots \quad \mathbf{c}_n \mid \mathbf{b}_1 \quad \cdots \quad \mathbf{b}_n] \sim [I \mid \underset{\mathcal{C}\leftarrow\mathcal{B}}{P}]$$

Equivalently, using the notation of Section 4.4,

$$[P_\mathcal{C} \quad P_\mathcal{B}] \sim [I \quad \underset{\mathcal{C}\leftarrow\mathcal{B}}{P}]$$

where $P_\mathcal{B}$ is the matrix $[\mathbf{b}_1 \quad \cdots \quad \mathbf{b}_n]$ that changes $\mathcal{B}$-coordinates to *standard coordinates*, and $P_\mathcal{C}$ is similarly defined. If you refer back to Exercise 12 of Section 2.2, you will see that $\underset{\mathcal{C}\leftarrow\mathcal{B}}{P}$ is the same as $(P_\mathcal{C})^{-1}P_\mathcal{B}$. Since $(P_\mathcal{C})^{-1}$ changes standard coordinates to $\mathcal{C}$-coordinates, you can obtain $[\mathbf{x}]_\mathcal{C}$ from $[\mathbf{x}]_\mathcal{B}$ as follows:

$$[\mathbf{x}]_\mathcal{B} \longrightarrow P_\mathcal{B}[\mathbf{x}]_\mathcal{B} \longrightarrow (P_\mathcal{C})^{-1}P_\mathcal{B}[\mathbf{x}]_\mathcal{B} = \underset{\mathcal{C}\leftarrow\mathcal{B}}{P}[\mathbf{x}]_\mathcal{B} = [\mathbf{x}]_\mathcal{C}$$

$\mathcal{B}$-coordinates   standard $\mathcal{C}$-coordinates
                              coordinates

This diagram provides another way of viewing the change of coordinates.

SOLUTIONS TO EXERCISES

1. a. From $\mathbf{b}_1 = 6\mathbf{c}_1 - 2\mathbf{c}_2$ and $\mathbf{b}_2 = 9\mathbf{c}_1 - 4\mathbf{c}_2$, write

$$[\mathbf{b}_1]_\mathscr{C} = \begin{bmatrix} 6 \\ -2 \end{bmatrix}, \mathbf{b}_2 = \begin{bmatrix} 9 \\ -4 \end{bmatrix}, \text{ and } \underset{\mathcal{C}\leftarrow\mathcal{B}}{P} = \begin{bmatrix} 6 & 9 \\ -2 & -4 \end{bmatrix}$$

b. Since $\mathbf{x} = -3\mathbf{b}_1 + 2\mathbf{b}_2$,

$$[\mathbf{x}]_\mathcal{B} = \begin{bmatrix} -3 \\ 2 \end{bmatrix} \text{ and } [\mathbf{x}]_\mathcal{C} = \begin{bmatrix} 6 & 9 \\ -2 & -4 \end{bmatrix}\begin{bmatrix} -3 \\ 2 \end{bmatrix} = \begin{bmatrix} 0 \\ -2 \end{bmatrix}$$

7. Unlike Exercise 1, you do not have direct information from which you can write $[\mathbf{b}_1]_\mathcal{C}$ and $[\mathbf{b}_2]_\mathcal{C}$. Rather than compute these two coordinate vectors separately, use the algorithm from Example 2:

$$[\mathbf{c}_1 \quad \mathbf{c}_2 \quad \mathbf{b}_1 \quad \mathbf{b}_2] = \begin{bmatrix} 1 & -2 & 7 & -3 \\ -5 & 2 & 5 & -1 \end{bmatrix} \sim \begin{bmatrix} 1 & -2 & 7 & -3 \\ 0 & -8 & 40 & -16 \end{bmatrix}$$

$$\sim \begin{bmatrix} 1 & -2 & 7 & -3 \\ 0 & 1 & -5 & 2 \end{bmatrix} \sim \begin{bmatrix} 1 & 0 & -3 & 1 \\ 0 & 1 & -5 & 2 \end{bmatrix}$$

Thus $\underset{C \leftarrow B}{P} = \begin{bmatrix} -3 & 1 \\ -5 & 2 \end{bmatrix}$. The change-of-coordinates matrix from C to B is

$$\underset{B \leftarrow C}{P} = (\underset{C \leftarrow B}{P})^{-1} = \begin{bmatrix} -3 & 1 \\ -5 & 2 \end{bmatrix}^{-1} = \frac{1}{-1}\begin{bmatrix} 2 & -1 \\ 5 & -3 \end{bmatrix} = \begin{bmatrix} -2 & 1 \\ -5 & 3 \end{bmatrix}$$

11. a. See Theorem 15.

   b. See the first paragraph in the subsection on *Change of Basis* in $\mathbb{R}^n$.

13. Let b_1 represent the polynomial $1 - 2t + t^2$, let b_2 be $3 - 5t + 4t^2$, let b_3 be $2t + 3t^2$, and let C be the standard basis $\{1, t, t^2\}$ for $\mathbb{P}^2$. The C-coordinate vectors of the vectors b_1, b_2, b_3 are

$$[b_1]_C = \begin{bmatrix} 1 \\ -2 \\ 1 \end{bmatrix}, [b_2]_C = \begin{bmatrix} 3 \\ -5 \\ 4 \end{bmatrix}, [b_3]_C = \begin{bmatrix} 0 \\ 2 \\ 3 \end{bmatrix}, \text{ and}$$

$$\underset{C \leftarrow B}{P} = \begin{bmatrix} 1 & 3 & 0 \\ -2 & -5 & 2 \\ 1 & 4 & 3 \end{bmatrix}$$

The coordinate vector $[-1 + 2t]_B$ satisfies

$$\underset{C \leftarrow B}{P}[-1+2t]_B = [-1+2t]_C = \begin{bmatrix} -1 \\ 2 \\ 0 \end{bmatrix}$$

This equation can be solved by row reduction:

$$\begin{bmatrix} 1 & 3 & 0 & -1 \\ -2 & -5 & 2 & 2 \\ 1 & 4 & 3 & 0 \end{bmatrix} \sim \cdots \sim \begin{bmatrix} 1 & 0 & 0 & 5 \\ 0 & 1 & 0 & -2 \\ 0 & 0 & 1 & 1 \end{bmatrix}; \quad [-1+2t]_B = \begin{bmatrix} 5 \\ -2 \\ 1 \end{bmatrix}$$

19. a. If P is to be the change-of-coordinates matrix from $\{u_1, u_2, u_3\}$ to $\{v_1, v_2, v_3\}$, then the columns of P should be C-coordinate vectors, where $C = \{v_1, v_2, v_3\}$. That is, the columns of P should be $[u_1]_C$, $[u_2]_C$, and $[u_3]_C$. You know P, but you do not know u_1, u_2, or u_3. Ask yourself, for example, what is the meaning of $[u_1]_C$? By definition, a C-coordinate vector tells how to build a vector out of the C-basis vectors, $v_1, v_2,$ and v_3. So, for $j = 1, 2, 3,$

$$u_j = [v_1 \quad v_2 \quad v_3] [u_j]_C = V[u_j]_C$$

where $V = [v_1 \quad v_2 \quad v_3]$. Then, by the definition of matrix multiplication,

$$[u_1 \quad u_2 \quad u_3] = [V[u_j]_C \quad V[u_j]_C \quad V[u_j]_C] = V[[u_j]_C \quad [u_j]_C \quad [u_j]_C] = VP$$

You know V and P, so you can compute $u_1, u_2,$ and u_3.

$$VP = \begin{bmatrix} -2 & -8 & -7 \\ 2 & 5 & 2 \\ 3 & 2 & 6 \end{bmatrix} \begin{bmatrix} 1 & 2 & -1 \\ -3 & -5 & 0 \\ 4 & 6 & 1 \end{bmatrix} = \begin{bmatrix} -6 & -6 & -5 \\ -5 & -9 & 0 \\ 21 & 32 & 3 \end{bmatrix}$$

$$\text{Thus, } \mathbf{u}_1 = \begin{bmatrix} -6 \\ -5 \\ 21 \end{bmatrix}, \mathbf{u}_2 = \begin{bmatrix} -6 \\ -9 \\ 32 \end{bmatrix}, \mathbf{u}_3 = \begin{bmatrix} -5 \\ 0 \\ 3 \end{bmatrix}.$$

b. Stop here! Part (a) was fairly difficult. If you were not able to work it by yourself and you have read the solution, then you should try part (b) by yourself. Here are the steps of the solution:

(i) Write in symbols what the columns of P should be.

(ii) Decide how these columns are related to the matrix $W = [\mathbf{w}_1 \quad \mathbf{w}_2 \quad \mathbf{w}_3]$.

(iii) Obtain a matrix equation that involves W in some way.

(iv) Compute W, and list its columns as the answer to the problem.

Study the solution to (a), close the *Study Guide*, and work on (b) as if it were a new problem. The solution of (b) is at the end of the solutions for Section 4.8.

MATLAB   Change-of-Coordinates Matrix

The Laydata Toolbox has data for Exercises 7–10 and 17–19. The command `ref(M)` row reduces a matrix such as $[\mathbf{c}_1 \quad \mathbf{c}_2 \quad \mathbf{b}_1 \quad \mathbf{b}_2]$ to the desired form.

4.8   APPLICATIONS TO DIFFERENCE EQUATIONS _____

Difference equations are the discrete analogues of differential equations. Both are important in science and engineering. The discrete and continuous theories are remarkably parallel, and linear algebra is applied in similar ways, although the calculations are somewhat easier for difference equations. A variety of examples and exercises here illustrate some difference equations you may encounter later in your work.

KEY IDEAS

Each signal in $\mathbb{S}$ is an infinite list of numbers. Linear independence of a set of signals can often be demonstrated by looking at short segments of the signals, that is, by showing that a Casorati matrix is invertible. A Casorati matrix cannot be used in general to demonstrate linear *dependence* of a set. However, see the appendix at the end of this *Study Guide* section.

The main focus of the section is on difference equations. Given a homogeneous difference equation, you should be able to:

- determine whether a specified signal is a solution of the equation;

- find solutions of the equation, using the auxiliary equation;

- give the general solution (when the auxiliary equation has no multiple roots and no complex roots).

Theorem 17 is the key result that enables you to write the general solution. Just finding some specific solutions is not enough; you must show that they *span* the set of all solutions. But Theorem 17 and the Basis Theorem in Section 4.5 together show that for an *n*th order equation, you only need to find *n linearly independent* solutions. (See Example 5.) The same principle apples to an *n*th order differential equation (discussed later in Section 5.7). This principle is one of the most powerful applications of linear algebra in the text.

The subsection on nonhomogenous equations is optional. If this is covered, you should be able to work Exercises 25–28. The general principle is illustrated in Figure 4, page 284:

$$\left\{\begin{array}{l}\text{General solution of} \\ \text{nonhomogeneous eqn.}\end{array}\right\} = \left\{\begin{array}{l}\text{Particular solution of} \\ \text{nonhomogeneous eqn.}\end{array}\right\} + \left\{\begin{array}{l}\text{General solution of} \\ \text{homogeneous eqn.}\end{array}\right\}$$

The final subsection shows the modern way to study an *n*th order linear difference equation, rewriting it as a first order system $\mathbf{x}_{k+1} = A\mathbf{x}_k$ ($k = 1, 2, \ldots$). Such systems were introduced in Section 1.10 and they will be discussed further in Sections 4.9 and 5.6.

SOLUTIONS TO EXERCISES

1. If $y_k = (-4)^k$, then $y_{k+1} = (-4)^{k+1}$ and $y_{k+2} = (-4)^{k+2}$. Substitute these formulas into the left side of the equation:

$$y_{k+2} + 2y_{k+1} - 8y_k = (-4)^{k+2} + 2(-4)^{k+1} - 8(-4)^k$$

$$= (-4)^k [(-4)^2 + 2(-4) - 8]$$

$$= (-4)^k [16 - 8 - 8] = 0 \text{ for all } k$$

Since the difference equation holds for all k, $(-4)^k$ is a solution. The text answer displays the similar calculations for $y_k = 2^k$.

7. Compute the Casorati matrix for the signals 1^k, 2^k, and $(-2)^k$, setting $k = 0$ for convenience:

$$\begin{bmatrix} 1^0 & 2^0 & (-2)^0 \\ 1^1 & 2^1 & (-2)^1 \\ 1^2 & 2^2 & (-2)^2 \end{bmatrix} = \begin{bmatrix} 1 & 1 & 1 \\ 1 & 2 & -2 \\ 1 & 4 & 4 \end{bmatrix} \sim \begin{bmatrix} 1 & 1 & 1 \\ 0 & 1 & -3 \\ 0 & 3 & 3 \end{bmatrix} \sim \begin{bmatrix} 1 & 1 & 1 \\ 0 & 1 & -3 \\ 0 & 0 & 12 \end{bmatrix}$$

This Casorati matrix has three pivots and hence is invertible, by the IMT. Hence the set of signals $\{1^k, 2^k, (-2)^k\}$ is linearly independent in $\mathbb{S}$. We know (from the text) that these signals are in the solution space H of a third-order difference equation. By Theorem 17, dim $H = 3$. Since the three signals are linearly independent, they form a basis for H, by the Basis Theorem in Section 4.5.

Warning: Many student papers for Exercise 7 suffer from a lack of precision, often confusing linear independence of the columns of the Casorati matrix with linear independence of the signals in $\mathbb{S}$. There is no need to discuss the columns of the Casorati matrix—just observe that the matrix is invertible. But you must point out that the three *signals* are linearly independent, in order to apply the Basis Theorem to the vector space H of solutions to the difference equation.

13. The auxiliary equation for $y_{k+2} - y_{k+1} + \dfrac{2}{9} y_k = 0$ is $r^2 - r + \dfrac{2}{9} = 0$. By the quadratic formula,

$$r = \frac{1 \pm \sqrt{1 - 8/9}}{2} = \frac{1 \pm 1/3}{2} = \frac{2}{3} \text{ or } \frac{1}{3}$$

Two solutions of the difference equation are $\left(\dfrac{2}{3}\right)^k$ and $\left(\dfrac{1}{3}\right)^k$. These signals are obviously linearly independent because neither is a multiple of the other. Since the solution space is two-dimensional (Theorem 17), the two signals form a basis for the solution space, by the Basis Theorem.

Study Tip: I think Exercises 7–19 (and 25–28) make good test questions because they illustrate how important Theorem 17 and the Basis Theorem really are. Probably, you do not have to remember the specific number of Theorem 17, but your discussion should show that you have it in mind and know how to use it with the Basis Theorem. (Check with your instructor.)

19. The auxiliary equation for $y_{k+2} + 4y_{k+1} + y_k = 0$ is $r^2 + 4r + 1 = 0$. By the quadratic formula,

$$r = \frac{-4 \pm \sqrt{16 - 4}}{2} = \frac{-4 \pm 2\sqrt{3}}{2} = -2 \pm \sqrt{3}$$

Two solutions of the difference equation are $(-2 + \sqrt{3})^k$ and $(-2 - \sqrt{3})^k$. They are obviously linearly independent because neither is a multiple of the other. Since the solution space is two-dimensional (Theorem 17), the two signals form a fundamental set of solutions by the Basis Theorem, and the general solution has the form $c_1(-2 + \sqrt{3})^k + c_2(-2 - \sqrt{3})^k$.

25. To prove that $y_k = k^2$ is a solution of

$$y_{k+2} + 3y_{k+1} - 4y_k = 10k + 7 \tag{1}$$

show that when k^2, $(k + 1)^2$, and $(k + 2)^2$ are substituted for y_k, y_{k+1}, and y_{k+2}, respectively, the resulting equation is true for all k:

$$(k+2)^2 + 3(k+1)^2 - 4k^2 = (k^2 + 4k + 4) + 3(k^2 + 2k + 1) - 4k^2$$

$$= (1 + 3 - 4)k^2 + (4 + 6)k + (4 + 3)$$

$$= 10k + 7 \quad \text{for all } k$$

So k^2 is a solution of (1). The auxiliary equation for the homogeneous difference equation

$$y_{k+2} + 3y_{k+1} - 4y_k = 0 \quad \text{for all } k \tag{2}$$

is $r^2 + 3r - 4 = 0$, which factors as $(r - 1)(r + 4) = 0$, so $r = 1, -4$. Thus 1^k and $(-4)^k$ are solutions of (2). The signals are linearly independent (for neither is a multiple of the other), so they form a basis for the two-dimensional solution space. The general solution of (2) is $c_1 \cdot 1^k + c_2(-4)^k$. Add this to a particular solution of (1) and obtain the general solution $k^2 + c_1 + c_2(-4)^k$ of (1).

31. The full explanation is in the text's answer section.

35. For $\{y_k\}$ and $\{z_k\}$ in $\mathbb{S}$, the kth term of $\{y_k\} + \{z_k\}$ is $y_k + z_k$. Hence

$$T(\{y_k\} + \{z_k\}) = \{(y_{k+2} + z_{k+2}) + a(y_{k+1} + z_{k+1}) + b(y_k + z_k)\}$$

$$= \{y_{k+2} + ay_{k+1} + by_k\} + \{z_{k+2} + az_{k+1} + bz_k\}$$

$$= T\{y_k\} + T\{z_k\}$$

For any scalar r, the kth term of $r\{y_k\}$ is ry_k, and so

$$T(r\{y_k\}) = \{ry_{k+2} + a(ry_{k+1}) + b(ry_k)\}$$
$$= r\{y_{k+2} + ay_{k+1} + by_k\} = rT\{y_k\}$$

Thus T has the two properties that define a linear transformation.

19. b. (*This solution is for Section 4.7.*) The columns of a change-of-coordinates matrix P from $\{v_1, v_2, v_3\}$ to $\{w_1, w_2, w_3\}$ are $\mathcal{D}$-coordinate vectors, where $\mathcal{D} = \{w_1, w_2, w_3\}$. That is, the columns of P are $[v_1]_\mathcal{D}$, $[v_2]_\mathcal{D}$, and $[v_3]_\mathcal{D}$. How are these columns related to the matrix $W = [w_1 \quad w_2 \quad w_3]$? A $\mathcal{D}$-coordinate vector tells how to build a vector out of the $\mathcal{D}$-basis vectors (the columns of the matrix W). For $j = 1, 2, 3$,

$$v_j = [w_1 \quad w_2 \quad w_3][v_j]_\mathcal{D} = W[v_j]_\mathcal{D}$$

By definition of matrix multiplication,

$$V = [v_1 \quad v_2 \quad v_3] = [W[v_j]_\mathcal{D} \quad W[v_j]_\mathcal{D} \quad W[v_j]_\mathcal{D}] = W[[v_j]_\mathcal{D} \quad [v_j]_\mathcal{D} \quad [v_j]_\mathcal{D}] = WP$$

You know V and P, so compute W from $VP^{-1} = W$. Use MATLAB or other matrix program to compute P^{-1}. Then

$$W = VP^{-1} = \begin{bmatrix} -2 & -8 & -7 \\ 2 & 5 & 2 \\ 3 & 2 & 6 \end{bmatrix} \begin{bmatrix} 5 & 8 & 5 \\ -3 & -5 & -3 \\ -2 & -2 & -1 \end{bmatrix} = \begin{bmatrix} 28 & 38 & 21 \\ -9 & -13 & -7 \\ -3 & 2 & 3 \end{bmatrix}$$

Thus, $\mathbf{w}_1 = \begin{bmatrix} 28 \\ -9 \\ -3 \end{bmatrix}$, $\mathbf{w}_2 = \begin{bmatrix} 38 \\ -13 \\ 2 \end{bmatrix}$, $\mathbf{w}_3 = \begin{bmatrix} 21 \\ -7 \\ 3 \end{bmatrix}$.

Appendix: The Casorati Test

Let $\{\mathbf{y}_1, \ldots, \mathbf{y}_n\}$ be a set of signals in $\mathbb{S}$. For $j = 1, \ldots, n$ and for any k, let $\mathbf{y}_j(k)$ denote the kth entry in the signal $\mathbf{y}_j$ and let

$$C(k) = \begin{bmatrix} \mathbf{y}_1(k) & \cdots & \mathbf{y}_n(k) \\ \mathbf{y}_1(k+1) & \cdots & \mathbf{y}_n(k+1) \\ \vdots & & \vdots \\ \mathbf{y}_1(k+n-1) & \cdots & \mathbf{y}_n(k+n-1) \end{bmatrix} \quad \text{The Casorati matrix}$$

a. If $C(k)$ is invertible for some k, $\{\mathbf{y}_1, \ldots, \mathbf{y}_n\}$ is linearly independent.

b. If $\mathbf{y}_1, \ldots, \mathbf{y}_n$ all satisfy a homogeneous difference equation of order n,

$$y_{k+n} + a_1 y_{k+n-1} + \cdots + a_n y_k = 0 \text{ for all } k \tag{*}$$

(with $a_n \neq 0$), and if the Casorati matrix $C(k)$ is not invertible for some k, then $\{\mathbf{y}_1, \ldots, \mathbf{y}_n\}$ is linearly dependent in $\mathbb{S}$, and for all k, $C(k)$ is not invertible.

Proof. (a) The argument given in the text (page 279) for a set of three signals generalizes immediately to n signals. (b) Suppose that $\mathbf{y}_1, \ldots, \mathbf{y}_n$ are in the set H of solutions of (*) and $C(k_o)$ is not invertible for some k_0. It is readily verified that if $T: H \to \mathbb{R}^n$ is defined by

$$T(\mathbf{y}) = \begin{bmatrix} \mathbf{y}(k_0) \\ \mathbf{y}(k_0 + 1) \\ \vdots \\ \mathbf{y}(k_0 + n-1) \end{bmatrix}$$

then T is a linear transformation. The proof of Theorem 16 is easily modified to show that (*) has a unique solution $\mathbf{y}$ whenever $\mathbf{y}(k_0), \ldots, \mathbf{y}(k_0 + n - 1)$ are specified. This means that T is a one-to-one mapping of H onto $\mathbb{R}^n$. Furthermore, the images $T(\mathbf{y}_1), \ldots, T(\mathbf{y}_n)$ form the columns of the Casorati matrix $C(k_0)$ and hence are linearly dependent, because $C(k_0)$ is not invertible. Since T is one-to-one, $\{\mathbf{y}_1, \ldots, \mathbf{y}_n\}$ is linearly dependent, by Exercise 32 in Section 4.3. This proves the first statement in (b). The second statement follows immediately from part (a), because if $C(k)$ were invertible for some k, then $\{\mathbf{y}_1, \ldots, \mathbf{y}_n\}$ would be linearly independent, which is not true. So $C(k)$ is not invertible for each k.

MATLAB   roots

In Exercises 7–16 and 25–28, the polynomial in the auxiliary equation is stored in a row vector **p**, with coefficients in descending order. For instance, if the auxiliary equation is $r^2 + 6r + 9 = 0$, then **p** = [1 6 9]

The MATLAB command **roots(p)** produces a column vector whose entries are the roots of the polynomial described by **p**.

4.9   APPLICATIONS TO MARKOV CHAINS _____

This section builds on the population movement example in Section 1.9. You should review that example now. Markov chains are widely used in applications and there is a rich theory connected with them. The simple examples and exercises in this section provide a basic foundation on which you can build later as needed. Two of the examples here will be analyzed from a different point of view in Section 5.2.

KEY IDEAS

A probability vector is a list of nonnegative numbers that sum to one. A Markov chain is a sequence of probability vectors $\{x_k\}$ that satisfy a difference equation $x_{k+1} = Px_k$ ($k = 0, 1, \ldots$) for some stochastic matrix P (whose columns are themselves probability vectors).

The theory of this chapter can be used to show that $P - I$ always has a nontrivial null space when P is a stochastic matrix (Exercise 17). Advanced texts show that the null space of $P - I$ always includes at least one probability vector, which then is a steady-state vector for P, because the equation $(P - I)q = 0$ is equivalent to $Pq = q$. (Also, see Exercise 18.)

Our main interest is in a *regular* stochastic matrix P. In this case the steady-state vector is unique, according to Theorem 18. The key to predicting the distant future for a Markov chain associated with such a P is to find the steady-state vector **q**, since the sequence $\{x_k\}$ converges to **q** no matter what the initial state.

SOLUTIONS TO EXERCISES

1. **a.** To set up the stochastic matrix P, label the columns N (for news) and M (for music) in some order; use the *same* order for the rows. (Failure to keep the same order is a common source of error in this type of problem.) The data should be arranged so you read *down* a column and then to the right along a row.

From:

$$\begin{array}{cc} N & M \end{array} \quad \underline{\text{To:}}$$

$$\begin{bmatrix} .7 & .6 \\ .3 & .4 \end{bmatrix} \begin{array}{l} \text{News} \\ \text{Music} \end{array}$$

b. You are told that 100% of the listeners are listening to the news at 8:15 a.m., so start the Markov chain then, with $\mathbf{x}_0 = \begin{bmatrix} 1 \\ 0 \end{bmatrix}$.

c. There are two breaks between 8:15 and 9:25, so you need $\mathbf{x}_2$.

$$\mathbf{x}_1 = P\mathbf{x}_0 = \begin{bmatrix} .7 & .6 \\ .3 & .4 \end{bmatrix} \begin{bmatrix} 1 \\ 0 \end{bmatrix} = \begin{bmatrix} .7 \\ .3 \end{bmatrix}$$

$$\mathbf{x}_2 = P\mathbf{x}_1 = \begin{bmatrix} .7 & .6 \\ .3 & .4 \end{bmatrix} \begin{bmatrix} .7 \\ .3 \end{bmatrix} = \begin{bmatrix} .67 \\ .33 \end{bmatrix}$$

The entries in $\mathbf{x}_2$ show that after two station breaks, 67% of the audience is listening to the news and 33% is listening to music.

Study Tip: When you compute a typical probability vector $P\mathbf{x}$, be sure to *compute all* of the entries in the product $P\mathbf{x}$. Then check your work by verifying that the entries sum to 1.

7. To find the steady state vector for a regular stochastic matrix P:

   (i)   set up the matrix $P - I$;

   (ii)   find the general solution of $(P - I)\mathbf{x} = \mathbf{0}$;

   (iii) choose a basis vector for Nul$(P - I)$ whose entries sum to 1.

$$P = \begin{bmatrix} .7 & .1 & .1 \\ .2 & .8 & .2 \\ .1 & .1 & .7 \end{bmatrix}, P - I = \begin{bmatrix} .7 & .1 & .1 \\ .2 & .8 & .2 \\ .1 & .1 & .7 \end{bmatrix} - \begin{bmatrix} 1 & 0 & 0 \\ 0 & 1 & 0 \\ 0 & 0 & 1 \end{bmatrix} = \begin{bmatrix} -.3 & .1 & .1 \\ .2 & -.2 & .2 \\ .1 & .1 & -.3 \end{bmatrix}$$

Solve $(P - I)\mathbf{x} = \mathbf{0}$:

$$\begin{bmatrix} -.3 & .1 & .1 & 0 \\ .2 & -.2 & .2 & 0 \\ .1 & .1 & -.3 & 0 \end{bmatrix} \sim \begin{bmatrix} .1 & .1 & -.3 & 0 \\ .2 & -.2 & .2 & 0 \\ -.3 & .1 & .1 & 0 \end{bmatrix}$$

Interchange rows 1 and 3
Scale every row by 10

$$\sim \cdots \sim \begin{bmatrix} 1 & 0 & -1 & 0 \\ 0 & 1 & -2 & 0 \\ 0 & 0 & 0 & 0 \end{bmatrix} ; \begin{matrix} x_1 = x_3 \\ x_2 = 2x_3 ; \\ x_3 \text{ is free} \end{matrix} \begin{bmatrix} x_1 \\ x_2 \\ x_3 \end{bmatrix} = \begin{bmatrix} x_3 \\ 2x_3 \\ x_3 \end{bmatrix} = x_3 \begin{bmatrix} 1 \\ 2 \\ 1 \end{bmatrix}$$

The entries in $\begin{bmatrix} 1 \\ 2 \\ 1 \end{bmatrix}$ sum to 4, so $\mathbf{q} = \dfrac{1}{4} \begin{bmatrix} 1 \\ 2 \\ 1 \end{bmatrix} = \begin{bmatrix} 1/4 \\ 1/2 \\ 1/4 \end{bmatrix}$ or $\begin{bmatrix} .25 \\ .50 \\ .25 \end{bmatrix}$.

Study Tip: Notice that the column sums are all zero for the matrix $I - P$ of Exercise 7. This always happens (see Exercise 17), and so you have a fast way to check your arithmetic for the entries in $P - I$.

Warning: You may have noticed that in Exercise 7, I scaled rows by 10, to avoid decimals. A common mistake is to do this only to P, before forming $P - I$. That changes P drastically. The scaling I did was permissible because it was applied to all the coefficients in an *equation*.

13. a. From Exercise 3, $P = \begin{bmatrix} .95 & .45 \\ .05 & .55 \end{bmatrix}$. So $P - I = \begin{bmatrix} -.05 & .45 \\ .05 & -.45 \end{bmatrix}$. Solve $(P - I)\mathbf{x} = \mathbf{0}$:

$$\begin{bmatrix} -.05 & .45 & 0 \\ .05 & -.45 & 0 \end{bmatrix} \sim \begin{bmatrix} -.05 & .45 & 0 \\ 0 & 0 & 0 \end{bmatrix} \sim \begin{bmatrix} 1 & -9 & 0 \\ 0 & 0 & 0 \end{bmatrix} \quad \begin{matrix} x_1 = 9x_2 \\ x_2 \text{ is free} \end{matrix}$$

A basis for $\text{Nul}(P - I)$ is $\left\{ \begin{bmatrix} 9 \\ 1 \end{bmatrix} \right\}$; the steady-state vector is $\mathbf{q} = \begin{bmatrix} .9 \\ .1 \end{bmatrix}$.

b. The description in Exercise 3 may be interpreted as saying that the "state" of any specified person (in some group of students) on day k is predicted by a probability vector, say, $\mathbf{x}_k$. The second entry in $\mathbf{x}_k$ is the probability that the person is ill on day k. The starting vector for a specified person is $(1,0)$ if the person is well today, and $(0,1)$ if the person is ill. This situation applies to each person, because the exercise says, for example, that *every* healthy student has a 95% probability of being healthy the next day. That is, the stochastic matrix P applies to each person in the group.

   The question in part (b) is about $\mathbf{x}_k$ for a large value of k. By Theorem 18, $\mathbf{x}_k$ approaches $\mathbf{q}$, so it is reasonable to assume that $\mathbf{q}$ may be used to answer a question about $\mathbf{x}_k$. Thus, the probability is .10 that after many days a specific student is ill. The second question essentially asks, "If $\mathbf{x}_0 = (0,1)$, does this have any affect on $\mathbf{x}_k$ for large k?" No, by Theorem 18, because the sequence $\{\mathbf{x}_k\}$ approaches $\mathbf{q}$ no matter what $\mathbf{x}_0$ is.

19. a. The product $S\mathbf{x}$ equals the sum of the entries in $\mathbf{x}$. Thus, by definition, $\mathbf{x}$ is a probability vector if and only if its entries are nonnegative and $S\mathbf{x} = 1$.

b. Let $P = [\mathbf{p}_1 \ \mathbf{p}_2 \ \cdots \ \mathbf{p}_n]$, where the $\mathbf{p}_i$ are probability vectors. By matrix multiplication and part (a),

$$SP = [S\mathbf{p}_1 \quad S\mathbf{p}_2 \quad \cdots \quad S\mathbf{p}_n] = [1 \ \ 1 \ \cdots \ \ 1] = S$$

c. By part (b), $S(P\mathbf{x}) = (SP)\mathbf{x} = S\mathbf{x} = 1$. The entries in $P\mathbf{x}$ are obviously nonnegative, because P and $\mathbf{x}$ have only nonnegative entries. By (a), the condition $S(P\mathbf{x}) = 1$ shows that $P\mathbf{x}$ is a probability vector.

MATLAB `randomstoc`

The MATLAB box for Section 1.10 contains information that is useful here. The command `randomstoc(n)` produces a random $n \times n$ stochastic matrix.

Chapter 4   GLOSSARY CHECKLIST

Check your knowledge by attempting to write definitions of the terms below. Then compare your work with the definitions given in the text's Glossary. Ask your instructor which definitions, if any, might appear on a test.

auxiliary equation: A polynomial equation in a variable r, created from . . . .

basis (for a nonzero subspace H):   A set $\mathcal{B} = \{\mathbf{v}_1, ..., \mathbf{v}_p\}$ in V such that: . . . .

$\mathcal{B}$-coordinates of x: *See* coordinates of $\mathbf{x}$ relative to the basis $\mathcal{B}$.

change-of-coordinates matrix (from a basis $\mathcal{B}$ to a basis $\mathcal{C}$):   A matrix $\underset{\mathcal{C} \leftarrow \mathcal{B}}{P}$ that transforms . . . , namely, (equation). . . .

column space (of an $m \times n$ matrix A):   The set Col A of . . . . In set notation, Col $A = \{$: $\}$.

controllable (pair of matrices):   A matrix pair (A, B) where A is $n \times n$, B has n rows, and . . . .

coordinate mapping (determined by an ordered basis $\mathcal{B}$ in a vector space V):   A mapping that associates to each . . . .

coordinates of x relative to the basis $\mathcal{B} = \{b_1, ..., b_n\}$:   . . . .

coordinate vector of x relative to $\mathcal{B}$:   The vector $[\mathbf{x}]_\mathcal{B}$ whose entries . . . .

dimension (of a vector space V):   The number . . . .

explicit description (of a subspace W of $\mathbb{R}^n$):   A parametric representation of W as the set of . . . .

finite-dimensional (vector space):   A vector space that is . . . .

full rank (matrix):   An $m \times n$ matrix whose rank is . . . .

fundamental set of solutions: A . . . for the set of solutions of . . . .

fundamental subspaces (determined by A):   The . . . of A, ...

implicit description (of a subspace W of $\mathbb{R}^n$):   A set of one or more . . . .

infinite-dimensional (vector space):   A nonzero vector space V that . . . .

isomorphism: A . . . mapping from one vector space . . . .

kernel (of a linear transformation $T: V \rightarrow W$):   The set of . . . such that . . . .

linear combination: A sum . . . .

linear dependence relation: A . . . vector equation where . . . .

linear filter: A . . . equation used to transform . . . .

linearly dependent (vectors):   A set $\{v_1, \ldots, v_p\}$ with the property that . . . .

linearly independent (vectors):   A set $\{v_1, \ldots, v_p\}$ with the property . . . .

linear transformation T (from a vector space V into a vector space W):   A rule $T: V \rightarrow W$ that to each vector $\mathbf{x}$ in V assigns a unique vector $T(\mathbf{x})$ in W, such that: . . . .

Markov chain: A sequence of . . . vectors $v_0, v_1, v_2, \ldots$, together with a . . . matrix P such that . . . .

maximal linearly independent set (in V):   A linearly independent set $\mathcal{B}$ in V such that if . . . , then . . . .

minimal spanning set (for a subspace H):   A set $\mathcal{B}$ that spans H and has the property that if . . . , then . . . .

null space (of an $m \times n$ matrix A):   The set Nul A of all . . . . In set notation, Nul $A = \{ \ : \ \ \}$.

probability vector: A vector in $\mathbb{R}^n$ whose entries . . . .

proper subspace: Any subspace of a vector space V . . . .

range (of a linear transformation $T: V \rightarrow W$):   The set of all vectors . . . .

rank (of a matrix A):   . . . .

regular stochastic matrix: A stochastic matrix P such that . . . .

row space (of a matrix A):   The set Row A of all . . . ; also denoted by . . . .

signal (or **discrete-time signal**):   . . . .

Span$\{v_1, \ldots, v_p\}$:   The set of . . . . Also, the . . . *spanned* . . . .

spanning set (for a subspace H):   Any set $\{v_1, \ldots, v_p\}$ . . . such that . . . .

standard basis: The basis . . . for $\mathbb{R}^n$ consisting of . . . , or the basis . . . for $\mathbb{P}_n$.

state vector: A . . . vector. In general, a vector that describes . . . , often in connection with a difference equation . . . .

steady-state vector (for a stochastic matrix P):   A . . . vector $\mathbf{v}$ such that . . . .

stochastic matrix: A . . . matrix whose columns . . . .

submatrix (of A):   Any matrix obtained by . . . .

subspace: A subset H of some vector space V such that H has these properties . . . .

vector space: A set of objects, called vectors, on which . . . .

zero subspace: The subspace . . . consisting of . . . .

5 Eigenvalues and Eigenvectors

5.1 EIGENVECTORS AND EIGENVALUES ─────────────

This section introduces eigenvectors and eigenvalues. A hint about the connection with dynamical systems appears at the end of the section.

KEY IDEAS

In words, a nonzero vector $\mathbf{v}$ is an eigenvector of a matrix A if and only if the transformed vector $A\mathbf{x}$ points in the same or opposite direction of $\mathbf{v}$.

Notice that while an eigenvalue λ might be zero, an eigenvector is never zero (by definition). An eigenspace contains eigenvectors together with the zero vector.

The two equations $A\mathbf{x} = \lambda\mathbf{x}$ and $(A - \lambda I)\mathbf{x} = \mathbf{0}$ are equivalent. See Example 3. The first equation is useful for understanding what eigenvalues and eigenvectors are, and it shows the geometric effect of the linear transformation $\mathbf{x} \mapsto A\mathbf{x}$ on an eigenvector. The second equation shows that the eigenspace is a subspace (because it is the null space of the matrix $A - \lambda I$), and the equation is used to find a basis for the eigenspace, when λ is a known eigenvalue. The second equation will be used again in Section 5.2 for another purpose.

SOLUTIONS TO EXERCISES

1. The number 2 is an eigenvalue of A if and only if the equation $A\mathbf{x} = 2\mathbf{x}$ has a nontrivial solution. This equation is equivalent to $(A - 2I)\mathbf{x} = \mathbf{0}$. Compute

$$A - 2I = \begin{bmatrix} 3 & 2 \\ 3 & 8 \end{bmatrix} - \begin{bmatrix} 2 & 0 \\ 0 & 2 \end{bmatrix} = \begin{bmatrix} 1 & 2 \\ 3 & 6 \end{bmatrix}$$

The columns of A are obviously linearly dependent, so $(A - 2I)\mathbf{x} = \mathbf{0}$ has a nontrivial solution, and so 2 is an eigenvalue of A.

7. Proceed as in Exercise 1:

$$A - 4I = \begin{bmatrix} 3 & 0 & -1 \\ 2 & 3 & 1 \\ -3 & 4 & 5 \end{bmatrix} - \begin{bmatrix} 4 & 0 & 0 \\ 0 & 4 & 0 \\ 0 & 0 & 4 \end{bmatrix} = \begin{bmatrix} -1 & 0 & -1 \\ 2 & -1 & 1 \\ -3 & 4 & 1 \end{bmatrix}$$

You need to know whether $A - 4I$ is invertible. This could be checked in several ways, but since you are asked for an eigenvector, in the event that one exists, the best strategy is to row reduce the augmented matrix for $(A - 4I)\mathbf{x} = \mathbf{0}$:

$$\begin{bmatrix} -1 & 0 & -1 & 0 \\ 2 & -1 & 1 & 0 \\ -3 & 4 & 1 & 0 \end{bmatrix} \sim \begin{bmatrix} -1 & 0 & -1 & 0 \\ 0 & -1 & -1 & 0 \\ 0 & 4 & 4 & 0 \end{bmatrix} \sim \begin{bmatrix} -1 & 0 & -1 & 0 \\ 0 & -1 & 1 & 0 \\ 0 & 0 & 0 & 0 \end{bmatrix}$$

Now it is clear that 4 is an eigenvalue of A [because $(A - 4I)\mathbf{x} = \mathbf{0}$ has a nontrivial solution]. The coordinates of an eigenvector satisfy $-x_1 - x_3 = 0$ and $-x_2 - x_3 = 0$. The general solution is not requested, so take any nonzero value for x_3 to produce an eigenvector. If $x_3 = 1$, then $\mathbf{x} = (-1, -1, 1)$.

Checkpoint 1:   The answer in the text is different, namely, $(1, 1, -1)$. Why is this also correct?

Helpful Hint: Suppose you think that 4 is an eigenvalue of a matrix, as in Exercise 7, and you row reduce the augmented matrix for $(A - 4I)\mathbf{x} = \mathbf{0}$. If you discover that there are no free variables, then there are only two possibilities: (1) 4 is *not* an eigenvalue of A, or (2) you have made an arithmetic error.

13. <u>For $\lambda = 1$:</u>

$$A - 1I = \begin{bmatrix} 4 & 0 & 1 \\ -2 & 1 & 0 \\ -2 & 0 & 1 \end{bmatrix} - \begin{bmatrix} 1 & 0 & 0 \\ 0 & 1 & 0 \\ 0 & 0 & 1 \end{bmatrix} = \begin{bmatrix} 3 & 0 & 1 \\ -2 & 0 & 0 \\ -2 & 0 & 0 \end{bmatrix}$$

The equations for $(A - I)\mathbf{x} = \mathbf{0}$ are easy to solve: $\begin{cases} 3x_1 + x_3 = 0 \\ -2x_1 \quad\quad = 0 \end{cases}$

Row operations hardly seem necessary. Obviously x_1 is zero, and hence x_3 is also zero. There are three-variables, so x_2 is free. The general solution of $(A - I)\mathbf{x} = \mathbf{0}$ is $x_2\mathbf{e}_2$, where $\mathbf{e}_2 = (0,1,0)$, and so $\mathbf{e}_2$ provides a basis for the eigenspace.

<u>For $\lambda = 2$:</u>

$$A - 2I = \begin{bmatrix} 4 & 0 & 1 \\ -2 & 1 & 0 \\ -2 & 0 & 1 \end{bmatrix} - \begin{bmatrix} 2 & 0 & 0 \\ 0 & 2 & 0 \\ 0 & 0 & 2 \end{bmatrix} = \begin{bmatrix} 2 & 0 & 1 \\ -2 & -1 & 0 \\ -2 & 0 & 1 \end{bmatrix}$$

$$[(A-2I) \quad \mathbf{0}] = \begin{bmatrix} 2 & 0 & 1 & 0 \\ -2 & -1 & 0 & 0 \\ -2 & 0 & -1 & 0 \end{bmatrix} \sim \begin{bmatrix} 2 & 0 & 1 & 0 \\ 0 & -1 & 1 & 0 \\ 0 & 0 & 0 & 0 \end{bmatrix} \sim \begin{bmatrix} ① & 0 & 1/2 & 0 \\ 0 & ① & -1 & 0 \\ 0 & 0 & 0 & 0 \end{bmatrix}$$

So $x_1 = -(1/2)x_3$, $x_2 = x_3$, with x_3 free. The general solution of $(A-2I)\mathbf{x} = \mathbf{0}$ is $x_3 \begin{bmatrix} -1/2 \\ 1 \\ 1 \end{bmatrix}$. A

nice basis vector for the eigenspace is $\begin{bmatrix} -1 \\ 2 \\ 2 \end{bmatrix}$.

For $\lambda = 3$:

$$A - 3I = \begin{bmatrix} 4 & 0 & 1 \\ -2 & 1 & 0 \\ -2 & 0 & 1 \end{bmatrix} - \begin{bmatrix} 3 & 0 & 0 \\ 0 & 3 & 0 \\ 0 & 0 & 3 \end{bmatrix} = \begin{bmatrix} 1 & 0 & 1 \\ -2 & -2 & 0 \\ -2 & 0 & -2 \end{bmatrix}$$

$$[(A-3I) \quad \mathbf{0}] = \begin{bmatrix} 1 & 0 & 1 & 0 \\ -2 & -2 & 0 & 0 \\ -2 & 0 & -2 & 0 \end{bmatrix} \sim \begin{bmatrix} 1 & 0 & 1 & 0 \\ 0 & -2 & 2 & 0 \\ 0 & 0 & 0 & 0 \end{bmatrix} \sim \begin{bmatrix} ① & 0 & 1 & 0 \\ 0 & ① & -1 & 0 \\ 0 & 0 & 0 & 0 \end{bmatrix}$$

So $x_1 = -x_3$, $x_2 = x_3$, with x_3 free. A basis vector for the eigenspace is $\begin{bmatrix} -1 \\ 1 \\ 1 \end{bmatrix}$.

Study Tip: The text's answer to Exercise 15 is likely to be the same as yours, but there are many answers. What should you do if your vectors differ from those in the answer key? Example 2 gives the answer. Whenever you compute an $\mathbf{x}$ that you think is an eigenvector of A, you can check this simply by computing $A\mathbf{x}$. There is a little more to do in Exercise 15, however. The answer shows a basis of two eigenvectors, which means that the eigenspace is two-dimensional. So your answer must consist of two linearly independent eigenvectors. You can check that they are indeed eigenvectors, and then their linear independence can be checked by inspection.

19. The matrix $\begin{bmatrix} 1 & 2 & 3 \\ 1 & 2 & 3 \\ 1 & 2 & 3 \end{bmatrix}$ is not invertible because its columns are linearly dependent. So the

number 0 is an eigenvalue of the matrix. See the discussion following Example 5.

21. **a**. Carefully read the definition of an eigenvalue.

   b. See the paragraph before Example 5.

    c. See the discussion of equation (3).

    d. See Example 2 and the paragraph preceding it. Also, see the Numerical Note.

    e. See the Warning after Example 3.

23. If a 2×2 matrix A had three distinct eigenvalues, then by Theorem 2 there would correspond three linearly independent eigenvectors (one for each eigenvalue). This is impossible because the vectors all belong to a two-dimensional vector space in which any set of three vectors is linearly dependent. See Theorem 8 in Section 1.6. In general, if an $n \times n$ matrix has p distinct eigenvalues, then by Theorem 2 there would be a linearly independent set of p eigenvectors (one for each eigenvalue). Since these vectors belong to an n-dimensional vector space, p cannot exceed n.

25. Let $\mathbf{x}$ be a nonzero vector such that $A\mathbf{x} = \lambda\mathbf{x}$. Then $A^{-1}A\mathbf{x} = A^{-1}(\lambda\mathbf{x})$, and $\mathbf{x} = \lambda(A^{-1}\mathbf{x})$. Since $\mathbf{x} \neq \mathbf{0}$ (and since A is invertible), λ cannot be zero. Then $\lambda^{-1}\mathbf{x} = A^{-1}\mathbf{x}$, which shows that λ^{-1} is an eigenvalue of A^{-1}, because $\mathbf{x} \neq \mathbf{0}$.

Note: The relation between the eigenvalues of A and A^{-1} is important in the so-called *inverse power* method for estimating an eigenvalue of a matrix. (See Section 5.8.)

27. For any λ, $(A - \lambda I)^T = A^T - (\lambda I)^T = A^T - \lambda I$. Since $(A - \lambda I)^T$ is invertible if and only if $A - \lambda I$ is invertible (by Theorem 6(c) in Section 2.2), we conclude that $A^T - \lambda I$ is *not* invertible if and only if $A - \lambda I$ is *not* invertible. That is, λ is an eigenvalue of A^T if and only if λ is an eigenvalue of A.

33. The solution is given in the text. This exercise is important because it sets the stage for work later in the chapter. You ought to spend at least a little time on Exercise 34, too, even if that is not assigned.

Answer to Checkpoint: The answer in the text is also correct because it is a nonzero multiple of the eigenvector found in the solution to Exercise 7, and any nonzero multiple of an eigenvector is another eigenvector. (The eigenspace is a *subspace* and so is closed under scalar multiplication.)

MATLAB   Finding Eigenvectors

When you know an eigenvalue, MATLAB can produce a basis for the corresponding eigenspace. For example, if A is a 5×5 matrix with an eigenvalue 7, then the commands

```
C = A - 7*eye(5)
B = nulbasis(C)
```

or, simply, `B = nulbasis(A - 7*eye(5))`, produce a matrix B whose columns form a basis for the eigenspace for A corresponding to $\lambda = 7$. In general, `eye(k)` is the $k \times k$ identity matrix, and `nulbasis(C)` is a matrix whose columns form a basis for Nul C (the same basis you would get if you started with `ref(C)` and made the calculations by hand). The command `nulbasis` is in the Laydata Toolbox.

If the numbers in the basis matrix B are messy, use **format rat; B** , which will display the entries in B as rational numbers. (All eigenvectors calculated in this section have rational entries, so the rational format introduces no error.) To return to the usual decimal number display, enter **format short**.

For Exercises 37–40, you need the command **eig(A)**, which lists the eigenvalues of A accurately to many decimal places, in most cases. For example, enter

```
ev = eig(A)
B = nulbasis(A - ev(2)*eye(5))
```

to compute a basis for the eigenspace corresponding to the second eigenvalue listed in the vector **ev**. It is dangerous to type **eig(A)** and simply "look" at the list of eigenvalues to use in the **nulbasis** command. You may make a mistake when you type an eigenvalue, particularly when the eigenvalue has sixteen nonzero digits and MATLAB displays only six of them.

5.2   THE CHARACTERISTIC EQUATION

There are several equivalent definitions of the determinant of a matrix. The definition here in terms of the pivots in an echelon form has the advantage that it is easy to state and understand, and in most cases it provides the most efficient way to compute a determinant.

KEY IDEAS

When A is 3×3, the geometric interpretation of det A as a volume explains why det $A = 0$ if and only if A is not invertible:

The determinant of A is zero.
<=> The parallelepiped determined by the columns of A has zero volume.
<=> One column of A is in the subspace spanned by the other columns.
<=> The columns of A are linearly dependent.
<=> The matrix A is not invertible.

If A is $n \times n$, then $\det(A - \lambda I) = 0$ if and only if $A - \lambda I$ is not invertible, and this happens if and only if λ is an eigenvalue of A.

Exercises 1–14 are designed only to provide some basic familiarity with characteristic polynomials. The main use of $\det(A - \lambda I)$ is as a tool for *studying* eigenvalues rather than computing them.

Sometimes the characteristic polynomial is defined as $\det(\lambda I - A)$. A property of determinants implies that $\det(\lambda I - A) = (-1)^n \det(A - \lambda I)$, when A is $n \times n$, so the two polynomials are either the same (when n is even) or they are negatives of one another. The use of $\det(A - \lambda I)$ tends to make hand calculations easier and less prone to copying errors.

SOLUTIONS TO EXERCISES

1. $A = \begin{bmatrix} 2 & 7 \\ 7 & 2 \end{bmatrix}$, $A - \lambda I = \begin{bmatrix} 2 & 7 \\ 7 & 2 \end{bmatrix} - \begin{bmatrix} \lambda & 0 \\ 0 & \lambda \end{bmatrix} = \begin{bmatrix} 2-\lambda & 7 \\ 7 & 2-\lambda \end{bmatrix}$, the characteristic polynomial is

$$\det(A - \lambda I) = (2 - \lambda)^2 - 7^2 = 4 - 4\lambda + \lambda^2 - 49 = \lambda^2 - 4\lambda - 45$$

In factored form, the characteristic equation is $(\lambda - 9)(\lambda + 5) = 0$, so the eigenvalues of A are 9 and -5.

Warning: Don't row reduce a matrix A to find its eigenvalues. Row reduction preserves the null space of A but not the eigenvalues of A.

7. $A = \begin{bmatrix} 5 & 3 \\ -4 & 4 \end{bmatrix}$, $A - \lambda I = \begin{bmatrix} 5-\lambda & 3 \\ -4 & 4-\lambda \end{bmatrix}$, the characteristic polynomial is

$$\det(A - \lambda I) = (5 - \lambda)(4 - \lambda) - (3)(-4) = 20 - 9\lambda + \lambda^2 + 12$$
$$= \lambda^2 - 9\lambda + 32$$

The characteristic polynomial does not factor easily, but the quadratic formula provides the solutions of $\lambda^2 - 9\lambda + 32 = 0$.

$$\lambda = \frac{+9 \pm \sqrt{81 - 4(32)}}{2}$$

These values for λ are not real numbers, so A has no real eigenvalues. There is no nonzero vector $\mathbf{x}$ in $\mathbb{R}^2$ such that $A\mathbf{x} = \lambda\mathbf{x}$ for such a λ. (For any $\mathbf{x} \neq \mathbf{0}$ in $\mathbb{R}^2$, the vector $A\mathbf{x}$ has only real entries and thus could not equal a complex multiple of $\mathbf{x}$.)

Study Tip: If you are asked to work some of Exercises 9–14, you may be tested on them. This is one way of finding out if you know what the characteristic polynomial is and how it is connected with eigenvalues. Also, you can show that you know some elementary properties of determinants.

13. $A - \lambda I = \begin{bmatrix} 6-\lambda & -2 & 0 \\ -2 & 9-\lambda & 0 \\ 5 & 8 & 3-\lambda \end{bmatrix}$

The method using the special 3 × 3 formula will produce the characteristic polynomial $-\lambda^3 + 18\lambda^2 - 95\lambda + 150$. Factoring such a polynomial to find the eigenvalues requires a little experience. (See the appendix at the end of the exercise solutions.) However, if you use a cofactor expansion down the third column (see Section 3.1), you immediately obtain

$$\det(A - \lambda I) = (3 - \lambda) \cdot \det \begin{bmatrix} 6 - \lambda & -2 \\ -2 & 9 - \lambda \end{bmatrix}$$

$$= (3 - \lambda)[(6 - \lambda)(9 - \lambda) - 4]$$

$$= (3 - \lambda)(\lambda^2 - 15\lambda + 50)$$

The characteristic polynomial is already partially factored, and the remaining quadratic factor is itself easily factored. The factored characteristic polynomial is $(3 - \lambda)(\lambda - 10)(\lambda - 5)$ or, equivalently, $-(\lambda - 3)(\lambda - 5)(\lambda - 10)$.

Note: The solutions of Exercises 11–14 are similar to that of Exercise 13. These matrices have the property that if a cofactor expansion is chosen along a column or row that contains two zeros, then the characteristic polynomial appears in a partially factored form.

19. Since the equation $\det(A - \lambda I) = (\lambda_1 - \lambda)(\lambda_2 - \lambda) \cdots (\lambda_n - \lambda)$ holds for all λ, set $\lambda = 0$ and conclude that $\det A = \lambda_1 \lambda_2 \cdots \lambda_n$.

21. a. See Example 1.   **b.** See Theorem 3.
   c. See Theorem 3.   **d.** See the solution of Example 4.

23. If $A = QR$, with Q invertible, and if $A_1 = RQ$, then $A_1 = Q^{-1}QRQ = Q^{-1}AQ$, which shows that A_1 is similar to A.

25. $A = \begin{bmatrix} .6 & .3 \\ .4 & .7 \end{bmatrix}$, $\mathbf{v}_1 = \begin{bmatrix} 3/7 \\ 4/7 \end{bmatrix}$, $\mathbf{x}_0 = \begin{bmatrix} .5 \\ .5 \end{bmatrix}$

a. The problem statement implies that $\mathbf{v}_1$ is an eigenvector of A. This is readily verified:

$$A\mathbf{v}_1 = \begin{bmatrix} .6 & .3 \\ .4 & .7 \end{bmatrix} \begin{bmatrix} 3/7 \\ 4/7 \end{bmatrix} = \begin{bmatrix} (1.8 + 1.2)/7 \\ (1.2 + 2.8)/7 \end{bmatrix} = \begin{bmatrix} 3/7 \\ 4/7 \end{bmatrix} = 1 \cdot \mathbf{v}_1$$

So $\mathbf{v}_1$ is an eigenvector corresponding to the eigenvalue 1. To find another eigenvector, first compute the characteristic polynomial:

$$\det(A - \lambda I) = \det \begin{bmatrix} .6 - \lambda & .3 \\ .4 & .7 - \lambda \end{bmatrix} = (.6 - \lambda)(.7 - \lambda) - .12$$

$$= .42 - 1.3\lambda + \lambda^2 - .12 = \lambda^2 - 1.3\lambda + .3$$

Factoring might be a little difficult, but since 1 is an eigenvalue $\lambda - 1$ must be a factor of the polynomial. This helps to see that $\det(A - \lambda I) = (\lambda - 1)(\lambda - .3)$. So .3 is the other eigenvalue. For a corresponding eigenvector, solve $(A - .3)\mathbf{x} = \mathbf{0}$:

$$[(A - .3I) \quad \mathbf{0}] = \begin{bmatrix} .3 & .3 & 0 \\ .4 & .4 & 0 \end{bmatrix} \sim \begin{bmatrix} 1 & 1 & 0 \\ 0 & 0 & 0 \end{bmatrix}, \quad x_1 + x_2 = 0$$

So $x_1 = -x_2$, with x_2 free. An eigenvector is $\mathbf{v}_2 = \begin{bmatrix} -1 \\ 1 \end{bmatrix}$. Now, $\{\mathbf{v}_1, \mathbf{v}_2\}$ is linearly

independent because the eigenvectors correspond to different eigenvectors (also, they are not multiples). Thus, $\{\mathbf{v}_1, \mathbf{v}_2\}$ is basis for $\mathbb{R}^2$, by the Basis Theorem, because the set contains two vectors and $\mathbb{R}^2$ is two-dimensional. (You can use other arguments, too.)

b. To show that $\mathbf{x}_0 = \mathbf{v}_1 + c\mathbf{v}_2$, just compare $\mathbf{x}_0 - \mathbf{v}_1$ with $\mathbf{v}_2$, to find c:

$$\mathbf{x}_0 - \mathbf{v}_1 = \begin{bmatrix} 1/2 \\ 1/2 \end{bmatrix} - \begin{bmatrix} 3/7 \\ 4/7 \end{bmatrix} = \begin{bmatrix} 1/14 \\ -1/14 \end{bmatrix} = \frac{1}{14}\begin{bmatrix} -1 \\ 1 \end{bmatrix} = \frac{1}{14}\mathbf{v}_2$$

So $\mathbf{x}_0 = \mathbf{v}_1 + (1/14)\mathbf{v}_2$.

c. Compute $\mathbf{x}_1 = A\mathbf{x}_0 = A(\mathbf{v}_1 + (1/14)\mathbf{v}_2) = A\mathbf{v}_1 + (1/14)A\mathbf{v}_2 = \mathbf{v}_1 + (1/14)(.3)\mathbf{v}_2$, because $\mathbf{v}_1$ and $\mathbf{v}_2$ are eigenvectors of A corresponding to eigenvalues 1 and .3, respectively. To continue, recall from Practice Problem 2 in Section 5.1 that $A^k\mathbf{x} = \lambda^k\mathbf{x}$ when $\mathbf{x}$ is an eigenvector corresponding to an eigenvalue λ. Thus,

$$\mathbf{x}_2 = A^2\mathbf{x}_0 = A^2(\mathbf{v}_1 + (1/14)\mathbf{v}_2) = A^2\mathbf{v}_1 + (1/14)A^2\mathbf{v}_2 = \mathbf{v}_1 + (1/14)(.3)^2\mathbf{v}_2$$
$$\mathbf{x}_3 = A^3\mathbf{x}_0 = A^3(\mathbf{v}_1 + (1/14)\mathbf{v}_2) = A^3\mathbf{v}_1 + (1/14)A^3\mathbf{v}_2 = \mathbf{v}_1 + (1/14)(.3)^3\mathbf{v}_2$$

Note: Another way to find these formulas is to compute

$$\mathbf{x}_2 = A^2\mathbf{x}_0 = A(A\mathbf{x}_0) = A(\mathbf{v}_1 + (1/14)(.3)\mathbf{v}_2) = A\mathbf{v}_1 + (1/14)(.3)A\mathbf{v}_2 = \mathbf{v}_1 + (1/14)(.3)^2\mathbf{v}_2$$
$$\mathbf{x}_3 = A^3\mathbf{x}_0 = A(A^2\mathbf{x}_0) = A(\mathbf{v}_1 + (1/14)(.3)^2\mathbf{v}_2) = A\mathbf{v}_1 + (1/14)(.3)^2A\mathbf{v}_2$$
$$= \mathbf{v}_1 + (1/14)(.3)^3\mathbf{v}_2$$

By inspection, it seems that

$$\mathbf{x}_k = \mathbf{v}_1 + (1/14)(.3)^k\mathbf{v}_2 \tag{1}$$

As $k \to \infty$, the powers of .3 tend to 0, and $\mathbf{x}_k$ tends to $\mathbf{v}_1$.

Appendix: Factoring a Polynomial

In general it is difficult to factor a polynomial of degree 3 or higher (unless you have one of several powerful computer programs available). Fortunately, textbook examples and exercises tend to have small integer solutions. The following observation is helpful.

> Let $p(\lambda)$ be a polynomial with integer coefficients. If $p(c) = 0$ for some integer c, then $\lambda - c$ is a factor of $p(\lambda)$ and c is a divisor of the constant term of $p(\lambda)$.

EXAMPLE Find the eigenvalues of the matrix A whose characteristic polynomial is $p(\lambda) = -\lambda^3 + 18\lambda^2 - 96\lambda + 160$.

Solution By the observation above, any integer eigenvalue of A must be a divisor of the constant term 160 in the characteristic polynomial. There are twenty-four such divisors: 1, 2, 4, 5, 8, 10, 16, 20, 32, 40, 80, and 160, together with the negatives of these numbers. We let c be one of these numbers and try to divide $\lambda - c$ into the polynomial by long division. The terms $\lambda \pm 1$ and $\lambda \pm 2$ don't work, and the first successful division is

$$\require{enclose}\begin{array}{r}-\lambda^2+14\lambda-40\\[-2pt]\lambda-4\overline{\smash{\big)}\,-\lambda^3+18\lambda^2-96\lambda+160}\\[-2pt]\underline{-\lambda^3+4\lambda^2}\\[-2pt]14\lambda^2-96\lambda\\[-2pt]\underline{14\lambda^2-56\lambda}\\[-2pt]-40\lambda+160\\[-2pt]\underline{-40\lambda+160}\end{array}$$

Thus the characteristic polynomial is $(\lambda - 4)(-\lambda^2 + 14\lambda - 40)$. The quadratic polynomial factors easily, and the characteristic equation is $(\lambda - 4)(\lambda - 4)(10 - \lambda) = 0$. The eigenvalues of A are 4 (with multiplicity two) and 10.

MATLAB poly, plot

You can use the MATLAB command **poly(A)** to check your answers in Exercises 9–14. Note that if A is $n \times n$, this command lists the coefficients of the characteristic polynomial of A in order of decreasing powers of λ, beginning with λ^n. If the polynomial is of odd degree, the coefficients are multiplied by -1, to make $+1$ the coefficient of λ^n. This corresponds to using $\det(\lambda I - A)$ instead of $\det(A - \lambda I)$.

For Exercises 28 and 29, use **randomint(4)** to create a 4×4 matrix with random integer entries. For Exercise 29, use **gauss** and perhaps **swap** to create the echelon form without row scaling. See the MATLAB box for Section 1.4.

The following commands will produce the graph of the characteristic polynomial of the matrix A in Exercise 30 (with $a = 32$).

```
x = linspace(0,3)        Choose 100 points between 0 and 3
grid on                  Include a grid on the display
hold on                  Add the next graph to the display
A(3,2)= 32; p = poly(A); Compute the characteristic polynomial
v = polyval(p,x);        Evaluate it at the points in x
plot(x,v,'b')            Plot the graph in blue.
```

Edit line 4 to change the value of a (from 32 to another value). Edit line 6 to change the color of the graph. When the commands are run again, the old graph(s) will remain visible. If you do not specify the color of the graph, MATLAB will automatically cycle through a set of colors, one for each graph on the display. To create a fresh display, enter **hold off**.

5.3   DIAGONALIZATION

The factorization $A = PDP^{-1}$ is used to compute powers of A, decouple dynamical systems in Sections 5.6 and 5.7, and study symmetric matrices and quadratic forms in Chapter 7.

KEY IDEAS

Example 3 gives the algorithm for diagonalizing a matrix A. After you construct P and D, check your calculations:

1. Compute AP and PD, and check that $AP = PD$.

2. Make sure the columns of P are linearly independent. Use Theorem 7 to save time. You only have to verify that for each eigenvalue, the corresponding eigenvectors are linearly independent. That's easy if the dimension of the eigenspace is 2 or 1.

The key equation in this section is $AP = PD$. It will help you to keep the order of the factors correct when you write $A = PDP^{-1}$, and it also leads immediately to $P^{-1}AP = D$, which you may need occasionally. Possible test question: If $AP = PD$, explain why the first column of P is an eigenvector of A (if the column is nonzero). (Study the proof of the Diagonalization Theorem.)

Warning: Do not confuse the property of being diagonalizable with the property of being invertible. They are not connected. The matrix in Example 5 is diagonalizable, but it is not invertible because 0 is an eigenvalue. The matrix $\begin{bmatrix} 1 & -2 \\ 0 & 1 \end{bmatrix}$ is invertible, but it is not diagonalizable because the eigenspace for $\lambda = 1$ is only one-dimensional.

SOLUTIONS TO EXERCISES

1. $P = \begin{bmatrix} 5 & 7 \\ 2 & 3 \end{bmatrix}$, $D = \begin{bmatrix} 2 & 0 \\ 0 & 1 \end{bmatrix}$, $A = PDP^{-1}$, and $A^4 = PD^4P^{-1}$. Next compute $P^{-1} = \frac{1}{1}\begin{bmatrix} 3 & -7 \\ -2 & 5 \end{bmatrix}$,

$D^4 = \begin{bmatrix} 2^4 & 0 \\ 0 & 1 \end{bmatrix} = \begin{bmatrix} 16 & 0 \\ 0 & 1 \end{bmatrix}$. Putting this together,

$A^4 = \begin{bmatrix} 5 & 7 \\ 2 & 3 \end{bmatrix}\begin{bmatrix} 16 & 0 \\ 0 & 1 \end{bmatrix}\begin{bmatrix} 3 & -7 \\ -2 & 5 \end{bmatrix} = \begin{bmatrix} 80 & 7 \\ 32 & 3 \end{bmatrix}\begin{bmatrix} 3 & -7 \\ -2 & 5 \end{bmatrix} = \begin{bmatrix} 226 & -525 \\ 90 & -209 \end{bmatrix}$

7. $A = \begin{bmatrix} 1 & 0 \\ 6 & -1 \end{bmatrix}$. The eigenvalues are obviously ± 1 (since A is triangular).

<u>For $\lambda = 1$</u>: $A - 1I = \begin{bmatrix} 0 & 0 \\ 6 & -2 \end{bmatrix}$. The equation $(A - I)\mathbf{x} = \mathbf{0}$ amounts to $6x_1 - 2x_2 = 0$. So $x_1 = (1/3)x_2$, with x_2 free. The general solution is $x_2\begin{bmatrix} 1/3 \\ 1 \end{bmatrix}$. A nice basis vector for the eigenspace is $\mathbf{u}_1 = \begin{bmatrix} 1 \\ 3 \end{bmatrix}$.

For $\lambda = -1$: $A - (-1)I = \begin{bmatrix} 2 & 0 \\ 6 & 0 \end{bmatrix}$. The equation $(A + I)\mathbf{x} = \mathbf{0}$ amounts to $2x_1 = 0$, with x_2 free.

The general solution is $x_2 \begin{bmatrix} 0 \\ 1 \end{bmatrix}$. Take $\mathbf{u}_2 = \begin{bmatrix} 0 \\ 1 \end{bmatrix}$ as a basis vector for the eigenspace.

From $\mathbf{u}_1$ and $\mathbf{u}_2$, construct $P = [\mathbf{u}_1 \quad \mathbf{u}_2] = \begin{bmatrix} 1 & 0 \\ 3 & 1 \end{bmatrix}$. Then set $D = \begin{bmatrix} 1 & 0 \\ 0 & -1 \end{bmatrix}$, where the eigenvalues in D correspond to $\mathbf{u}_1$ and $\mathbf{u}_2$, respectively.

Warning: The 3×3 matrices in Exercises 12–18 may be diagonalizable even though each matrix has only two distinct eigenvalues. (Theorem 6 gives only a *sufficient* condition for diagonalizability.) You have to check for three linearly independent eigenvectors.

13. $A = \begin{bmatrix} 2 & 2 & -1 \\ 1 & 3 & -1 \\ -1 & -2 & 2 \end{bmatrix}$. The eigenvalues 5 and 1 are given. Because A is 3×3, you need three linearly independent eigenvectors.

For $\lambda = 5$: Solve $(A - 5I)\mathbf{x} = \mathbf{0}$. Form $A - 5I = \begin{bmatrix} -3 & 2 & -1 \\ 1 & -2 & -1 \\ -1 & -2 & -3 \end{bmatrix}$ and compute

$$\begin{bmatrix} -3 & 2 & -1 & 0 \\ 1 & -2 & -1 & 0 \\ -1 & -2 & -3 & 0 \end{bmatrix} \sim \begin{bmatrix} 1 & -2 & -1 & 0 \\ -3 & 2 & -1 & 0 \\ -1 & -2 & -3 & 0 \end{bmatrix} \sim \cdots \sim \begin{bmatrix} ① & 0 & 1 & 0 \\ 0 & ① & 1 & 0 \\ 0 & 0 & 0 & 0 \end{bmatrix}$$

So $x_1 = -x_3$, $x_2 = -x_3$, with x_3 free. Take $\mathbf{v}_1 = \begin{bmatrix} -1 \\ -1 \\ 1 \end{bmatrix}$, for instance, as a basis vector for the eigenspace. At this point, you don't know if A is diagonalizable. Your only hope is to find two linearly independent eigenvectors inside the eigenspace for $\lambda = 1$, because there are no other eigenspaces in which to look.

For $\lambda = 1$: Form $A - I = \begin{bmatrix} 1 & 2 & -1 \\ 1 & 2 & -1 \\ -1 & -2 & 1 \end{bmatrix}$. The equation $(A - I)\mathbf{x} = \mathbf{0}$ reduces to $x_1 + 2x_2 - x_3 = 0$. So $x_1 = -2x_2 + x_3$, with x_2 and x_3 free. At this point you know that the eigenspace is two-dimensional (because there are two free variables). So there are the necessary two linearly independent eigenvectors, and hence A is diagonalizable. To produce the eigenvectors, write the solution of $(A - I)\mathbf{x} = \mathbf{0}$ in the form

$$\begin{bmatrix} x_1 \\ x_2 \\ x_3 \end{bmatrix} = \begin{bmatrix} -2x_2 + x_3 \\ x_2 \\ x_3 \end{bmatrix} = x_2 \begin{bmatrix} -2 \\ 1 \\ 0 \end{bmatrix} + x_3 \begin{bmatrix} 1 \\ 0 \\ 1 \end{bmatrix}$$

Set $\mathbf{v}_2 = \begin{bmatrix} -2 \\ 1 \\ 0 \end{bmatrix}$, $\mathbf{v}_3 = \begin{bmatrix} 1 \\ 0 \\ 1 \end{bmatrix}$, and $P = [\mathbf{v}_1 \quad \mathbf{v}_2 \quad \mathbf{v}_3] = \begin{bmatrix} -1 & -2 & 1 \\ -1 & 1 & 0 \\ 1 & 0 & 1 \end{bmatrix}$.

The columns of P are linearly independent, by Theorem 7, because the eigenvectors form bases for their respective eigenspaces. So P is invertible. Since the first column of P corresponds to $\lambda = 5$, the first diagonal entry in D must be 5, which means that

$$D = \begin{bmatrix} 5 & 0 & 0 \\ 0 & 1 & 0 \\ 0 & 0 & 1 \end{bmatrix}$$

Warning: A common mistake in Exercise 13 is to build a 3×3 matrix P in the usual way and then take D to be a 2×2 matrix, such as

$$D = \begin{bmatrix} 5 & 0 \\ 0 & 1 \end{bmatrix}$$

Of course this doesn't make sense because PD isn't even defined. Another mistake is to make a 2×2 matrix D as above and build P with only two of the three eigenvectors, say one from each eigenspace. Now $AP = PD$, but P is a 3×2 matrix and is not invertible.

Study Tip: Exercise 18 is a good test question. Try it. Note: the eigenvector $(-2, 1, 2)$ does not correspond to the eigenvalue $\lambda = 5$. If you have trouble here, review Exercises 5–8 in Section 5.1.

19. $A = \begin{bmatrix} 5 & -3 & 0 & 9 \\ 0 & 3 & 1 & -2 \\ 0 & 0 & 2 & 0 \\ 0 & 0 & 0 & 2 \end{bmatrix}$. The eigenvalues are obviously 5, 3, and 2. (Why?)

For $\underline{\lambda = 2}$: To solve $(A - 2I)\mathbf{x} = \mathbf{0}$, completely reduce $[(A - 2I) \quad \mathbf{0}]$:

$$\begin{bmatrix} 3 & -3 & 0 & 9 & 0 \\ 0 & 1 & 1 & -2 & 0 \\ 0 & 0 & 0 & 0 & 0 \\ 0 & 0 & 0 & 0 & 0 \end{bmatrix} \sim \begin{bmatrix} 3 & 0 & 3 & 3 & 0 \\ 0 & 1 & 1 & -2 & 0 \\ 0 & 0 & 0 & 0 & 0 \\ 0 & 0 & 0 & 0 & 0 \end{bmatrix} \sim \begin{bmatrix} ① & 0 & 1 & 1 & 0 \\ 0 & ① & 1 & -2 & 0 \\ 0 & 0 & 0 & 0 & 0 \\ 0 & 0 & 0 & 0 & 0 \end{bmatrix}$$

So $x_1 = -x_3 - x_4$, $x_2 = -x_3 + 2x_4$, with x_3 and x_4 free. The usual calculations produce a basis for the eigenspace:

$$\begin{bmatrix} x_1 \\ x_2 \\ x_3 \\ x_4 \end{bmatrix} = \begin{bmatrix} -x_3 - x_4 \\ -x_3 + 2x_4 \\ x_3 \\ x_4 \end{bmatrix} = x_3 \begin{bmatrix} -1 \\ -1 \\ 1 \\ 0 \end{bmatrix} + x_4 \begin{bmatrix} -1 \\ 2 \\ 0 \\ 1 \end{bmatrix}. \quad \text{Basis: } \mathbf{v}_1 = \begin{bmatrix} -1 \\ -1 \\ 1 \\ 0 \end{bmatrix}, \ \mathbf{v}_2 = \begin{bmatrix} -1 \\ 2 \\ 0 \\ 1 \end{bmatrix}$$

Checkpoint: If you happened to choose $\lambda = 2$ first, as in this solution, would you have enough information at this point to determine whether A is diagonalizable?

For $\lambda = 3$: To solve $(A - 3I)\mathbf{x} = \mathbf{0}$, completely reduce $[(A - 3I) \ \mathbf{0}]$:

$$\begin{bmatrix} 2 & -3 & 0 & 9 & 0 \\ 0 & 0 & 1 & -2 & 0 \\ 0 & 0 & -1 & 0 & 0 \\ 0 & 0 & 0 & -1 & 0 \end{bmatrix} \sim \cdots \sim \begin{bmatrix} ① & -3/2 & 0 & 0 & 0 \\ 0 & 0 & ① & 0 & 0 \\ 0 & 0 & 0 & ① & 0 \\ 0 & 0 & 0 & 0 & 0 \end{bmatrix}, \quad \begin{cases} x_1 = (3/2)x_2 \\ x_2 \text{ is free} \\ x_3 = 0 \\ x_4 = 0 \end{cases}$$

Choosing $x_2 = 2$ produces the eigenvector $\mathbf{v}_3 = (3, 2, 0, 0)$.

For $\lambda = 5$: To solve $(A - 5I)\mathbf{x} = \mathbf{0}$, completely reduce $[(A - 5I) \ \mathbf{0}]$:

$$\begin{bmatrix} 0 & -3 & 0 & 9 & 0 \\ 0 & -2 & 1 & -2 & 0 \\ 0 & 0 & -3 & 0 & 0 \\ 0 & 0 & 0 & -3 & 0 \end{bmatrix} \sim \cdots \sim \begin{bmatrix} 0 & ① & 0 & 0 & 0 \\ 0 & 0 & ① & 0 & 0 \\ 0 & 0 & 0 & ① & 0 \\ 0 & 0 & 0 & 0 & 0 \end{bmatrix}, \quad \begin{cases} x_1 \text{ is free} \\ x_2 = 0 \\ x_3 = 0 \\ x_4 = 0 \end{cases}$$

A basis vector for the eigenspace is $\mathbf{v}_4 = (1, 0, 0, 0)$. Set

$$P = [\mathbf{v}_1 \quad \mathbf{v}_2 \quad \mathbf{v}_3 \quad \mathbf{v}_4] = \begin{bmatrix} -1 & -1 & 3 & 1 \\ -1 & 2 & 2 & 0 \\ 1 & 0 & 0 & 0 \\ 0 & 1 & 0 & 0 \end{bmatrix}, \quad D = \begin{bmatrix} 2 & 0 & 0 & 0 \\ 0 & 2 & 0 & 0 \\ 0 & 0 & 3 & 0 \\ 0 & 0 & 0 & 5 \end{bmatrix}$$

This answer differs from that in the text. There, $P = [\mathbf{v}_4 \ \mathbf{v}_3 \ \mathbf{v}_1 \ \mathbf{v}_2]$, and the entries in D are rearranged to match the new order of the eigenvectors. According to the Diagonalization Theorem, both answers are correct.

21. **a.** The symbol D does not automatically denote a diagonal matrix.

   b. See the remark after the statement of the Diagonalization Theorem.

   c. Check out the matrix in Example 4.

   d. See Example 4.

23. A is diagonalizable because you know that five linearly independent eigenvectors exist: three in the three-dimensional eigenspace and two in the two-dimensional eigenspace. Theorem 7 guarantees that the set of all five eigenvectors is linearly independent.

25. Let $\{\mathbf{v}_1\}$ be a basis for the one-dimensional eigenspace, let $\mathbf{v}_2$ and $\mathbf{v}_3$ form a basis for the two-dimensional eigenspace, and let $\mathbf{v}_4$ be any eigenvector in the remaining eigenspace. By Theorem 7, $\{\mathbf{v}_1, \mathbf{v}_2, \mathbf{v}_3, \mathbf{v}_4\}$ is linearly independent. Since A is 4×4, the Diagonalization Theorem shows that A is diagonalizable.

27. If A is diagonalizable, then $A = PDP^{-1}$ for some invertible P and diagonal D. Since A is invertible, 0 is not an eigenvalue of A. So the diagonal entries in D (which are eigenvalues of A) are not zero, and D is invertible. By the theorem on the inverse of a product,
$$A^{-1} = (PDP^{-1})^{-1} = (P^{-1})^{-1}D^{-1}P^{-1} = PD^{-1}P^{-1}$$
Since D^{-1} is obviously diagonal, A^{-1} is diagonalizable.

A Second Proof: If A is $n \times n$, it has n linearly independent eigenvectors, say, $\mathbf{v}_1, \ldots, \mathbf{v}_n$. Then $\mathbf{v}_1, \ldots, \mathbf{v}_n$ are also eigenvectors of A^{-1}, by the solution of Exercise 25 in Section 5.1. Hence A^{-1} is diagonalizable, by the Diagonalization Theorem.

29. The diagonal entries in D_1 are reversed from those in D. So interchange the (eigenvector) columns of P to make them correspond properly to the eigenvalues in D_1. In this case,
$$P_1 = \begin{bmatrix} 1 & 1 \\ -2 & -1 \end{bmatrix} \text{ and } D_1 = \begin{bmatrix} 3 & 0 \\ 0 & 5 \end{bmatrix}$$
Although the first column of P must be an eigenvector corresponding to the eigenvalue 3, there is nothing to prevent us from selecting some multiple of $\begin{bmatrix} 1 \\ -2 \end{bmatrix}$, say $\begin{bmatrix} -3 \\ 6 \end{bmatrix}$, and letting
$$P_2 = \begin{bmatrix} -3 & 1 \\ 6 & -1 \end{bmatrix}.$$ We now have three different factorizations or "diagonalizations" of A:
$$A = PDP^{-1} = P_1D_1P_1^{-1} = P_2D_1P_2^{-1}$$

Answer to Checkpoint: Yes. In Exercise 19, the fact that the eigenspace for $\lambda = 2$ is two-dimensional guarantees that A is diagonalizable, because each of the other two eigenvalues will produce at least one eigenvector, and the resulting set of four eigenvectors will be linearly independent, by Theorem 7. So A is diagonalizable, by the Diagonalization Theorem. (Note that since one eigenspace is two-dimensional, the other eigenspaces must be one-dimensional, because there could not possibly be *more* than four linearly independent eigenvectors in $\mathbb{R}^4$.)

Mastering Linear Algebra Concepts: Eigenvalue, Eigenvector, Eigenspace

I suggest that you take time now to prepare review sheets for the three terms listed above. Begin with their definitions, of course.

Eigenvalue:
- equivalent description   Sec. 5.1: Equation (3); Box on page 313
- geometric interpretation   Sec. 5.1: Fig. 2
- special cases   Theorems 1, 6
- typical computations   Sec. 5.1: Exer. 7, 19; Sec. 5.2: Exer. 7, 15
- connections with other concepts   The IMT; Theorem 4; Sec. 5.2: Exer. 19

Eigenvector:
- special cases   Theorem 2
- typical computations   Sec. 5.1: Examples 2, 4, Exer. 5, 15
- connections with other concepts   Sec. 5.1: Exer. 33; Sec. 5.2: Example 5

                                            Sec. 5.3: Theorem 5, Exer. 5, 18

Eigenspace:
- equivalent description   Sec. 5.1: Equation (3)
- geometric interpretation   Sec. 5.1: Fig. 2, 3
- typical computations   Sec. 5.1: Example 4
- connections with other concepts   Theorem 7

MATLAB   Diagonalization

To practice the diagonalization procedure in this section, you should use **nulbasis** to produce eigenvectors. For Exercises 33–36, first enter the command **ev = eig(A)**, to provide the eigenvalues. See the MATLAB box for Section 5.1.

   In later work, you may automate the diagonalization process. The command **[P D] = eig(A)** produces two square matrices P and D (or any other names you choose) such that $AP = PD$, with D diagonal. If P is invertible, then A is diagonalizable. Check whether **P*D*inv(P) - A** is the zero matrix. In any case, P is likely to be different from what you construct for the homework. The columns of P are scaled so they are "unit" vectors (studied in Chapter 6). Dividing P by one of its entries may make some of the entries recognizable, because the problems in the text usually involve simple numbers.

5.4   EIGENVALUES AND LINEAR TRANSFORMATIONS _____

This section introduces the matrix of a linear transformation relative to specified bases for vector spaces, and then uses this concept to give a new interpretation of the matrix factorization $A = PDP^{-1}$.

STUDY NOTES

The exercises will help you learn the definition of the matrix representation of a transformation relative to specified bases, say $\mathcal{B}$ and $\mathcal{C}$. Compare this definition with the standard matrix of a transformation from $\mathbb{R}^n$ into $\mathbb{R}^m$. What are $\mathcal{B}$ and $\mathcal{C}$ in this case?

After Example 1 (and Exercises 1–6 and 9–10), the section focuses on the case when $T:V \rightarrow V$ and $\mathcal{B} = \mathcal{C}$. The following algorithm and the solution to Exercise 7 describe two different ways to construct $[T]_\mathcal{B}$.

Algorithm for Finding the $\mathcal{B}$-matrix of $T : V \rightarrow V$

1. Compute the images of the basis vectors:
$$T(\mathbf{b}_1), \ldots, T(\mathbf{b}_n)$$

2. Convert these images into $\mathcal{B}$-coordinate vectors:
$$[T(\mathbf{b}_1)]_\mathcal{B}, \ldots, [T(\mathbf{b}_n)]_\mathcal{B}$$

3. Place the $\mathcal{B}$-coordinate vectors into the columns of $[T]_\mathcal{B}$.

The sentence before Theorem 8 summarizes the main idea of the theorem. Studying the proof should help you understand the theorem and review important concepts. The subsection on Similarity of Matrix Representations could have contained more ideas. Take notes carefully if your instructor decides to expand this material somewhat.

SOLUTIONS TO EXERCISES

1. $T(\mathbf{b}_1) = 3\mathbf{d}_1 - 5\mathbf{d}_2,$ $T(\mathbf{b}_2) = -\mathbf{d}_1 + 6\mathbf{d}_2,$ $T(\mathbf{b}_3) = 4\mathbf{d}_2$

$$[T(\mathbf{b}_1)]_\mathcal{D} = \begin{bmatrix} 3 \\ -5 \end{bmatrix} \qquad [T(\mathbf{b}_2)]_\mathcal{D} = \begin{bmatrix} -1 \\ 6 \end{bmatrix} \qquad [T(\mathbf{b}_3)]_\mathcal{D} = \begin{bmatrix} 0 \\ 4 \end{bmatrix}$$

Matrix for T
relative to $\mathcal{B}$ and $\mathcal{D}$: $\begin{bmatrix} 3 & -1 & 0 \\ -5 & 6 & 4 \end{bmatrix}$

7. The method of Example 2 works. But, perhaps a faster way is to realize that the information given provides the general form of $T(\mathbf{p})$ as shown in the figure below:

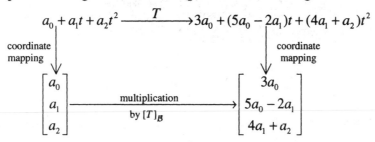

The matrix that implements the multiplication along the bottom of the figure is easily filled in by inspection:

$$\begin{bmatrix} ? & ? & ? \\ ? & ? & ? \\ ? & ? & ? \end{bmatrix} \begin{bmatrix} a_0 \\ a_1 \\ a_2 \end{bmatrix} = \begin{bmatrix} 3a_0 \\ 5a_0 - 2a_1 \\ 4a_1 + a_2 \end{bmatrix} \text{ implies that } [T]_B = \begin{bmatrix} 3 & 0 & 0 \\ 5 & -2 & 0 \\ 0 & 4 & 1 \end{bmatrix}$$

Study Tip: See the *Study Guide* notes for Section 1.9 (particularly for Exercise 19). This method allows you to find a matrix that implements the mapping without assuming that T is linear. In fact, this derivation *proves* that T is linear, because it is now represented as a matrix transformation. Why not try this method on Example 2?

9. b. The transformation T maps $\mathbb{P}_2$ into $\mathbb{R}^3$ by the formula $T(\mathbf{p}) = \begin{bmatrix} \mathbf{p}(-1) \\ \mathbf{p}(0) \\ \mathbf{p}(1) \end{bmatrix}$. Take any $\mathbf{p}$ and $\mathbf{q}$

in $\mathbb{P}_2$ and any scalar c. Then T is linear because

$$T(\mathbf{p}+\mathbf{q}) = \begin{bmatrix} (\mathbf{p}+\mathbf{q})(-1) \\ (\mathbf{p}+\mathbf{q})(0) \\ (\mathbf{p}+\mathbf{q})(1) \end{bmatrix} = \begin{bmatrix} \mathbf{p}(-1) \\ \mathbf{p}(0) \\ \mathbf{p}(1) \end{bmatrix} + \begin{bmatrix} \mathbf{q}(-1) \\ \mathbf{q}(0) \\ \mathbf{q}(1) \end{bmatrix} = T(\mathbf{p})+T(\mathbf{q}), \text{ and}$$

$$T(c\mathbf{p}) = \begin{bmatrix} (c\mathbf{p})(-1) \\ (c\mathbf{p})(0) \\ (c\mathbf{p})(1) \end{bmatrix} = c \begin{bmatrix} \mathbf{p}(-1) \\ \mathbf{p}(0) \\ \mathbf{p}(1) \end{bmatrix} = cT(\mathbf{p})$$

13. Start by diagonalizing $A = \begin{bmatrix} 0 & 1 \\ -3 & 4 \end{bmatrix}$. The characteristic polynomial is

$\lambda^2 - 4\lambda + 3 = (\lambda - 1)(\lambda - 3)$, so the eigenvalues are 1 and 3.

For $\underline{\lambda = 1}$: $A - I = \begin{bmatrix} -1 & 1 \\ -3 & 3 \end{bmatrix}$. The equation $(A - I)\mathbf{x} = \mathbf{0}$ amounts to $-x_1 + x_2 = 0$. So $x_1 = x_2$, with x_2 free. As a basis vector, take $\mathbf{u}_1 = \begin{bmatrix} 1 \\ 1 \end{bmatrix}$.

For $\underline{\lambda = 3}$: $A - 3I = \begin{bmatrix} -3 & 1 \\ -3 & 3 \end{bmatrix}$. The equation $(A - I)\mathbf{x} = \mathbf{0}$ amounts to $-3x_1 + x_2 = 0$. So $x_1 = x_2/3$, with x_2 free. Take $\mathbf{u}_2 = \begin{bmatrix} 1 \\ 3 \end{bmatrix}$ as a basis vector. The vectors $\mathbf{u}_1$, $\mathbf{u}_2$ can form the columns of a matrix P that diagonalizes A. By Theorem 8, the basis $\mathcal{B} = \{\mathbf{u}_1, \mathbf{u}_2\}$ has the property that the $\mathcal{B}$-matrix of the transformation $\mathbf{x} \mapsto A\mathbf{x}$ is a diagonal matrix.

19. If A is similar to B, then there exists an invertible matrix P such that $P^{-1}AP = B$. Thus B is invertible because it is the product of invertible matrices. By a theorem about inverses of products, $B^{-1} = P^{-1}A^{-1}(P^{-1})^{-1} = P^{-1}A^{-1}P$, which shows that A^{-1} is similar to B^{-1}.

21. By hypothesis, there exist invertible P and Q such that $P^{-1}BP = A$ and $Q^{-1}CQ = A$. Then $P^{-1}BP = Q^{-1}CQ$. Left-multiply by Q and right-multiply by Q^{-1} to obtain $QP^{-1}BPQ^{-1} = QQ^{-1}CQQ^{-1}$. So $C = QP^{-1}BPQ^{-1} = (PQ^{-1})^{-1}B(PQ^{-1})$, which shows that B is similar to C.

23. If $A\mathbf{x} = \lambda\mathbf{x}$, $\mathbf{x} \neq \mathbf{0}$, then $P^{-1}A\mathbf{x} = \lambda P^{-1}\mathbf{x}$. If $B = P^{-1}AP$, then

$$B(P^{-1}\mathbf{x}) = P^{-1}AP(P^{-1}\mathbf{x}) = P^{-1}A\mathbf{x} = \lambda P^{-1}\mathbf{x} \qquad (*)$$

by the first calculation. Note that $P^{-1}\mathbf{x} \neq \mathbf{0}$, because $\mathbf{x} \neq \mathbf{0}$ and P^{-1} is invertible. Hence $(*)$ shows that $P^{-1}\mathbf{x}$ is an eigenvector of B corresponding to λ. (Of course, λ is an eigenvalue of both A and B because the matrices are similar, by Theorem 4 in Section 5.2.)

25. If $A = PBP^{-1}$, then

$$\begin{aligned} \text{tr}(A) &= \text{tr}((PB)P^{-1}) = \text{tr}(P^{-1}(PB)) \qquad \text{By the trace property} \\ &= \text{tr}(P^{-1}PB) = \text{tr}(IB) = \text{tr}(B) \end{aligned}$$

If B is diagonal, then the diagonal entries of B must be the eigenvalues of A, by the Diagonalization Theorem (Theorem 5 in Section 5.3). So tr A = tr B = {sum of the eigenvalues of A}.

Study Tip: It can be shown that for *any* square matrix A, the trace of A is the sum of the eigenvalues of A, counted according to multiplicities. You can use this fact to provide a quick check on your eigenvalue calculations. The *sum* of the eigenvalues must match the *sum* of the diagonal entries in A (even if A is not diagonalizable).

29. If $\mathcal{B} = \{\mathbf{b}_1, ..., \mathbf{b}_n\}$, then the $\mathcal{B}$-coordinate vector of $\mathbf{b}_j$ is $\mathbf{e}_j$, the standard basis vector for $\mathbb{R}^n$. For instance,

$$\mathbf{b}_1 = 1\cdot\mathbf{b}_1 + 0\cdot\mathbf{b}_2 + \cdots + 0\cdot\mathbf{b}_n$$

Thus $[I(\mathbf{b}_j)]_\mathcal{B} = [\mathbf{b}_j]_\mathcal{B} = \mathbf{e}_j$, and

$$[I]_\mathcal{B} = \big[[I(\mathbf{b}_1)]_\mathcal{B} \quad \cdots \quad [I(\mathbf{b}_n)]_\mathcal{B}\big] = [\mathbf{e}_1 \quad \cdots \quad \mathbf{e}_n] = I$$

31. See the subsection, "Similarity of Matrix Representations," which points out that the matrix D in Theorem 8 need not be diagonal. If P is the matrix whose columns come from $\mathcal{B}$, then the $\mathcal{B}$-matrix of the transformation $\mathbf{x} \mapsto A\mathbf{x}$ is $D = P^{-1}AP$. From the data in the text,

$$A = \begin{bmatrix} -7 & -48 & -16 \\ 1 & 14 & 6 \\ -3 & -45 & -19 \end{bmatrix}, \quad P = [\mathbf{b}_1 \ \mathbf{b}_2 \ \mathbf{b}_3] = \begin{bmatrix} -3 & -2 & 3 \\ 1 & 1 & -1 \\ -3 & -3 & 0 \end{bmatrix}$$

$$D = \begin{bmatrix} -1 & -3 & -1/3 \\ 1 & 3 & 0 \\ 0 & -1 & -1/3 \end{bmatrix}\begin{bmatrix} -7 & -48 & -16 \\ 1 & 14 & 6 \\ -3 & -45 & -19 \end{bmatrix}\begin{bmatrix} -3 & -2 & 3 \\ 1 & 1 & -1 \\ -3 & -3 & 0 \end{bmatrix} = \begin{bmatrix} -7 & -2 & -6 \\ 0 & -4 & -6 \\ 0 & 0 & -1 \end{bmatrix}$$

5.5 COMPLEX EIGENVALUES

If the characteristic equation of an $n \times n$ real matrix A has a complex eigenvalue λ and if $\mathbf{v}$ is a nonzero vector in $\mathbb{C}_n$ such that $A\mathbf{v} = \lambda\mathbf{v}$, then both λ and $\mathbf{v}$ provide useful information about A.

STUDY NOTES

Only matrices with real entries are considered here. If λ is a complex eigenvalue of A, with $\mathbf{v}$ a corresponding eigenvector, then $\bar{\lambda}$ is also an eigenvalue of A, with $\bar{\mathbf{v}}$ an eigenvector. Find out if you should know how to prove this fact.

Example 6 describes the prototype for all 2×2 matrices with a complex eigenvalue λ. Only λ is needed if you want to know the angle φ of rotation and the scale factor $|\lambda|$, but an associated eigenvector $\mathbf{v}$ is also needed if you want to factor A as PCP^{-1}, as in Example 7.

SOLUTIONS TO EXERCISES

1. $A = \begin{bmatrix} 1 & -2 \\ 1 & 3 \end{bmatrix}$, $A - \lambda I = \begin{bmatrix} 1-\lambda & -2 \\ 1 & 3-\lambda \end{bmatrix}$

$\det(A - \lambda I) = (1 - \lambda)(3 - \lambda) - (-2) = \lambda^2 - 4\lambda + 5$

Use the quadratic formula to find the eigenvalues: $\lambda = \dfrac{4 \pm \sqrt{16 - 20}}{2} = 2 \pm i$. Example 2

gives a shortcut for finding one eigenvector, and Example 5 shows how to write the other eigenvector with no effort.

For $\lambda = 2 + i$: $A - (2+i)I = \begin{bmatrix} -1-i & -2 \\ 1 & 1-i \end{bmatrix}$. The equation $(A - \lambda I)\mathbf{x} = \mathbf{0}$ gives

$$
\begin{aligned}
(-1-i)x_1 - \quad 2x_2 &= 0 \\
x_1 + (1-i)x_2 &= 0
\end{aligned}
$$

As in Example 2, the two equations are equivalent—each determines the same relation between x_1 and x_2. So use the second equation to obtain $x_1 = -(1-i)x_2$, with x_2 free. The

general solution is $x_2 \begin{bmatrix} -1+i \\ 1 \end{bmatrix}$, and the vector $\mathbf{v}_1 = \begin{bmatrix} -1+i \\ 1 \end{bmatrix}$ provides a basis for the

eigenspace.

For ~$\lambda = 2 - i$: Let $\mathbf{v}_2 = \overline{\mathbf{v}}_1 = \begin{bmatrix} -1-i \\ 1 \end{bmatrix}$. The remark prior to Example 5 shows that $\mathbf{v}_2$ is

automatically an eigenvector for $\overline{2+i}$. In fact, calculations similar to those above would show that $\{\mathbf{v}_2\}$ is a basis for the eigenspace. (In general, for a real matrix A, it can be shown that the set of complex conjugates of the vectors in a basis of the eigenspace for λ is a basis of the eigenspace for $\overline{\lambda}$.)

7. $A = \begin{bmatrix} \sqrt{3} & -1 \\ 1 & \sqrt{3} \end{bmatrix}$. The eigenvalues are $\sqrt{3} \pm i$. Ask your instructor if you are permitted to

write down the eigenvalues of $\begin{bmatrix} a & -b \\ b & a \end{bmatrix}$ from memory, or if you are expected to find them

via the characteristic equation. Note that the eigenvectors are easy to remember, too. See the Practice Problem.

The scale factor associated with the transformation $\mathbf{x} \mapsto A\mathbf{x}$ is simply

$r = |\lambda| = \left((\sqrt{3})^2 + 1^2 \right)^{1/2} = 2$. For the angle of the rotation, plot the point $(a, b) = (\sqrt{3}, 1)$ in

the xy-plane and use trigonometry.

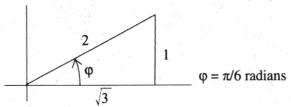

$\varphi = \pi/6$ radians

13. From Exercise 1, $\lambda = 2 \pm i$, and the eigenvector $\mathbf{v} = \begin{bmatrix} -1-i \\ 1 \end{bmatrix}$ corresponds to $\lambda = 2 - i$. Since

$\operatorname{Re}\mathbf{v} = \begin{bmatrix} -1 \\ 1 \end{bmatrix}$ and $\operatorname{Im}\mathbf{v} = \begin{bmatrix} -1 \\ 0 \end{bmatrix}$, take $P = \begin{bmatrix} -1 & -1 \\ 1 & 0 \end{bmatrix}$. Then compute

$$C = P^{-1}AP = \begin{bmatrix} 0 & 1 \\ -1 & -1 \end{bmatrix} \begin{bmatrix} 1 & -2 \\ 1 & 3 \end{bmatrix} \begin{bmatrix} -1 & -1 \\ 1 & 0 \end{bmatrix} = \begin{bmatrix} 0 & 1 \\ -1 & -1 \end{bmatrix} \begin{bmatrix} -3 & -1 \\ 2 & -1 \end{bmatrix} = \begin{bmatrix} 2 & -1 \\ 1 & 2 \end{bmatrix}$$

Actually, Theorem 9 gives the formula for C. Note that the eigenvector $\mathbf{v}$ corresponds to $a - bi$ instead of $a + bi$. If, for instance, you use the eigenvector for $2 + i$, your C will be $\begin{bmatrix} 2 & 1 \\ -1 & 2 \end{bmatrix}$. The imaginary part of the eigenvalue is the $(1, 2)$-entry in C.

So, there are two possible choices for C (depending on the vector used to produce P). On an exam, if you are not sure of the form of C, you can always compute it quickly from the formula $C = P^{-1}AP$, as in the solution above.

Note: Because there are two possibilities for C in the factorization of a 2×2 matrix as in Exercise 13, the measure of rotation φ associated with the transformation $\mathbf{x} \mapsto A\mathbf{x}$ is determined only up to a change of sign. The "orientation" of the angle is determined by the change of variable $\mathbf{x} = P\mathbf{u}$. See Fig. 4 in the text.

19. $A = \begin{bmatrix} 1.52 & -.7 \\ .56 & .4 \end{bmatrix}$, $\det(A - \lambda I) = \lambda^2 - 1.92\lambda + 1$

Use the quadratic formula to solve $\lambda^2 - 1.92\lambda + 1 = 0$:

$$\lambda = \frac{1.92 \pm \sqrt{-.3136}}{2} = .96 \pm .28i$$

To find the eigenvector for $\lambda = .96 - .28i$, solve

$$(1.52 - .96 + .28i)x_1 \qquad\qquad -.7x_2 = 0$$
$$.56x_1 + (.4 - .96 + .28i)x_2 = 0$$

There is a nonzero solution (because $.96 - .28i$ is an eigenvalue), so you can use either equation to find the solution. From the second equation,

$$x_1 = \frac{.56 - .28i}{.56}x_2 = (1 - .5i)x_2$$

Setting $x_2 = 2$ produces the (complex) eigenvector $\mathbf{v} = \begin{bmatrix} 2-i \\ 2 \end{bmatrix}$.

Since Re $\mathbf{v} = \begin{bmatrix} 2 \\ 2 \end{bmatrix}$ and Im $\mathbf{v} = \begin{bmatrix} -1 \\ 0 \end{bmatrix}$, take $P = \begin{bmatrix} 2 & -1 \\ 2 & 0 \end{bmatrix}$. Finally, compute

$$P^{-1}AP = \frac{1}{2}\begin{bmatrix} 0 & 1 \\ -2 & 2 \end{bmatrix}\begin{bmatrix} 1.52 & -.7 \\ .56 & .4 \end{bmatrix}\begin{bmatrix} 2 & -1 \\ 2 & 0 \end{bmatrix} = \begin{bmatrix} .96 & -.28 \\ .28 & .96 \end{bmatrix}$$

This final matrix, which has the proper form, is C.

25. Write $\mathbf{x} = $ Re $\mathbf{x} + i($Im $\mathbf{x})$, so that $A\mathbf{x} = A($Re $\mathbf{x}) + i\,A($Im $\mathbf{x})$. Since A is real, so are $A($Re $\mathbf{x})$ and $A($Im $\mathbf{x})$. Thus $A($Re $\mathbf{x})$ is the real part of $A\mathbf{x}$ and $A($Im $\mathbf{x})$ is the imaginary part of $A\mathbf{x}$.

MATLAB   Complex Eigenvalues

The command **[V D] = eig(A)** (mentioned in Section 5.3) works for matrices with complex eigenvalues. In this case V and the diagonal matrix D have some complex entries. For a 2×2 real matrix with a complex eigenvalue, MATLAB tends to place the eigenvalue $a - bi$ (where $b > 0$) as the (2, 2)-entry of D. MATLAB does not produce matrices for a factorization $A = PCP^{-1}$ of the sort described in this section.

For any matrix V, the commands **real(V)** and **imag(V)** produce the real and imaginary parts of the entries in V, displayed as matrices the same size as V.

5.6   DISCRETE DYNAMICAL SYSTEMS

This section presents the climax to a crescendo of ideas that began in Section 1.10 and flowed through parts of Chapters 4 and 5. You need not have read all the material to appreciate the interesting applications in this section, but you will profit from a review of page 307 and Example 5 in Section 5.2.

KEY IDEAS

A solution of a first order homogeneous difference equation

$$\mathbf{x}_{k+1} = A\mathbf{x}_k \qquad (k = 0, 1, 2, \ldots) \tag{1}$$

is a sequence $\{\mathbf{x}_k\}$ that satisfies (1) and is described by a formula for each $\mathbf{x}_k$ that does not depend on the preceding terms in the sequence other than the initial term $\mathbf{x}_0$. In Section 5.1, you saw how a solution can be constructed when $\mathbf{x}_0$ is an eigenvector. When $\mathbf{x}_0$ is *not* an eigenvector, look for an eigenvector decomposition of $\mathbf{x}_0$:

$$\mathbf{x}_0 = c_1\mathbf{v}_1 + \cdots + c_n\mathbf{v}_n \qquad \text{Each } \mathbf{v}_i \text{ is an eigenvector.} \tag{2}$$

To make (2) possible for any $\mathbf{x}_0$ in $\mathbb{R}^n$, the section assumes that the $n \times n$ matrix A has n linearly independent eigenvectors. If $\mathbf{x}_k = A^k \mathbf{x}_0$, then

$$\mathbf{x}_k = c_1(\lambda_1)^k \mathbf{v}_1 + \cdots + c_n(\lambda_n)^k \mathbf{v}_n$$

When $\{\mathbf{x}_k\}$ describes the "state" of a system at discrete times (denoted by $k = 0, 1, 2, \ldots$), the *long-term behavior* of this dynamical system is a description of what happens to $\mathbf{x}_k$ as $k \to \infty$. The text focuses on the following important situation.

Let A be an $n \times n$ matrix with n linearly independent eigenvectors, corresponding to eigenvalues such that $|\lambda_1| \geq 1 > |\lambda_j|$ for $j = 2, \ldots, n$. If $\mathbf{x}_0$ is given by (2) with $c_1 \neq 0$, then for all sufficiently large k,

$\mathbf{x}_{k+1} \approx \lambda_1 \mathbf{x}_k$	Each entry in $\mathbf{x}_k$ grows by a factor of λ_1.
$\mathbf{x}_k \approx c_1(\lambda_1)^k \mathbf{v}_1$	$\mathbf{x}_k$ is approximately a multiple of $\mathbf{v}_1$, and so the ratio between any two entries in $\mathbf{x}_k$ is nearly the same as the corresponding ratio for $\mathbf{v}_1$.

STUDY NOTES

When the two approximations above are true in an application, the eigenvalue λ_1 and the eigenvector $\mathbf{v}_1$ have interesting physical interpretations. Make sure you can describe these on an exam. (See the last four sentences in the solution of Example 1, for instance.)

The predator-prey model is rather primitive and provides only a starting point for more refined models. Still, you might enjoy considering what the model in Example 1 predicts if $\mathbf{x}_0$ happens to be a multiple of $\mathbf{v}_2 = (5, 1)$, or if initially there are *more* than 5 owls for every 1 thousand rats, assuming $p = .104$.

The graphical descriptions of solutions to difference equations should help you understand what can happen to $\mathbf{x}_k$ as $k \to \infty$. I hope you enjoy studying the figures even if your class does not have time to cover this part of the section. Only the simplest cases are shown, but these cases form the foundation for studying *nonlinear* dynamical systems which are widely used (but require calculus techniques not covered here). Even for nonlinear systems, eigenvalues and eigenvectors of certain matrices play an important role.

SOLUTIONS TO EXERCISES

1. a. The eigenvectors $\mathbf{v}_1 = \begin{bmatrix} 1 \\ 1 \end{bmatrix}$ and $\mathbf{v}_2 = \begin{bmatrix} -1 \\ 1 \end{bmatrix}$ form a basis for $\mathbb{R}^2$. To find the action of A on

$\mathbf{x}_0 = \begin{bmatrix} 9 \\ 1 \end{bmatrix}$ express $\mathbf{x}_0$ in terms of $\mathbf{v}_1$ and $\mathbf{v}_2$. That is, find c_1, c_2 such that $\mathbf{x}_0 = c_1 \mathbf{v}_1 + c_2 \mathbf{v}_2$:

$$\begin{bmatrix} 1 & -1 & 9 \\ 1 & 1 & 1 \end{bmatrix} \sim \begin{bmatrix} 1 & 0 & 5 \\ 0 & 1 & -4 \end{bmatrix} \Rightarrow \mathbf{x}_0 = 5\mathbf{v}_1 - 4\mathbf{v}_2$$

Since v_1, v_2 are eigenvectors (for the eigenvalues 3 and 1/3):

$$x_1 = Ax_0 = 5Av_1 - 4Av_2 = 5 \cdot 3v_1 - 4 \cdot (1/3)v_2$$

$$= \begin{bmatrix} 15 \\ 15 \end{bmatrix} - \begin{bmatrix} -4/3 \\ 4/3 \end{bmatrix} = \begin{bmatrix} 49/3 \\ 41/4 \end{bmatrix}$$

b. Each time A acts on a linear combination of v_1 and v_2, the v_1 term is multiplied by the eigenvalue 3 and the v_2 term is multiplied by the eigenvalue 1/3.

$$x_2 = Ax_1 = A[5(3)v_1 - 4(1/3)v_2] = 5(3)^2v_1 - 4(1/3)^2v_2$$

In general, $x_k = 5(3)^k v_1 - 4(1/3)^k v_2$ for $k \geq 0$.

7. a. The matrix A in Exercise 1 has eigenvalues 3 and 1/3. Since $|3| > 1$ and $|1/3| < 1$, the origin is a saddle point.

b. The direction of greatest attraction is determined by $v_2 = \begin{bmatrix} -1 \\ 1 \end{bmatrix}$, the eigenvector

corresponding to the eigenvalue with absolute value less than 1. The direction of greatest

repulsion is determined by $v_1 = \begin{bmatrix} 1 \\ 1 \end{bmatrix}$, the eigenvector corresponding to the eigenvalue

greater than 1.

c. The drawing below shows: (1) lines through the eigenvectors and the origin, (2) arrows toward the origin (showing attraction) on the line through v_2 and arrows away from the origin (showing repulsion) on the line through v_1, (3) several typical trajectories (with arrows) that show the general flow of points. No specific points other than v_1 and v_2 were computed. This type of drawing is about all that one can make without using a computer to plot points.

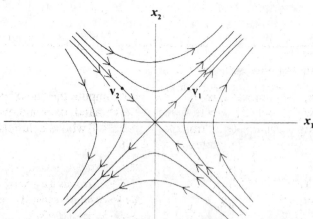

Remark: Sketching trajectories for a dynamical system in which the origin is an attractor or a repellor is more difficult than the sketch in Exercise 7. There has been no discussion of the direction in which the trajectories "bend" as they move toward or away from the origin. For instance, if you rotate Figure 1 of Section 5.6 through a quarter-turn and relabel the axes so that x_1 is on the horizontal axis, then the new figure corresponds to the matrix A with the diagonal entries .8 and .64 interchanged. In general, if A is a diagonal matrix, with positive diagonal entries a and d, unequal to 1, then the trajectories lie on the axes or on curves whose equations have the form $x_2 = r(x_1)^s$, where $s = (\ln d)/(\ln a)$ and r depends on the initial point x_0. (See *Encounters with Chaos*, by Denny Gulick, New York: McGraw-Hill, 1992, pp. 147–150.)

Study Tip: If your instructor wants you to graph trajectories when the origin is an attractor or repellor, there will need to be some class discussion of exactly how to do this.

13. $A = \begin{bmatrix} .8 & .3 \\ -.4 & 1.5 \end{bmatrix}$. First find the eigenvalues:

$$\det(A - \lambda I) = (.8 - \lambda)(1.5 - \lambda) - (.3)(-.4) = \lambda^2 - 2.3 + 1.32$$
$$= (\lambda - 1.2)(\lambda - 1.1) \qquad \text{Use the quadratic formula, if needed.}$$
$$= 0$$

Since both eigenvalues, 1.2 and 1.1, are greater than 1, the origin is a repellor. For the direction of greatest repulsion, find the eigenvector for the larger eigenvalue, 1.2:

$$[(A - 1.2I) \quad \mathbf{0}] = \begin{bmatrix} -.4 & .3 & 0 \\ -.4 & .3 & 0 \end{bmatrix} \sim \begin{bmatrix} 1 & -3/4 & 0 \\ 0 & 0 & 0 \end{bmatrix}, \quad \mathbf{x} = x_2 \begin{bmatrix} 3/4 \\ 1 \end{bmatrix}$$

Any multiple of $\begin{bmatrix} 3/4 \\ 1 \end{bmatrix}$, such as $\begin{bmatrix} 3 \\ 4 \end{bmatrix}$, determines the direction of greatest repulsion.

MATLAB Plotting Trajectories

Given a vector **x**, the command **x = A*x** will compute the "next" point on the trajectory. Use the up-arrow (↑) and <Enter> to repeat the command, over and over.

The following steps create a "trajectory" matrix T whose columns are the points **x**, $A\mathbf{x}$, $A^2\mathbf{x}$, ..., $A^{15}\mathbf{x}$. (Change 15 to any integer you wish.)

`T = x`	Put **x** in the first column of T.
`for j=1:15`	This loop repeats the next two lines 15 times.
`  x = A*x;`	Compute the next point on the trajectory.
`  T = [T x]`	Store the new point in T.
`end`	End of the loop.

After you type the line beginning with "for," MATLAB will suspend all calculations (while you type additional lines) until you type "end" and press <Enter>.

If you want MATLAB itself to plot the points in *T*, use the commands:

```
plot(T(1,:),T(2,:),'ow'), grid
```

The 'o' produces a small circle at each point on the trajectory. The 'w' makes the circle white. If you have the data for another trajectory stored in a matrix *S*, you can plot both trajectories on the same graph:

```
plot(Ttl,:),T(2,:),'ow',S(1,:),S(2,:),'*g'),grid      g is for green.
```

Each new **plot** command erases the previous graph. If you want the new graph added to the previous graph, issue the command **hold on** before the next **plot** command.

For Exercise 17, the following commands will produce a graph of the first entry in each of the first nine columns of the matrix *T* constructed above.

```
k = 0:8
plot(k,T(1,1:9),'-'), grid
```

To graph the sum of the two entries in the first nine columns of *T*, with *k* already defined, enter

```
plot(k,T(1,1:9)+T(2,1:9),'-'), grid
```

The command **./** between two vectors of equal lengths divides each entry in the first vector by the corresponding entry in the second vector. Thus, to plot the quotient of the two entries in each column of *T*, use

```
plot(k,T(1,1:9)./ T(2,1:9),'-'), grid
```

5.7   APPLICATIONS TO DIFFERENTIAL EQUATIONS _____

If you plan to take a course in differential equations, the material in this section will be a valuable reference. If you have already studied differential equations, you may gain new understanding as you work through this section.

KEY IDEAS

A basic solution of the differential equation $\mathbf{x}' = A\mathbf{x}$ is an *eigenfunction* $\mathbf{x}(t) = \mathbf{v}e^{\lambda t}$, where λ is an eigenvalue of *A* and $\mathbf{v}$ is a corresponding eigenvector. In all examples and exercises in this section, every solution of $\mathbf{x}' = A\mathbf{x}$ is a linear combination of eigenfunctions. (This is because *A* is diagonalizable. The general case is usually handled in a full course in differential equations.) An initial condition, $\mathbf{x}(0) = \mathbf{x}_0$, determines the weights for the linear combination of eigenfunctions.

The eigenvalues of A determine the nature of the origin for the dynamical system described by $\mathbf{x}' = A\mathbf{x}$. Most of the discussion involves the case when A is 2×2. If one eigenvalue is negative and one is positive, the origin is a saddle point. If the real parts of the eigenvalues are negative (this includes the case when both eigenvalues are real and negative), the origin is an *attractor* of the dynamical system. If the real parts are positive, the origin is a *repellor*. If an eigenvalue is complex, then the trajectory of a corresponding eigenfunction forms a *spiral*— either toward the origin, away from the origin, or on an ellipse around the origin, depending on the real part of the eigenvalue.

Study Tip: Note that conditions on eigenvalues here differ from those in Section 5.6. For differential equations, the real parts of the eigenvalues determine the nature of the trajectories; for difference equations, the absolute values of the eigenvalues are important. You can remember this if you note that a basic solution $\mathbf{v}e^{\lambda t}$ of $\mathbf{x}' = A\mathbf{x}$ tends to $\mathbf{0}$ (as $t \to \infty$) only if the real part of λ is negative. In contrast, a basic solution $\lambda^k \mathbf{v}$ of $\mathbf{x}_{k+1} = A\mathbf{x}_k$ tends to $\mathbf{0}$ (as $k \to \infty$) only if the absolute value of λ is less than 1.

SOLUTIONS TO EXERCISES

1. The eigenfunctions for $\mathbf{x}' = A\mathbf{x}$ are $\mathbf{v}_1 e^{4t}$ and $\mathbf{v}_2 e^{2t}$. The general solution of $\mathbf{x}' = A\mathbf{x}$ has the form

$$c_1 \begin{bmatrix} -3 \\ 1 \end{bmatrix} e^{4t} + c_2 \begin{bmatrix} -1 \\ 1 \end{bmatrix} e^{2t}$$

The initial condition $\mathbf{x}(0) = (-6, 1)$ determines c_1 and c_2:

$$c_1 \begin{bmatrix} -3 \\ 1 \end{bmatrix} e^{4(0)} + c_2 \begin{bmatrix} -1 \\ 1 \end{bmatrix} e^{2(0)} = \begin{bmatrix} -6 \\ 1 \end{bmatrix}$$

$$\begin{bmatrix} -3 & -1 & -6 \\ 1 & 1 & 1 \end{bmatrix} \sim \begin{bmatrix} 1 & 1 & 1 \\ -3 & -1 & -6 \end{bmatrix} \sim \cdots \sim \begin{bmatrix} 1 & 0 & 5/2 \\ 0 & 1 & -3/2 \end{bmatrix}$$

Thus $c_1 = 5/2$, $c_2 = -3/2$, and $\mathbf{x}(t) = \dfrac{5}{2}\begin{bmatrix} -3 \\ 1 \end{bmatrix} e^{4t} - \dfrac{3}{2}\begin{bmatrix} -1 \\ 1 \end{bmatrix} e^{2t}$.

Checkpoint: Let A be the matrix $\begin{bmatrix} 1 & 3/2 \\ -1/2 & -1 \end{bmatrix}$, obtained by dividing the matrix in Exercise 3 by 2. Now the eigenvalues of A are .5 and $-.5$. Is the origin an attractor or a saddle point for the equation $\mathbf{x}' = A\mathbf{x}$?

7. Use the eigenvectors $\mathbf{v}_1 = \begin{bmatrix} 1 \\ 3 \end{bmatrix}$ and $\mathbf{v}_2 = \begin{bmatrix} 1 \\ 1 \end{bmatrix}$ (found in Exercise 5) to create $P = [\mathbf{v}_1 \quad \mathbf{v}_2]$.

Match the eigenvectors with the eigenvalues in $D = \begin{bmatrix} 4 & 0 \\ 0 & 6 \end{bmatrix}$. The details of the substitution

of $\mathbf{x} = P\mathbf{y}$ into $\mathbf{x}' = A\mathbf{x}$ are given in the answer section of the text.

Helpful Hint: The idea of changing variables to uncouple a differential equation is fairly common in engineering texts. Exercises 7 and 8 test your understanding of the value of a diagonalization. (You might see such a question on an exam.)

13. An eigenvalue of A is $\lambda = 1 + 3i$, with eigenvector $\mathbf{v} = (1 + i, 2)$. The complex eigenfunctions $\mathbf{v}e^{\lambda t}$ and $\overline{\mathbf{v}}e^{\overline{\lambda} t}$ provide a basis for the solution space of all complex solutions of $\mathbf{x}' = A\mathbf{x}$. The general (complex) solution is

$$c_1 \begin{bmatrix} 1+i \\ 2 \end{bmatrix} e^{(1+3i)t} + c_2 \begin{bmatrix} 1-i \\ 2 \end{bmatrix} e^{(1-3i)t} \qquad (c_1 \text{ and } c_2 \text{ are complex})$$

Use the real and imaginary parts of $\mathbf{v}e^{(1+3i)t}$ to build the general real solution. Rewrite $\mathbf{v}e^{(1+3i)t}$ as:

$$\begin{bmatrix} 1+i \\ 2 \end{bmatrix} e^{(1+3i)t} = \begin{bmatrix} 1+i \\ 2 \end{bmatrix} (\cos 3t + i \sin 3t) e^t$$

$$= \begin{bmatrix} \cos 3t - \sin 3t \\ 2\cos 3t \end{bmatrix} e^t + i \begin{bmatrix} \sin 3t + \cos 3t \\ 2\sin 3t \end{bmatrix} e^t$$

The general real solution has the form

$$c_1 \begin{bmatrix} \cos 3t - \sin 3t \\ 2\cos 3t \end{bmatrix} e^t + c_2 \begin{bmatrix} \sin 3t + \cos 3t \\ 2\sin 3t \end{bmatrix} e^t \qquad (c_1 \text{ and } c_2 \text{ are real})$$

The trajectories are spirals because the eigenvalues are complex. The spirals tend away from the origin because the real parts of the eigenvalues are positive.

19. Substitute $R_1 = 1/5$, $R_2 = 1/3$, $C_1 = 4$, and $C_2 = 3$ into the formula for A given in Example 1, and use a matrix program to find the eigenvalues and eigenvectors:

$$A = \begin{bmatrix} -2 & 3/4 \\ 1 & -1 \end{bmatrix}, \qquad \lambda_1 = -.5\text{: } \mathbf{v}_1 = \begin{bmatrix} 1 \\ 2 \end{bmatrix}, \qquad \lambda_2 = -2.5\text{: } \mathbf{v}_2 = \begin{bmatrix} -3 \\ 2 \end{bmatrix}$$

General solution: $\mathbf{x}(t) = c_1 \begin{bmatrix} 1 \\ 2 \end{bmatrix} e^{-.5t} + c_2 \begin{bmatrix} -3 \\ 2 \end{bmatrix} e^{-2.5t}$.

The condition $\mathbf{x}(0) = \begin{bmatrix} 4 \\ 4 \end{bmatrix}$ implies that $\begin{bmatrix} 1 & -3 \\ 2 & 2 \end{bmatrix}\begin{bmatrix} c_1 \\ c_2 \end{bmatrix} = \begin{bmatrix} 4 \\ 4 \end{bmatrix}$. By a matrix program, $c_1 = 5/2$ and $c_2 = -1/2$, so that

$$\begin{bmatrix} v_1(t) \\ v_2(t) \end{bmatrix} = \mathbf{x}(t) = \frac{5}{2}\begin{bmatrix} 1 \\ 2 \end{bmatrix} e^{-.5t} - \frac{1}{2}\begin{bmatrix} -3 \\ 2 \end{bmatrix} e^{-2.5t}$$

Helpful Hint: (for 21 and 22) Find the general real solution before you use the initial condition to find the constants c_1 and c_2. Otherwise, your c_1 and c_2 will probably be complex, and you will have to do unnecessary complex arithmetic to write the solution using only real scalars.

Answer to Checkpoint: One eigenvalue is positive and one is negative, so the origin is a saddle point. If you consider the *difference* equation $\mathbf{x}_{k+1} = A\mathbf{x}_k$ (with the same matrix A), then the origin is an attractor, because both eigenvalues are less than 1 in absolute value. Be careful if you have a test that covers both Sections 5.6 and 5.7.

MATLAB   Solutions of Differential Equations

For the eigenvalues of A, use **ev = eig(A)**. The eigenvectors shown in the text's answers were produced using commands such as

$$\texttt{v = nulbasis(A-ev(1)*eye(3))}$$

If the eigenvalue **ev(1)** is complex, the eigenvector v will be complex. The real and imaginary parts of v are **real(v)** and **imag(v)**.

   If you use the command **[P D] = eig(A)**, your eigenvectors should be multiples of those in the text's answer (when the eigenspaces are one-dimensional). To test whether a vector v is a multiple of a vector w, compute **v./w**. This divides each entry in v by the corresponding entry in w. If v is a multiple of w, the result of **v./w** should be a vector whose entries are all equal.

5.8   ITERATIVE ESTIMATES FOR EIGENVALUES _____

The algorithms in this section illustrate another use of the eigenvector decomposition described in Section 5.6 (on page 343). Other methods for eigenvalue estimation were mentioned in Section 5.2 (on page 317).

STUDY NOTES

Throughout the section, we suppose that the initial vector x_0 can be written as $x_0 = c_1 v_1 + \cdots + c_n v_n$, where $v_1, \ldots, v_n$ are eigenvectors of A and $c_1 \neq 0$. (In practice, you will not know $c_1 v_1, \ldots, c_n v_n$. This eigenvector decomposition is used only to explain why the power method works.)

The Power Method: Assume the eigenvalue λ_1 for v_1 is a strictly dominant eigenvalue (so that $|\lambda_1| > |\lambda_j|$ for $j = 2, \ldots, n$). Then, for large k, the line through $A^k x_0$ and 0 nearly coincides with the line through v_1 and 0. The vector $A^k x_0$ itself may never approach a multiple of v_1 (see Exercise 21), but if each $A^k x_0$ is scaled so its largest entry is 1, then the scaled vectors approach an eigenvector (a multiple of v_1) as $k \to \infty$.

The Inverse Power Method: You must start with an initial estimate α for a particular eigenvalue, say λ_2, and α must be closer to λ_2 than to any other eigenvalue of A. In this case, $1/(\lambda_2 - \alpha)$ is a strictly dominant eigenvalue of the matrix $B = (A - \alpha I)^{-1}$. The inverse power method avoids computing B. Instead of multiplying x_k by B to get x_{k+1} (suitably scaled), you solve the equation $(A - \alpha I)y_k = x_k$ for y_k and then scale y_k to produce x_{k+1}.

SOLUTIONS TO EXERCISES

1. The vectors in the sequence $\begin{bmatrix} 1 \\ 0 \end{bmatrix}, \begin{bmatrix} 1 \\ .25 \end{bmatrix}, \begin{bmatrix} 1 \\ .3158 \end{bmatrix}, \begin{bmatrix} 1 \\ .3298 \end{bmatrix}, \begin{bmatrix} 1 \\ .3326 \end{bmatrix}$ approach an

eigenvector v_1. Of these vectors, the last one, x_4, is probably the best estimate of v_1. To compute an estimate of λ_1, multiply one of the vectors by A and examine its entries. Again,

the best information probably comes from $A x_4 = \begin{bmatrix} 4.9978 \\ 1.6652 \end{bmatrix}$ whose entries are approximately

λ_1 times the entries in x_4. From the first entry, the estimate of λ_1 is 4.9978.

The computed value of $A x_4$ can be used as an estimate of the direction of the eigenspace.

Study Tip: Exercises 1–6 make good exam questions because they test your understanding of the power method without requiring extensive calculation.

7. The data in the table below and the tables in Exercise 19 were produced by MATLAB, which carried more decimal places than shown here.

k	0	1	2	3	4	5
$\mathbf{x}_k$	$\begin{bmatrix} 1 \\ 0 \end{bmatrix}$	$\begin{bmatrix} .75 \\ 1 \end{bmatrix}$	$\begin{bmatrix} 1 \\ .9565 \end{bmatrix}$	$\begin{bmatrix} .9932 \\ 1 \end{bmatrix}$	$\begin{bmatrix} 1 \\ .9990 \end{bmatrix}$	$\begin{bmatrix} .9998 \\ 1 \end{bmatrix}$
$A\mathbf{x}_k$	$\begin{bmatrix} 6 \\ 8 \end{bmatrix}$	$\begin{bmatrix} 11.5 \\ 11.0 \end{bmatrix}$	$\begin{bmatrix} 12.70 \\ 12.78 \end{bmatrix}$	$\begin{bmatrix} 12.959 \\ 12.946 \end{bmatrix}$	$\begin{bmatrix} 12.9927 \\ 12.9948 \end{bmatrix}$	$\begin{bmatrix} 12.9990 \\ 12.9987 \end{bmatrix}$
μ_k	8	11.5	12.78	12.959	12.9948	12.9990

The exact eigenvalues are 13 and –2. The subspaces determined by $A^k\mathbf{x}$ are lines whose slopes alternate above and below the slope of the eigenspace. (The eigenspace is the line $x_2 = x_1$.)

13. If the eigenvalues close to 4 and –4 have different absolute values, then one of these eigenvalues is a strictly dominant eigenvalue, so the power method will work. But the power method depends on powers of the quotients λ_2/λ_1 and λ_3/λ_1 going to zero. If $|\lambda_2/\lambda_1|$ is close to 1, its powers will go to zero slowly.

15. Suppose $A\mathbf{x} = \lambda\mathbf{x}$, with $\mathbf{x} \neq 0$. For any α, $A\mathbf{x} - \alpha I\mathbf{x} = (\lambda - \alpha)\mathbf{x}$, and $(A - \alpha I)\mathbf{x} = (\lambda - \alpha)\mathbf{x}$. If α is *not* an eigenvalue of A, then $A - \alpha I$ is invertible and $\lambda - \alpha$ is not 0; hence

$$\mathbf{x} = (A - \alpha I)^{-1}(\lambda - \alpha)\mathbf{x} \text{ and } (\lambda - \alpha)^{-1}\mathbf{x} = (A - \alpha I)^{-1}\mathbf{x}$$

This last equation shows that $\mathbf{x}$ is an eigenvector of $(A - \alpha I)^{-1}$ corresponding to the eigenvalue $(\lambda - \alpha)^{-1}$.

19. a. The data in the table below show that $\mu_6 = 30.2887 = \mu_7$ to four decimal places. Actually, to six places, the largest eigenvalue is 30.288685, with eigenvector (.957629, .688937, 1, .943782).

k	0	1	2	3	4	5	6	7
$\mathbf{x}_k$	$\begin{bmatrix} 1 \\ 0 \\ 0 \\ 0 \end{bmatrix}$	$\begin{bmatrix} 1 \\ .7 \\ .8 \\ .7 \end{bmatrix}$	$\begin{bmatrix} .99 \\ .71 \\ 1 \\ .93 \end{bmatrix}$	$\begin{bmatrix} .961 \\ .691 \\ 1 \\ .942 \end{bmatrix}$	$\begin{bmatrix} .9581 \\ .6893 \\ 1 \\ .9436 \end{bmatrix}$	$\begin{bmatrix} .9577 \\ .6890 \\ 1 \\ .9438 \end{bmatrix}$	$\begin{bmatrix} .957637 \\ .688942 \\ 1 \\ .943778 \end{bmatrix}$	$\begin{bmatrix} .957630 \\ .688938 \\ 1 \\ .943781 \end{bmatrix}$
$A\mathbf{x}_k$	$\begin{bmatrix} 10 \\ 7 \\ 8 \\ 7 \end{bmatrix}$	$\begin{bmatrix} 26.2 \\ 18.8 \\ 26.5 \\ 24.7 \end{bmatrix}$	$\begin{bmatrix} 29.4 \\ 21.1 \\ 30.6 \\ 28.8 \end{bmatrix}$	$\begin{bmatrix} 29.05 \\ 20.90 \\ 30.32 \\ 28.61 \end{bmatrix}$	$\begin{bmatrix} 29.01 \\ 20.87 \\ 30.29 \\ 28.59 \end{bmatrix}$	$\begin{bmatrix} 29.006 \\ 20.868 \\ 30.289 \\ 28.586 \end{bmatrix}$	$\begin{bmatrix} 29.0054 \\ 20.8671 \\ 30.2887 \\ 28.5859 \end{bmatrix}$	$\begin{bmatrix} 29.0053 \\ 20.8670 \\ 30.2887 \\ 28.5859 \end{bmatrix}$
μ_k	10	26.5	30.6	30.32	30.29	30.2892	30.2887	30.2887

b. The inverse power method (with $\alpha = 0$) produces $v_1 = \mu_1^{-1} = .010141$, and $v_2 = .0101501$, which seems to be accurate to at least four places. Actually, v_2 is accurate to six places, v_3 is accurate to eight places, and v_4 is accurate to ten places. The convergence is so rapid because the next-to-smallest eigenvalue is near .85, which is much farther away from 0 than .0101501. The vector $\mathbf{x}_4$ gives an estimate for the eigenvector that is accurate to seven places in each entry.

k	0	1	2	3	4
$\mathbf{x}_k$	$\begin{bmatrix} 1 \\ 0 \\ 0 \\ 0 \end{bmatrix}$	$\begin{bmatrix} -.6098 \\ 1 \\ -.2439 \\ .1463 \end{bmatrix}$	$\begin{bmatrix} -.60401 \\ 1 \\ -.25105 \\ .14890 \end{bmatrix}$	$\begin{bmatrix} -.603973 \\ 1 \\ -.251134 \\ .148953 \end{bmatrix}$	$\begin{bmatrix} -.6039723 \\ 1 \\ -.2511351 \\ .1489534 \end{bmatrix}$
$A\mathbf{x}_k$	$\begin{bmatrix} 25 \\ -41 \\ 10 \\ -6 \end{bmatrix}$	$\begin{bmatrix} -59.56 \\ 98.61 \\ -24.76 \\ 14.68 \end{bmatrix}$	$\begin{bmatrix} -59.5041 \\ 98.5211 \\ -24.7420 \\ 14.6750 \end{bmatrix}$	$\begin{bmatrix} -59.5044 \\ 98.5217 \\ -24.7423 \\ 14.6751 \end{bmatrix}$	$\begin{bmatrix} -59.50438 \\ 98.52170 \\ -24.74226 \\ 14.67515 \end{bmatrix}$
μ_k	-.024	.010141	.0101501	.010150059	.0101500484

MATLAB   Power Method and Inverse Power Method

Use **format long** to display 15 decimal digits in your data. The algorithms below assume that A has a strictly dominant eigenvalue, and the initial vector is **x**, with largest entry 1 (in magnitude). (If your initial vector is called **x0**, rename it by entering **x = x0**.)

    The Power Method   When the following sequence of commands is performed over and over, the values of **x** approach (in many cases) an eigenvector for a strictly dominant eigenvalue:

```
y = A*x                                                    (1)
[t r] = max(abs(y)); mu = y(r)    mu = estimate for eigenvalue    (2)
x = y/y(r)                        Estimate for the eigenvector     (3)
```

In (2), t is the absolute value of the largest entry in **y** and r is the index of that entry. As these commands are repeated, the numbers that appear in **y**(r) are the μ_k that approach the dominant eigenvalue.

    Recall that MATLAB commands can be recalled by the up-arrow key ($\uparrow$). After entering (1) – (3), your keystrokes can be

      $\uparrow\uparrow\uparrow$ <Enter> $\uparrow\uparrow\uparrow$ <Enter> $\uparrow\uparrow\uparrow$ <Enter>

and so on. Alternatively, you could enclose lines (1) – (3) in a loop (see page 5-25).

    The Inverse Power Method   Store the initial estimate of the eigenvalue in the variable **a**, and enter the command **C = A - a*eye(n)**, where n is the number of columns of A. Then enter the commands

```
y = C\x                              Solves (A – aI)y = x            (1)
[t r] = max(abs(y)); nu = a + 1/y(r)  nu = estimated eigenvalue      (2)
x = y/y(r)                           Estimate for the eigenvector    (3)
```

As these commands are repeated (using $\uparrow\uparrow\uparrow$ <Enter> each time), lines (2) and (3) produce the sequences $\{v_k\}$ and $\{x_k\}$ described in the text.

    Displaying Data   If your computer screen displays only 24 or 25 lines, vectors in the sequence $\{x_k\}$ tend to scroll off the screen soon after you compute them. To see more vectors at once, and to compare their entries more easily, you can *display* them as row vectors. Change (1) to **y = A*x; y'** (power method) or **y = C\x; y'** (inverse power), and for both methods, change (3) to **x = y/y(r); x'**.

    For even more data on your screen, use the command **format compact**, which removes extra lines between data displays. The simple command **format** returns everything to normal.

Chapter 5   GLOSSARY CHECKLIST _____

Check your knowledge by attempting to write definitions of the terms below. Then compare your work with the definitions given in the text's Glossary. Ask your instructor which definitions, if any, might appear on a test.

algebraic multiplicity: The multiplicity of an eigenvalue as . . . .

attractor (of a dynamical system in $\mathbb{R}^2$):   The origin of $\mathbb{R}^2$ when all trajectories tend . . . .

$\mathcal{B}$-matrix (for T):   A matrix $[T]_{\mathcal{B}}$ for a linear transformation $T{:}V \to V$ relative to a basis $\mathcal{B}$ for V, with the property that . . . .

characteristic equation (of A):   . . . .

characteristic polynomial (of A):   . . . .

companion matrix: A special form of matrix whose characteristic . . . is . . . .

complex eigenvalue: A nonreal root of the characteristic equation of an $n \times n$ matrix A, when . . . .

complex eigenvector: A nonzero vector $\mathbf{x}$ in $\mathbb{C}^n$ such that . . . , where . . . .

decoupled system: A difference equation $\mathbf{y}_{k+1} = A\mathbf{y}_k$, or a differential equation, in which A is a . . . .

determinant (of a square matrix A):   A number det A computed from A; equal to . . . .

diagonalizable (matrix):   A matrix that may be written in factored form as . . . .

difference equation (or **linear recurrence relation**):   An equation of the form . . . whose solution is . . . .

discrete linear dynamical system (or briefly, a **dynamical system**):   A difference equation of the form . . . that describes . . . .

eigenfunction (of the equation $\mathbf{x}'(t) = A\mathbf{x}(t)$):   A function of the form . . . .

eigenspace (of A corresponding to λ):   The set of . . . solutions of . . . .

eigenvalue (of A):   A scalar λ such that . . . .

eigenvector (of A):   A . . . vector $\mathbf{x}$ such that . . . .

eigenvector basis: A basis consisting entirely of . . . .

eigenvector decomposition (of $\mathbf{x}$):   An equation $\mathbf{x} = $ . . . .

fundamental set of solutions (for $\mathbf{x}' = A\mathbf{x}$):   A basis for . . . .

Im x: The vector in $\mathbb{R}^n$ formed from . . . .

invariant subspace (for A):   A subspace H such that . . . .

inverse power method: An algorithm for estimating . . . .

matrix for T relative to bases $\mathcal{B}$ and $\mathcal{C}$:   A matrix M for a linear transformation $T:V \rightarrow W$ with the property that . . . . When $W = V$ and $\mathcal{C} = \mathcal{B}$, the matrix M is called . . . and is denoted by . . . .

power method: An algorithm for estimating . . . .

repellor (of a dynamical system in $\mathbb{R}^2$):   The origin in $\mathbb{R}^2$ when all trajectories . . . tend . . . .

Rayleigh quotient: $R(\mathbf{x}) = $ . . . . An estimate of . . . .

Re x: The vector in $\mathbb{R}^n$ formed from . . . .

saddle point (of a dynamical system in $\mathbb{R}^2$):   The origin in $\mathbb{R}^2$ when . . . .

similar (matrices):   Matrices A and B such that . . . .

spiral point (of a dynamical system in $\mathbb{R}^2$):   The origin in $\mathbb{R}^2$ when . . . .

stage-matrix model: A difference equation $\mathbf{x}_{k+1} = A\mathbf{x}_k$ where $\mathbf{x}_k$ lists . . . .

strictly dominant eigenvalue: An eigenvalue λ_1 of a matrix A with the property that . . . .

trace (of a square matrix A):   The . . . , denoted by tr A.

trajectory: The graph of a solution $\{\mathbf{x}_0, \mathbf{x}_1, \mathbf{x}_2, \ldots\}$ of a . . . . Also, the graph of $\mathbf{x}(t)$ for $t \geq 0$, when . . . .

6 Orthogonality and Least Squares

6.1 INNER PRODUCT, LENGTH, AND ORTHOGONALITY ____

The concepts of length, distance, and orthogonality introduced in this section are essential for many geometric descriptions in the rest of the text.

STUDY NOTES

The first half of the section is computational and easily learned. The second half, however, requires more attention. Read it carefully. The concepts of orthogonality and orthogonal complements are the foundation for the rest of the chapter. In fact, **Theorem 3** is sometimes called the Fundamental Theorem of Linear Algebra.

SOLUTIONS TO EXERCISES

1.
$$\mathbf{u} = \begin{bmatrix} -1 \\ 2 \end{bmatrix}, \mathbf{v} = \begin{bmatrix} 4 \\ 6 \end{bmatrix}, \mathbf{u} \cdot \mathbf{u} = (-1)^2 + 2^2 = 5, \quad \mathbf{v} \cdot \mathbf{u} = 4(-1) + 6(2) = 8, \quad \frac{\mathbf{v} \cdot \mathbf{u}}{\mathbf{u} \cdot \mathbf{u}} = \frac{8}{5}$$

7. $\mathbf{w} = (3, -1, -5), \quad \|\mathbf{w}\|^2 = \mathbf{w} \cdot \mathbf{w} = 3^2 + (-1)^2 + (-5)^2 = 35.$ So $\|\mathbf{w}\| = \sqrt{35}$.

13. $\mathbf{x} = (10, -3), \mathbf{y} = (-1, -5)$

$$\|\mathbf{x} - \mathbf{y}\|^2 = [10 - (-1)]^2 + [-3 - (-5)]^2 = 121 + 4 = 125$$

$$\text{dist}(\mathbf{x}, \mathbf{y}) = \|\mathbf{x} - \mathbf{y}\| = \sqrt{125}, \text{ or } 5\sqrt{5}$$

19. a. See the definition of $\|\mathbf{v}\|$.

   b. See Theorem 1(c).

   c. See the discussion of Fig. 5.

d. Think about a 2×2 matrix of zeros and ones, and see Theorem 3 for statements that are always true.

e. See the box following Example 6.

21. Theorem l(b): $(\mathbf{u} + \mathbf{v}) \cdot \mathbf{w} = (\mathbf{u} + \mathbf{v})^T \mathbf{w} = (\mathbf{u}^T + \mathbf{v}^T)\mathbf{w} = \mathbf{u}^T\mathbf{w} + \mathbf{v}^T\mathbf{w} = \mathbf{u} \cdot \mathbf{w} + \mathbf{v} \cdot \mathbf{w}$. The second and third equalities used Theorems 3(b) and 2(c), respectively, from Section 2.1. Theorem l(c): $(c\mathbf{u}) \cdot \mathbf{v} = (c\mathbf{u})^T \mathbf{v} = (c\mathbf{u}^T)\mathbf{v} = c(\mathbf{u}^T\mathbf{v}) = c(\mathbf{u} \cdot \mathbf{v})$, by Theorems 3(c) and 2(d) in Section 2.1. Also, $\mathbf{u} \cdot (c\mathbf{v}) = \mathbf{u}^T(c\mathbf{v}) = c\mathbf{u}^T\mathbf{v} = c(\mathbf{u} \cdot \mathbf{v})$.

25. When $\mathbf{v} = \begin{bmatrix} a \\ b \end{bmatrix}$, the set H of all vectors $\begin{bmatrix} x \\ y \end{bmatrix}$ that are orthogonal to $\mathbf{v}$ is the subspace of vectors whose entries satisfy $ax + by = 0$. If $a \neq 0$, then $x = -(b/a)y$, with y a free variable. Then H is a line through the origin. A natural choice for a basis for this subspace is $\begin{bmatrix} -b \\ a \end{bmatrix}$.

If $a = 0$ and $b \neq 0$, then the vectors in H satisfy $by = 0$. Since b is nonzero, $y = 0$ and x is free. A basis for H is $\begin{bmatrix} 1 \\ 0 \end{bmatrix}$. Note, however, that $\begin{bmatrix} -b \\ a \end{bmatrix}$ is also a basis, since $a = 0$ and $b \neq 0$.

Finally, if a and b are both zero, then H is $\mathbb{R}^2$ itself, because the equation $0x + 0y = 0$ places no restrictions on x or y.

27. If $\mathbf{y}$ is orthogonal to $\mathbf{u}$ and $\mathbf{v}$, then $\mathbf{y} \cdot \mathbf{u} = 0$ and $\mathbf{y} \cdot \mathbf{v} = 0$, and hence by a property of the inner product, $\mathbf{y} \cdot (\mathbf{u} + \mathbf{v}) = \mathbf{y} \cdot \mathbf{u} + \mathbf{y} \cdot \mathbf{v} = 0 + 0 = 0$. So $\mathbf{y}$ is orthogonal to $\mathbf{u} + \mathbf{v}$.

29. Take a typical vector $\mathbf{w} = c_1\mathbf{v}_1 + \cdots + c_p\mathbf{v}_p$ in W. If $\mathbf{x}$ is orthogonal to each $\mathbf{v}_j$, then using the linearity of the inner product (Theorem l(b) and l(c)), $\mathbf{w} \cdot \mathbf{x} = (c_1\mathbf{v}_1 + \cdots + c_p\mathbf{v}_p) \cdot \mathbf{x} = c_1\mathbf{v}_1 \cdot \mathbf{x} + \cdots + c_p\mathbf{v}_p \cdot \mathbf{x} = 0$. So $\mathbf{x}$ is orthogonal to each $\mathbf{w}$ in W.

31. Suppose $\mathbf{x}$ is in W and $W^\perp$. Then, since $\mathbf{x}$ is in $W^\perp$, $\mathbf{x}$ is orthogonal to every vector in W, including $\mathbf{x}$ itself. So $\mathbf{x} \cdot \mathbf{x} = 0$. This is true only if $\mathbf{x} = 0$. This problem shows that $W \cap W^\perp$ is the zero subspace. (See Exercise 32 in Section 4.1.)

MATLAB   The inner product of real column vectors $\mathbf{u}$ and $\mathbf{v}$ is $\mathtt{u'*v}$ (and $\mathtt{v'*u}$); the length of $\mathbf{v}$ is $\mathtt{norm(v)}$. See the MATLAB note for Section 2.1.

6.2   ORTHOGONAL SETS

Orthogonal sets and orthogonal bases are used throughout the chapter. The "orthogonal projection" discussed in this section is an important special case of the orthogonal projections studied in Section 6.3.

STUDY NOTES

The proofs of Theorems 4 and 5 are worth studying because they involve a calculation you will see and use several times.

The subsection entitled *An Orthogonal Projection* is simple but extremely important. Also, the geometric interpretation of Theorem 5 on page 388 will be helpful when you study Theorem 8 in the next section.

The attention paid to Theorems 6 and 7 will depend on what your instructor plans to do later in the chapter. In some cases, an instructor may discuss Theorems 6 and 7 only for square matrices. The $m \times n$ case is needed later, for Theorems 10, 12 and 15. Remember: the term *orthogonal matrix* applies *only to a square matrix*. Also, the columns of an orthogonal matrix must be *orthonormal*, not simply orthogonal.

SOLUTIONS TO EXERCISES

1. $\mathbf{u} = \begin{bmatrix} -1 \\ 4 \\ -3 \end{bmatrix}$, $\mathbf{v} = \begin{bmatrix} 5 \\ 2 \\ 1 \end{bmatrix}$, $\mathbf{w} = \begin{bmatrix} 3 \\ -4 \\ -7 \end{bmatrix}$, $\mathbf{u} \cdot \mathbf{v} = -5 + 8 - 3 = 0$, $\mathbf{u} \cdot \mathbf{w} = -3 - 16 + 21 = 2 \neq 0$.

The set $\{\mathbf{u}, \mathbf{v}, \mathbf{w}\}$ is not orthogonal. There is no need to check $\mathbf{v} \cdot \mathbf{w}$.

7. $\mathbf{u}_1 = \begin{bmatrix} 2 \\ -3 \end{bmatrix}$, $\mathbf{u}_2 = \begin{bmatrix} 6 \\ 4 \end{bmatrix}$, $\mathbf{x} = \begin{bmatrix} 9 \\ -7 \end{bmatrix}$, $\mathbf{u}_1 \cdot \mathbf{u}_2 = 12 - 12 = 0$, so $\{\mathbf{u}_1, \mathbf{u}_2\}$ is an orthogonal set.

Since the vectors are nonzero, $\mathbf{u}_1$ and $\mathbf{u}_2$ are linearly independent, by Theorem 4. But two such vectors in $\mathbb{R}^2$ automatically form a basis for $\mathbb{R}^2$. So $\{\mathbf{u}_1, \mathbf{u}_2\}$ is an orthogonal basis for $\mathbb{R}^2$. By Theorem 5,

$$\mathbf{x} = \frac{\mathbf{x} \cdot \mathbf{u}_1}{\mathbf{u}_1 \cdot \mathbf{u}_1} \mathbf{u}_1 + \frac{\mathbf{x} \cdot \mathbf{u}_2}{\mathbf{u}_2 \cdot \mathbf{u}_2} \mathbf{u}_2 = \frac{18 + 21}{4 + 9} \mathbf{u}_1 + \frac{54 - 28}{36 + 16} \mathbf{u}_2 = 3 \begin{bmatrix} 2 \\ -3 \end{bmatrix} + \frac{1}{2} \begin{bmatrix} 6 \\ 4 \end{bmatrix}$$

13. $\mathbf{y} = \begin{bmatrix} 2 \\ 3 \end{bmatrix}$, $\mathbf{u} = \begin{bmatrix} 4 \\ -7 \end{bmatrix}$. The orthogonal projection of $\mathbf{y}$ onto $\mathbf{u}$ is

$$\hat{\mathbf{y}} = \frac{\mathbf{y} \cdot \mathbf{u}}{\mathbf{u} \cdot \mathbf{u}} \mathbf{u} = \frac{8 - 21}{16 + 49} \mathbf{u} = \frac{-13}{65} \mathbf{u} = \frac{-1}{5} \begin{bmatrix} 4 \\ -7 \end{bmatrix} = \begin{bmatrix} -4/5 \\ 7/5 \end{bmatrix}$$

The component of $\mathbf{y}$ orthogonal to $\mathbf{u}$ is $\mathbf{y} - \hat{\mathbf{y}} = \begin{bmatrix} 2 \\ 3 \end{bmatrix} - \begin{bmatrix} -4/5 \\ 7/5 \end{bmatrix} = \begin{bmatrix} 14/5 \\ 8/5 \end{bmatrix}$.

Thus, $\mathbf{y} = \hat{\mathbf{y}} + (\mathbf{y} - \hat{\mathbf{y}}) = \begin{bmatrix} -4/5 \\ 7/5 \end{bmatrix} + \begin{bmatrix} 14/5 \\ 8/5 \end{bmatrix}$.

19. $\mathbf{u} = \begin{bmatrix} -.6 \\ .8 \end{bmatrix}$, $\mathbf{v} = \begin{bmatrix} .8 \\ .6 \end{bmatrix}$, $\mathbf{u} \cdot \mathbf{v} = -.48 + .48 = 0$, so $\{\mathbf{u}, \mathbf{v}\}$ is an orthogonal set. Also,

$\|\mathbf{u}\|^2 = \mathbf{u} \cdot \mathbf{u} = (-.6)^2 + (.8)^2 = .36 + .64 = 1$. Similarly, $\|\mathbf{v}\|^2 = \mathbf{v} \cdot \mathbf{v} = 1$. Thus $\{\mathbf{u}, \mathbf{v}\}$ is an orthonormal set.

23. **a.** See Example 3, for instance.   **b.** See Theorem 5.

c. See the paragraph after Example 5.   **d.** See the paragraph before Example 7.

e. See Example 4.

25. $(U\mathbf{x}) \cdot (U\mathbf{y}) = (U\mathbf{x})^T(U\mathbf{y}) = \mathbf{x}^T U^T U\mathbf{y} = \mathbf{x}^T\mathbf{y} = \mathbf{x} \cdot \mathbf{y}$ (because $U^T U = I$). If $\mathbf{y} = \mathbf{x}$, Theorem 7(b) says that $\|U\mathbf{x}\|^2 = \|\mathbf{x}\|^2$, which implies part (a). Part (c) of Theorem 7 follows immediately from part (b).

Study Tip: If your instructor emphasizes orthogonal matrices, work Exercises 27–29. (They make good test questions.) In each case, mention explicitly how you use the fact that the matrices are square. Don't read the solutions below until you have first *written* your own solution.

27. If U has orthonormal columns, then $U^T U = I$, by Theorem 6. If U is also *square*, then the equation $U^T U = I$ implies that U is invertible, by the Invertible Matrix Theorem.

29. Since U and V are orthogonal, each is invertible. By Theorem 6 in Section 2.2, UV is invertible and $(UV)^{-1} = V^{-1}U^{-1} = V^T U^T = (UV)^T$ (by Theorem 3 in Section 2.1). Thus UV is an orthogonal matrix.

31. The full solution is in the text.

Mastering Linear Algebra Concepts: Orthogonal Basis

To the review sheet(s) you have on "basis", add the concepts of an orthogonal basis and an orthonormal basis for a subspace. You need to know what special properties they possess.

• basic definitions	Pages 385 and 389
• equivalent descriptions	Theorems 4 and 6
• geometric interpretation	Figs. 1 and 6
• special cases	Matrix with orthonormal columns
• examples and counterexamples	Examples 2 and 5
• algorithms and computations	Example 2
• connections with other concepts	Orthogonal matrix

MATLAB   Orthogonality

In Exercises 1–10 and 17–22, the fastest way (counting the keystrokes) in MATLAB to test a set such as $\{\mathbf{u}_1, \mathbf{u}_2, \mathbf{u}_3\}$ for orthogonality is to use a matrix $\mathbf{U} = [\mathbf{u1} \ \mathbf{u2} \ \mathbf{u3}]$ whose columns are the vectors from the set, and test whether $\mathbf{U'*U}$ is a diagonal matrix. See the proof of Theorem 6.

For column vectors $\mathbf{y}$ and $\mathbf{u}$, the orthogonal projection of $\mathbf{y}$ onto $\mathbf{u}$ is

$$(\mathbf{y'*u})/(\mathbf{u'*u})*\mathbf{u}$$

The parentheses (and the final *) are essential. MATLAB computes the scalar quotient $(\mathbf{y}^T\mathbf{u})/(\mathbf{u}^T\mathbf{u})$ and then multiplies $\mathbf{u}$ by this scalar.

6.3   ORTHOGONAL PROJECTIONS

A familiar idea in Euclidean geometry is to construct a line segment perpendicular to a line or plane. This section treats an analogous situation in $\mathbb{R}^n$, namely, the orthogonal projection of a vector (a point in $\mathbb{R}^n$) onto a subspace. The case when the subspace is a line through the origin was already examined in Section 6.2.

KEY IDEAS

If $\mathbf{y}$ is in $\mathbb{R}^n$ and if W is a subspace of $\mathbb{R}^n$, then the orthogonal projection of $\mathbf{y}$ onto W, denoted by $\hat{\mathbf{y}}$ or $\text{proj}_W \mathbf{y}$, has two important properties:

(i) $\mathbf{y} - \hat{\mathbf{y}}$ is orthogonal to W (so $\mathbf{y}$ is the sum of a vector $\hat{\mathbf{y}}$ in W and a vector $\mathbf{y} - \hat{\mathbf{y}}$ in $W^{\perp}$), and

(ii) $\hat{\mathbf{y}}$ is the closest point in W to $\mathbf{y}$.

Properties (i) and (ii) are described in the Orthogonal Decomposition Theorem and the Best Approximation Theorem. You should learn the statements of both theorems. (By now you probably know that whenever a theorem has an official name, an instructor has an easy time asking test questions about it.) When you need one of these theorems in a discussion (homework or test question), you should mention the theorem by name.

If your class covers Theorem 10, then the paragraph following the theorem will help you understand the difference between an *orthogonal matrix* (which must be square) and a rectangular matrix with orthonormal columns.

SOLUTIONS TO EXERCISES

1. $\mathbf{u}_1 = \begin{bmatrix} 0 \\ 1 \\ -4 \\ -1 \end{bmatrix}$, $\mathbf{u}_2 = \begin{bmatrix} 3 \\ 5 \\ 1 \\ 1 \end{bmatrix}$, $\mathbf{u}_3 = \begin{bmatrix} 1 \\ 0 \\ 1 \\ -4 \end{bmatrix}$, $\mathbf{u}_4 = \begin{bmatrix} 5 \\ -3 \\ -1 \\ 1 \end{bmatrix}$, $\mathbf{x} = \begin{bmatrix} 10 \\ -8 \\ 2 \\ 0 \end{bmatrix}$. You could calculate all the inner

products in the decomposition:

$$\mathbf{x} = \underbrace{\frac{\mathbf{x} \cdot \mathbf{u}_1}{\mathbf{u}_1 \cdot \mathbf{u}_1}\mathbf{u}_1 + \frac{\mathbf{x} \cdot \mathbf{u}_2}{\mathbf{u}_2 \cdot \mathbf{u}_2}\mathbf{u}_2 + \frac{\mathbf{x} \cdot \mathbf{u}_3}{\mathbf{u}_3 \cdot \mathbf{u}_3}\mathbf{u}_3}_{\text{in Span}\{\mathbf{u}_1,\,\mathbf{u}_2,\,\mathbf{u}_3\}} + \underbrace{\frac{\mathbf{x} \cdot \mathbf{u}_4}{\mathbf{u}_4 \cdot \mathbf{u}_4}\mathbf{u}_4}_{\text{in Span}\{\mathbf{u}_4\}} \qquad (1)$$

However, once you know the vector in Span$\{\mathbf{u}_4\}$, the vector in Span$\{\mathbf{u}_1, \mathbf{u}_2, \mathbf{u}_3\}$ is determined completely by (1). So all you need is

$$\frac{\mathbf{x} \cdot \mathbf{u}_4}{\mathbf{u}_4 \cdot \mathbf{u}_4}\mathbf{u}_4 = \frac{50 + 24 - 2 + 0}{25 + 9 + 1 + 1}\mathbf{u}_4 = 2\mathbf{u}_4 = \begin{bmatrix} 10 \\ -6 \\ -2 \\ 2 \end{bmatrix}$$

The vector in Span$\{\mathbf{u}_1, \mathbf{u}_2, \mathbf{u}_3\}$ is $\mathbf{x} - 2\mathbf{u}_4 = \begin{bmatrix} 10 \\ -8 \\ 2 \\ 0 \end{bmatrix} - \begin{bmatrix} 10 \\ -6 \\ -2 \\ 2 \end{bmatrix} = \begin{bmatrix} 0 \\ -2 \\ 4 \\ -2 \end{bmatrix}$.

Study Tip: One way to check whether $\text{proj}_W \mathbf{y}$ is computed correctly is to verify that $\mathbf{y} - \text{proj}_W \mathbf{y}$ is orthogonal to each vector in the orthogonal basis $\{\mathbf{u}_1, ..., \mathbf{u}_p\}$ for W. A faster check that will catch most errors (but not all) is to verify that $\mathbf{y} - \text{proj}_W \mathbf{y}$ is orthogonal to $\text{proj}_W \mathbf{y}$.

7. $\mathbf{y} = \begin{bmatrix} 1 \\ 3 \\ 5 \end{bmatrix}$, $\mathbf{u}_1 = \begin{bmatrix} 1 \\ 3 \\ -2 \end{bmatrix}$, $\mathbf{u}_2 = \begin{bmatrix} 5 \\ 1 \\ 4 \end{bmatrix}$. First, make sure that $\{\mathbf{u}_1, \mathbf{u}_2\}$ is an orthogonal basis for

Span$\{\mathbf{u}_1, \mathbf{u}_2\}$. This is easy, since $\mathbf{u}_1$ and $\mathbf{u}_2$ are nonzero and $\mathbf{u}_1 \cdot \mathbf{u}_2 = 0$. Next, by the Orthogonal Decomposition Theorem, $\mathbf{y}$ is the sum of $\text{proj}_W \mathbf{y}$ and $\mathbf{y} - \text{proj}_W \mathbf{y}$, where $W = \text{Span}\{\mathbf{u}_1, \mathbf{u}_2\}$.

$$\text{proj}_W \mathbf{y} = \frac{\mathbf{y} \cdot \mathbf{u}_1}{\mathbf{u}_1 \cdot \mathbf{u}_1} \mathbf{u}_1 + \frac{\mathbf{y} \cdot \mathbf{u}_2}{\mathbf{u}_2 \cdot \mathbf{u}_2} \mathbf{u}_2 = \frac{1+9-10}{1+9+4} \mathbf{u}_1 + \frac{5+3+20}{25+1+16} \mathbf{u}_2$$

$$= 0\mathbf{u}_1 + \frac{2}{3} \mathbf{u}_2 \begin{bmatrix} 10/3 \\ 2/3 \\ 8/3 \end{bmatrix}$$

and

$$\mathbf{y} - \text{proj}_W \mathbf{y} = \begin{bmatrix} 1 \\ 3 \\ 5 \end{bmatrix} - \begin{bmatrix} 10/3 \\ 2/3 \\ 8/3 \end{bmatrix} = \begin{bmatrix} -7/3 \\ 7/3 \\ 7/3 \end{bmatrix}$$

As a check, scale $\mathbf{y} - \text{proj}_W \mathbf{y} = \begin{bmatrix} -1 \\ 1 \\ 1 \end{bmatrix}$, and observe that the scaled vector is obviously

orthogonal to $\mathbf{u}_1$ and $\mathbf{u}_2$. Thus $\mathbf{y} - \text{proj}_W \mathbf{y}$ is in $W^\perp$, as it should be.

Warning: The formula for $\text{proj}_W \mathbf{y}$ applies only if $\{\mathbf{u}_1, \ldots, \mathbf{u}_p\}$ is an *orthogonal* basis for W. That's why you should check orthogonality, as in Exercise 7, if you are not sure that the basis is orthogonal. If an orthogonal basis is not available, then other methods can be used to compute $\hat{\mathbf{y}}$. (See Exercise 23 in Section 6.5, for example.)

13. $\mathbf{z} = \begin{bmatrix} 3 \\ -7 \\ 2 \\ 3 \end{bmatrix}$, $\mathbf{v}_1 = \begin{bmatrix} 2 \\ -1 \\ -3 \\ 1 \end{bmatrix}$, $\mathbf{v}_2 = \begin{bmatrix} 1 \\ 1 \\ 0 \\ -1 \end{bmatrix}$. Note that $\mathbf{v}_1$ and $\mathbf{v}_2$ are orthogonal. By the Best

Approximation Theorem, the closest point in $\text{Span}\{\mathbf{v}_1, \mathbf{v}_2\}$ to $\mathbf{z}$ is the orthogonal projection $\hat{\mathbf{z}}$, where

$$\hat{\mathbf{z}} = \frac{\mathbf{z} \cdot \mathbf{v}_1}{\mathbf{v}_1 \cdot \mathbf{v}_1} \mathbf{v}_1 + \frac{\mathbf{z} \cdot \mathbf{v}_2}{\mathbf{v}_2 \cdot \mathbf{v}_2} \mathbf{v}_2 = \frac{10}{15} \mathbf{v}_1 + \frac{-7}{3} \mathbf{v}_2 = \frac{2}{3} \begin{bmatrix} 2 \\ -1 \\ -3 \\ 1 \end{bmatrix} - \frac{7}{3} \begin{bmatrix} 1 \\ 1 \\ 0 \\ -1 \end{bmatrix} = \begin{bmatrix} -1 \\ -3 \\ -2 \\ 3 \end{bmatrix}$$

Check: $\mathbf{z} - \hat{\mathbf{z}} = \begin{bmatrix} 4 \\ -4 \\ 4 \\ 0 \end{bmatrix}$. The vector $\begin{bmatrix} 1 \\ -1 \\ 1 \\ 0 \end{bmatrix}$ is orthogonal to both $\mathbf{v}_1$ and $\mathbf{v}_2$.

19. By the Orthogonal Decomposition Theorem, $\mathbf{u}_3$ is the sum of a vector in $W = \text{Span}\{\mathbf{u}_1, \mathbf{u}_2\}$ and a vector $\mathbf{v}$ orthogonal to W. First,

$$\text{proj}_W \mathbf{u}_3 = \frac{-2}{6}\mathbf{u}_1 + \frac{2}{30}\mathbf{u}_2 = \begin{bmatrix} -2/6 \\ -2/6 \\ 4/6 \end{bmatrix} + \begin{bmatrix} 10/30 \\ -2/30 \\ 4/30 \end{bmatrix} = \begin{bmatrix} 0 \\ -2/5 \\ 4/5 \end{bmatrix}$$

Then

$$\mathbf{v} = \mathbf{u}_3 - \text{proj}_W \mathbf{u}_3 = \begin{bmatrix} 0 \\ 0 \\ 1 \end{bmatrix} - \begin{bmatrix} 0 \\ -2/5 \\ 4/5 \end{bmatrix} = \begin{bmatrix} 0 \\ 2/5 \\ 1/5 \end{bmatrix}$$

Not only is $\mathbf{v}$ orthogonal to W, but also any multiple of $\mathbf{v}$ is in $W^\perp$.

Study Tip: It would be a good idea to try Exercise 20 and compare the result with Exercise 19. Then think about the following problem: Suppose that $\{\mathbf{u}_1, \mathbf{u}_2\}$ is an orthogonal set of nonzero vectors in $\mathbb{R}^3$. How would you find an orthogonal basis of $\mathbb{R}^3$ that contains $\mathbf{u}_1$ and $\mathbf{u}_2$? You might discuss this with your instructor.

21. a. See the calculations for $\mathbf{z}_2$ in Example 1 or the box after Example 6 in Section 6.1.

    b. See the Orthogonal Decomposition Theorem.

    c. See the second paragraph after the statement of Theorem 9.

    d. See the box before the Best Approximation Theorem.

    e. See the paragraph after Theorem 10.

23. By the Orthogonal Decomposition Theorem, each $\mathbf{x}$ in $\mathbb{R}^n$ can be written uniquely as $\mathbf{x} = \mathbf{p} + \mathbf{u}$, with $\mathbf{p}$ in Row A and $\mathbf{u}$ in $(\text{Row } A)^\perp$. By Theorem 3 in Section 6.1, $\mathbf{u}$ is in Nul A.

    Next, suppose that $A\mathbf{x} = \mathbf{b}$ is consistent. Let $\mathbf{x}$ be a solution, and write $\mathbf{x} = \mathbf{p} + \mathbf{u}$, as above. Then $A\mathbf{p} = A(\mathbf{x} - \mathbf{u}) = A\mathbf{x} - A\mathbf{u} = \mathbf{b} - \mathbf{0} = \mathbf{b}$. So the equation $A\mathbf{x} = \mathbf{b}$ has at least one solution $\mathbf{p}$ in Row A.

    Finally, suppose that $\mathbf{p}$ and $\mathbf{p}_1$ are both in Row A and satisfy $A\mathbf{x} = \mathbf{b}$. Then $\mathbf{p} - \mathbf{p}_1$ is in Nul A because

$$A\,(\mathbf{p} - \mathbf{p}_1) = A\mathbf{p} - A\mathbf{p}_1 = \mathbf{b} - \mathbf{b} = \mathbf{0}$$

The equations $\mathbf{p} = \mathbf{p}_1 + (\mathbf{p} - \mathbf{p}_1)$ and $\mathbf{p} = \mathbf{p} + \mathbf{0}$ both decompose $\mathbf{p}$ as the sum of a vector in Row A and a vector in $(\text{Row } A)^\perp$. By the uniqueness of the orthogonal decomposition (Theorem 8), $\mathbf{p}_1 = \mathbf{p}$, so $\mathbf{p}$ is unique.

25. From Exercise 36 of Section 6.2, U should have orthonormal columns, because U is formed by normalizing the columns of the matrix A in Exercise 35 whose columns are orthogonal. Verify this by computing $U^T U$. The result should be the 4×4 identity matrix.

    The closest point to $\mathbf{y}$ in Col U is the orthogonal projection of $\mathbf{y}$ onto Col U. By Theorem 10, this closest point is $UU^T\mathbf{y}$. The MATLAB command is   `U*U' *y`. The result of this computation should be the (column) vector $(1.2, .4, 1.2, 1.2, .4, 1.2, .4, .4)$.

Warning: I had to work hard to make the arithmetic simple in the exercises for this section, to avoid distractions for you and to save you time. You might not be so lucky on an exam. Even if a problem is designed to be numerically simple, there is always a chance that a minor error will make the calculations messy. In such a case, don't despair. Carry out the arithmetic as best you can, showing the details of your work (patterned after the solutions in this *Study Guide*). Chances are that you will get substantial credit for showing that you understand the concepts.

MATLAB   Orthogonal Projections

The orthogonal projection of **y** onto a single vector was described in the MATLAB note for Section 6.2. The orthogonal projection onto the set spanned by an orthogonal set of vectors is the sum of the one-dimensional projections. Another way to construct this projection is to normalize the orthogonal vectors, place them in the columns of a matrix U, and use Theorem 10. For instance, if $\{\mathbf{y}_1, \mathbf{y}_2, \mathbf{y}_3\}$ is an orthogonal set of nonzero vectors, then the matrix

```
U = [y1/norm(y1) y2/norm(y2) y3/norm(y3)]
```

has orthonormal columns, and `U*(U'*y)` produces the orthogonal projection of **y** onto the subspace spanned by $\{\mathbf{y}_1, \mathbf{y}_2, \mathbf{y}_3\}$. (The parentheses around `U'*y` speed up the computation of `U*U'*y` by avoiding a matrix-matrix product.)

6.4   THE GRAM-SCHMIDT PROCESS

This section has a nice geometric appeal. The Gram-Schmidt process is well-liked by students and faculty because it is easily learned. Although the process is seldom used in practical computations, it has important generalizations to spaces other than $\mathbb{R}^n$ (to be discussed briefly in Section 6.7).

KEY IDEAS

When the Gram-Schmidt process is applied to $\{\mathbf{x}_1, \ldots, \mathbf{x}_p\}$, the first step is to set $\mathbf{v}_1 = \mathbf{x}_1$. For $k = 2, \ldots, n$, the kth step consists of subtracting from $\mathbf{x}_k$ its projection onto the subspace spanned by the previous **x**'s. At each step the projection is easy to compute because an orthogonal basis for the appropriate subspace has already been constructed.

   The QR factorization of a matrix A encapsulates the result of applying the Gram-Schmidt process to the columns of A, just as the LU factorization of a matrix encodes the row operations that reduce a matrix to echelon form. Also, just as the LU factorization can be implemented via multiplication by elementary matrices, so can the QR factorization be constructed via multiplication by orthogonal matrices.

SOLUTIONS TO EXERCISES

1. $\mathbf{x}_1 = \begin{bmatrix} 3 \\ 0 \\ -1 \end{bmatrix}$, $\mathbf{x}_2 = \begin{bmatrix} 8 \\ 5 \\ -6 \end{bmatrix}$. Set $\mathbf{v}_1 = \mathbf{x}_1$ and compute

$$\mathbf{v}_2 = \mathbf{x}_2 - \frac{\mathbf{x}_2 \cdot \mathbf{v}_1}{\mathbf{v}_1 \cdot \mathbf{v}_1} \mathbf{v}_1 = \begin{bmatrix} 8 \\ 5 \\ -6 \end{bmatrix} - \frac{30}{10} \begin{bmatrix} 3 \\ 0 \\ -1 \end{bmatrix} = \begin{bmatrix} -1 \\ 5 \\ -3 \end{bmatrix}$$

Check: $\mathbf{v}_2 \cdot \mathbf{v}_1 = -3 + 0 + 3 = 0$. So an orthogonal basis is $\left\{ \begin{bmatrix} 3 \\ 0 \\ -1 \end{bmatrix}, \begin{bmatrix} -1 \\ 5 \\ -3 \end{bmatrix} \right\}$.

7. $\mathbf{x}_1 = \begin{bmatrix} 2 \\ -5 \\ 1 \end{bmatrix}$, $\mathbf{x}_2 = \begin{bmatrix} 4 \\ -1 \\ 2 \end{bmatrix}$. From Exercise 3, use $\mathbf{v}_1 = \begin{bmatrix} 2 \\ -5 \\ 1 \end{bmatrix}$ and $\mathbf{v}_2 = \begin{bmatrix} 3 \\ 3/2 \\ 3/2 \end{bmatrix}$ as an orthogonal

basis for $W = \text{Span}\{\mathbf{x}_1, \mathbf{x}_2\}$. Scale $\mathbf{v}_2$ to $(2, 1, 1)$ before normalizing, and then obtain

$$\mathbf{u}_1 = \frac{1}{\sqrt{30}} \begin{bmatrix} 2 \\ -5 \\ 1 \end{bmatrix} = \begin{bmatrix} 2/\sqrt{30} \\ -5/\sqrt{30} \\ 1/\sqrt{30} \end{bmatrix}, \quad \mathbf{u}_2 = \frac{1}{\sqrt{6}} \begin{bmatrix} 2 \\ 1 \\ 1 \end{bmatrix} = \begin{bmatrix} 2/\sqrt{6} \\ 1/\sqrt{6} \\ 1/\sqrt{6} \end{bmatrix}$$

Study Tip: If you need to normalize a vector by hand, first consider scaling the entries in the vector to make them small integers, if possible.

13. $A = \begin{bmatrix} 5 & 9 \\ 1 & 7 \\ -3 & -5 \\ 1 & 5 \end{bmatrix}$, $Q = \begin{bmatrix} 5/6 & -1/6 \\ 1/6 & 5/6 \\ -3/6 & 1/6 \\ 1/6 & 3/6 \end{bmatrix}$. Let

$$R = Q^T A = \begin{bmatrix} 5/6 & 1/6 & -3/6 & 1/6 \\ -1/6 & 5/6 & 1/6 & 3/6 \end{bmatrix} \begin{bmatrix} 5 & 9 \\ 1 & 7 \\ -3 & -5 \\ 1 & 5 \end{bmatrix} = \begin{bmatrix} 36/6 & 72/6 \\ 0 & 36/6 \end{bmatrix} = \begin{bmatrix} 6 & 12 \\ 0 & 6 \end{bmatrix}$$

As a check, compute $QR = \begin{bmatrix} 5/6 & -1/6 \\ 1/6 & 5/6 \\ -3/6 & 1/6 \\ 1/6 & 3/6 \end{bmatrix} \begin{bmatrix} 6 & 12 \\ 0 & 6 \end{bmatrix} = \begin{bmatrix} 5 & 54/6 \\ 1 & 42/6 \\ -3 & -30/6 \\ 1 & 30/6 \end{bmatrix} = A$.

Remark: The reason the R in Exercise 13 works is that the columns of Q form an orthonormal basis for Col A (since they were obtained by the Gram-Schmidt process). Thus $QQ^T\mathbf{y} = \mathbf{y}$ for all $\mathbf{y}$ in Col A, by Theorem 10 in Section 6.3. In particular, $QQ^TA = A$. So if R is Q^TA, then $QR = Q(Q^TA) = A$.

17. a. See the remark after Example 5 in Section 6.2, and the reference there to Exercise 32.
   b. See (1) in the statement of Theorem 11.
   c. See the solution of Example 4.

19. The full solution is in the text.

21. The solution in the text is complete, except for the details of extending an orthonormal basis for Span$\{\mathbf{q}_1, \dots, \mathbf{q}_n\}$ to an orthonormal basis for $\mathbb{R}^m$. Here is one algorithm. Let $\{\mathbf{e}_1, \dots, \mathbf{e}_m\}$ be the standard basis for $\mathbb{R}^m$. Let $\mathbf{f}_1$ be the first vector in this basis that is *not* in $W_n = \text{Span}\{\mathbf{q}_1, \dots, \mathbf{q}_n\}$, and let $\mathbf{u}_1 = \mathbf{f}_1 - \text{proj}_W \mathbf{f}_1$. Then $\{\mathbf{q}_1, \dots, \mathbf{q}_n, \mathbf{u}_1\}$ is an orthogonal basis for $W_{n+1} = \text{Span}\{\mathbf{q}_1, \dots, \mathbf{q}_n, \mathbf{u}_1\}$. Let $\mathbf{f}_2$ be the first vector in $\{\mathbf{e}_1, \dots, \mathbf{e}_m\}$ that is not in W_{n+1}. (Of course $\mathbf{f}_2$ occurs later than $\mathbf{f}_1$ in the list $\mathbf{e}_1, \dots, \mathbf{e}_m$.) Form $\mathbf{u}_2 = \mathbf{f}_2 - \text{proj}_{W_{n+1}} \mathbf{f}_2$ and $W_{n+2} = \text{Span}\{\mathbf{q}_1, \dots, \mathbf{q}_n, \mathbf{u}_1, \mathbf{u}_2\}$. This process will continue until $m - n$ vectors have been added to the original n vectors. Normalizing the new vectors produces an orthonormal basis for $\mathbb{R}^m$.

23. Use the definition of matrix multiplication. When $A = QR$, the first p columns of A are determined by the action of Q on the first p columns of R. So, if $A = [A_1 \ \ A_2]$, make the same column-partition of R as $[R_1 \ \ R_2]$, where R_1 has p columns. Then

$$A = Q[R_1 \ \ R_2] = [QR_1 \ \ QR_2] = [A_1 \ \ A_2]$$

Is QR_1 a QR factorization of A_1? Unfortunately, no. The second factor in a QR factorization should be square and upper triangular with positive entries on the diagonal. Since R has those properties, its first p columns have zeros in rows $p + 1$ to n. (This is a key observation.) So, partition R_1 into two blocks, $R_1 = \begin{bmatrix} R_{11} \\ 0 \end{bmatrix}$, where R_{11} is square and upper triangular. The entries on the diagonal of R_{11} are positive because they come from R. Then $A_1 = QR_1 = Q\begin{bmatrix} R_{11} \\ 0 \end{bmatrix}$. You might consider left-multiplying R_{11} by Q, but partitioned matrix multiplication does not work that way. QR_{11} is not defined, because Q has more columns than R_{11} has rows. (Why?)

   The final idea needed is to view the product QR_1 as a product of block matrices, by partitioning the *columns* of Q to match the row partition of R_1. Write $Q = [Q_1 \ \ Q_2]$, where Q_1 consists of the first p columns of Q. The matrix Q_1 has orthonormal columns, because the columns come from Q and so the columns are unit vectors and are pairwise orthogonal. Finally,

$$A_1 = QR_1 = [Q_1 \ \ Q_2]\begin{bmatrix} R_{11} \\ 0 \end{bmatrix} = Q_1 R_{11} + Q_2 0 = Q_1 R_{11}$$

This concludes the construction, because the properties of Q_1 and R_{11} have already been discussed. This solution has followed a path that a good student might find. Now that you see how things fit together, you should be able to write a short proof that begins with appropriate partitions of Q and R. I encourage you to try it. The shorter proof is hidden at the end of the solutions for Section 6.5.

MATLAB   The Gram-Schmidt Process

If A has only two columns, then the Gram-Schmidt process is

```
v1 = A(:,1)
v2 = A(:,2) - (A(:,2)'*v1)/(v1'*v1)*v1
```

If A has three columns, add the command

```
v3 = A(:,3) - (A(:,3)'*v1)/(v1'*v1)*v1 - (A(:,3)'*v2)/(v2'*v2)*v2
```

You should use these commands for a while, to learn the general procedure. After that, you can use the Laydata command **proj(x,V)**, which computes the projection of a vector **x** onto the subspace spanned by the columns of a matrix (or vector) V. For example,

```
v2 = A(:,2) - proj(A(:,2),v1)        V = v1
v3 = A(:,3) - proj(A(:,3),[v1 v2])   V = [v1  v2]
```

The columns of V in **proj(x,V)** need not be orthogonal for the command to work, but if they are, the entries in **proj(x,V)** will usually agree with those computed via Theorem 10 in Section 6.3, to twelve or more decimal places. Enter **help proj** to learn more about **proj**.

To check your work or to save time, enter **Q = gs(A)**, which uses the Gram-Schmidt process to construct the columns of Q. See **help gs**.

Although you should construct the QR factorization of a matrix using the approach in the text, you might like to see what MATLAB does. The command **[Q1 R1] = qr(A)** creates a modified QR factorization of an $m \times n$ matrix A as described in Exercise 21 in the text. If rank $A = r$, then the first r rows of R_1 are nonzero and the first r columns of Q_1 form an orthonormal basis for Col A.

6.5   LEAST SQUARES PROBLEMS

The basic geometric principles in this section provide the foundation for all the applications in Sections 6.6–6.8.

KEY IDEAS

A least-squares solution of $A\mathbf{x} = \mathbf{b}$ is any vector $\mathbf{x}$ that makes $A\mathbf{x}$ as close as possible to $\mathbf{b}$. (Learn the formal definition, too.) If the columns of A are linearly dependent, then there are many least-squares solutions of $A\mathbf{x} = \mathbf{b}$. They can all be found by row reducing the augmented matrix for the normal equations $A^TA\mathbf{x} = A^T\mathbf{b}$ (Theorem 13).

If the columns of A are linearly independent, then there is only one least-squares solution. To find it, solve the normal equations or compute $(A^TA)^{-1}A^T\mathbf{b}$ (Theorem 14). If a QR factorization of A is available, say $A = QR$, solve the equation $R\mathbf{x} = Q^T\mathbf{b}$ (Theorem 15 and the Numerical Note).

STUDY NOTES

The material up to and including Figure 2 needs to be read carefully several times, so you understand what the term "least-squares solution" means. Be careful to distinguish between $\hat{\mathbf{x}}$ and $\hat{\mathbf{b}}$. A common mistake is to think that $\hat{\mathbf{x}}$ itself somehow has the least-squares norm or is the closest point to $\mathbf{b}$. Look at Fig. 2 again. The vector closest to $\mathbf{b}$ is $A\hat{\mathbf{x}}$, not $\hat{\mathbf{x}}$.

Theorem 13 provides a common way to find least-squares solutions. One way to remember the normal equations is to observe that they look the same as $A\mathbf{x} = \mathbf{b}$ with A^T left-multiplied on each side of the equation. It is *completely wrong*, however, to try to *derive* the normal equations from $A\mathbf{x} = \mathbf{b}$ via left-multiplication by A^T. If the equation $A\mathbf{x} = \mathbf{b}$ has no solution, then the equation itself is a false statement about every vector $\mathbf{x}$. Matrix algebra on such a false statement is meaningless.

SOLUTIONS TO EXERCISES

1. $A = \begin{bmatrix} -1 & 2 \\ 2 & -3 \\ -1 & 3 \end{bmatrix}$, $\mathbf{b} = \begin{bmatrix} 4 \\ 1 \\ 2 \end{bmatrix}$

$$A^TA = \begin{bmatrix} -1 & 2 & -1 \\ 2 & -3 & 3 \end{bmatrix}\begin{bmatrix} -1 & 2 \\ 2 & -3 \\ -1 & 3 \end{bmatrix} = \begin{bmatrix} 6 & -11 \\ -11 & 22 \end{bmatrix}, \quad A^T\mathbf{b} = \begin{bmatrix} -1 & 2 & -1 \\ 2 & -3 & 3 \end{bmatrix}\begin{bmatrix} 4 \\ 1 \\ 2 \end{bmatrix} = \begin{bmatrix} -4 \\ 11 \end{bmatrix}$$

a. The normal equations: $\begin{bmatrix} 6 & -11 \\ -11 & 22 \end{bmatrix}\begin{bmatrix} x_1 \\ x_2 \end{bmatrix} = \begin{bmatrix} -4 \\ 11 \end{bmatrix}$

b. Since A^TA is only 2×2, $(A^TA)^{-1}$ is easy to compute, and

$$\hat{\mathbf{x}} = \begin{bmatrix} 6 & -11 \\ -11 & 22 \end{bmatrix}^{-1}\begin{bmatrix} -4 \\ 11 \end{bmatrix} = \frac{1}{11}\begin{bmatrix} 22 & 11 \\ 11 & 6 \end{bmatrix}\begin{bmatrix} -4 \\ 11 \end{bmatrix} = \frac{1}{11}\begin{bmatrix} 33 \\ 22 \end{bmatrix} = \begin{bmatrix} 3 \\ 2 \end{bmatrix}$$

Warning: It is important to distinguish between the normal equations $A^T A\hat{x} = A^T b$ and the formula $\hat{x} = (A^T A)^{-1} A^T b$. Both equations describe $\hat{x}$ (implicitly or explicitly), but the formula for $\hat{x}$ holds only when A has linearly independent columns. Note that the expression $(A^T A)^{-1} A^T$ cannot be simplified when A is not invertible.

7. $A = \begin{bmatrix} 1 & -2 \\ -1 & 2 \\ 0 & 3 \\ 2 & 5 \end{bmatrix}$, $b = \begin{bmatrix} 3 \\ 1 \\ -4 \\ 2 \end{bmatrix}$, $A^T A = \begin{bmatrix} 1 & -1 & 0 & 2 \\ -2 & 2 & 3 & 5 \end{bmatrix} \begin{bmatrix} 1 & -2 \\ -1 & 2 \\ 0 & 3 \\ 2 & 5 \end{bmatrix} = \begin{bmatrix} 6 & 6 \\ 6 & 42 \end{bmatrix}$

$$A^T b = \begin{bmatrix} 1 & -1 & 0 & 2 \\ -2 & 2 & 3 & 5 \end{bmatrix} \begin{bmatrix} 3 \\ 1 \\ -4 \\ 2 \end{bmatrix} = \begin{bmatrix} 6 \\ -6 \end{bmatrix}$$

The normal equations: $\begin{bmatrix} 6 & 6 \\ 6 & 42 \end{bmatrix} \begin{bmatrix} x_1 \\ x_2 \end{bmatrix} = \begin{bmatrix} 6 \\ -6 \end{bmatrix}$

The particular numbers in $A^T A$ suggest that the normal equations might be solved easily via row operations:

$$\begin{bmatrix} 6 & 6 & 6 \\ 6 & 42 & -6 \end{bmatrix} \sim \begin{bmatrix} 6 & 6 & 6 \\ 0 & 36 & -12 \end{bmatrix} \sim \begin{bmatrix} 1 & 1 & 1 \\ 0 & 1 & -1/3 \end{bmatrix} \sim \begin{bmatrix} 1 & 0 & 4/3 \\ 0 & 1 & -1/3 \end{bmatrix}$$

Thus $\hat{x} = \begin{bmatrix} 4/3 \\ -1/3 \end{bmatrix}$. The least-squares error is $\| A\hat{x} - b \|$, so compute

$$A\hat{x} - b = \begin{bmatrix} 1 & -2 \\ -1 & 2 \\ 0 & 3 \\ 2 & 5 \end{bmatrix} \begin{bmatrix} 4/3 \\ -1/3 \end{bmatrix} - \begin{bmatrix} 3 \\ 1 \\ -4 \\ 2 \end{bmatrix} = \begin{bmatrix} 2 \\ -2 \\ -1 \\ 1 \end{bmatrix} - \begin{bmatrix} 3 \\ 1 \\ -4 \\ 2 \end{bmatrix} = \begin{bmatrix} -1 \\ -3 \\ 3 \\ -1 \end{bmatrix}$$

$\| A\hat{x} - b \|^2 = 1 + 9 + 9 + 1 = 20$, and $\| A\hat{x} - b \| = \sqrt{20} = 2\sqrt{5}$

Study Tip: A good way to check your work in Exercises 1–8 is to verify, that $A\hat{x} - b$ is orthogonal to each column of A.

Warning: The matrices in Exercises 9–12 are special—their columns are orthogonal. That is why these exercises are not difficult. See Example 4. In general, if the columns of A are not orthogonal, finding the orthogonal projection of b onto $\text{Col}(A)$ takes more work. See Exercise 23.

13. $A\mathbf{u} = \begin{bmatrix} 3 & 4 \\ -2 & 1 \\ 3 & 4 \end{bmatrix} \begin{bmatrix} 5 \\ -1 \end{bmatrix} = \begin{bmatrix} 11 \\ -11 \\ 11 \end{bmatrix}$, $\mathbf{b} - A\mathbf{u} = \begin{bmatrix} 11 \\ -9 \\ 5 \end{bmatrix} - \begin{bmatrix} 11 \\ -11 \\ 11 \end{bmatrix} = \begin{bmatrix} 0 \\ 2 \\ -6 \end{bmatrix}$, $\| \mathbf{b} - A\mathbf{u} \| = \sqrt{40}$

$A\mathbf{v} = \begin{bmatrix} 3 & 4 \\ -2 & 1 \\ 3 & 4 \end{bmatrix} \begin{bmatrix} 5 \\ -2 \end{bmatrix} = \begin{bmatrix} 7 \\ -12 \\ 7 \end{bmatrix}$, $\mathbf{b} - A\mathbf{v} = \begin{bmatrix} 11 \\ -9 \\ 5 \end{bmatrix} - \begin{bmatrix} 7 \\ -12 \\ 7 \end{bmatrix} = \begin{bmatrix} 4 \\ 3 \\ -2 \end{bmatrix}$, $\| \mathbf{b} - A\mathbf{u} \| = \sqrt{29}$

Obviously, $A\mathbf{u}$ is not the closest point of Col A to $\mathbf{b}$, because $A\mathbf{v}$ is closer. Hence $\mathbf{u}$ is *not* the least-squares solution of $A\mathbf{x} = \mathbf{b}$.

17. a. See the beginning of the section.

    b. See the comments about equation (1).

    c. Read the definition of a least-squares solution.

    d. See Theorem 13.

    e. See Theorem 14.

19. The full solution is in the text.

21. a. If A has linearly independent columns, then the equation $A\mathbf{x} = \mathbf{0}$ has only the trivial solution. By Exercise 17, $A^T A \mathbf{x} = \mathbf{0}$ also has only the trivial solution. Since $A^T A$ is *square*, it must be invertible, by the Invertible Matrix Theorem.

    b. Since the n linearly independent columns of A belong to $\mathbb{R}^m$, m could not be less than n.

    c. The n linearly independent columns of A form a basis for Col A, so the rank of A is n.

23. (*This solution is for Section 6.4.*) Given $A = QR$, partition $A = [A_1 \quad A_2]$, where A_1 has p columns. Consider the following partitions of Q and R, which are possible because R is upper triangular:

$$A = [A_1 \quad A_2] = QR = [Q_1 \quad Q_2] \begin{bmatrix} R_{11} & R_{12} \\ 0 & R_{22} \end{bmatrix} = [Q_1 R_{11} \quad Q_1 R_{12} + Q_2 R_{22}]$$

where Q_1 has p columns and R_{11} is a $p \times p$ matrix. Then $A_1 = Q_1 R_{11}$. The matrix Q_1 has orthonormal columns, because the columns come from Q and so the columns are unit vectors and pairwise orthogonal. The matrix R_{11} is square and upper triangular, because of its position inside R. The diagonal entries of R_{11} are positive because they are diagonal entries of R. Thus $Q_1 R_{11}$ is a QR factorization of A_1.

MATLAB   The Backslash Command

When A has linearly dependent columns, you can write the general description of all least-squares solutions on paper after you row reduce the augmented matrix for the normal equations: `ref([A'*A   A'*b])`. When A has linearly independent columns, enter the MATLAB "backslash" command

`x = (A'*A)\(A'*b)`

to solve the normal equations. You can also enter `inv(A'*A)*(A'*b)`; or use `ref`, as above. For Exercises 15 and 16, see the Numerical Note on page 415 in the text and use the backslash command `R\(Q'*b)` to solve $Rx = Q^T\mathbf{b}$.

When A is not square but has linearly independent columns, this procedure of first forming Q and R and then solving $Rx = Q^T\mathbf{b}$ is exactly what MATLAB does (internally) when the MATLAB command `A\b` is used to solve $A\mathbf{x} = \mathbf{b}$. You should use the normal equations or QR for computations here, instead of just using the backslash. This will give you a solid conceptual background for applying least-squares techniques later in your career. For more about the backslash, see the MATLAB box for Section 2.5.

For Exercise 26, the command `A = [A1; A2]` creates a (partitioned) matrix whose top block is A1 and bottom block is A2. Of course, A1 and A2 must have the same number of columns.

6.6   APPLICATIONS TO LINEAR MODELS _____

This section of the text will be a valuable reference for any person who works with data that require statistical analysis. Many graduate fields require such work, often in connection with doctoral research. Even most undergraduates will take a course where least-squares lines are used.

KEY IDEA

Linear algebra unifies the study of many problems in statistics and data analysis. All the examples in this section, from ordinary linear regression (using a least-squares line) to multiple regression, concern just one idea: find a least-squares solution of $X\beta = \mathbf{y}$. Only the design matrix X varies. The exercises help you practice choosing X. The least-squares solution $\hat{\beta}$ always satisfies the normal equations $X^TX\hat{\beta} = X^T\mathbf{y}$.

STUDY NOTES

Don't confuse the least-squares line in Fig.1 with the lines and planes in Section 6.5 onto which we projected various vectors **b**. The line is nothing more than a special case of the curves in Figures 2–5. In each case, the "linearity" of the model lies not in the curve, but rather in the fact that the unknown parameters (or *weights*) β_0, β_1, ... occur linearly in the formula for the curve, just as the variables x_1, x_2, ... occur in an ordinary linear equation.

Any 4×4 submatrix of the design matrix in Example 3 is called a Vandermonde matrix. Using Exercise 11 on page 184, one can show that if at least four of the values x_1, ..., x_n are distinct, then the least-squares solution $\hat{\beta}$ will be unique, by Theorem 14 in Section 6.5.

FURTHER READING

An important generalization of the discussion here is to *multivariate* analysis, which involves several **y** vectors rather than just one. In this case the basic equation is $XB = Y$, where each column of Y is a data set for one dependent variable, and each column of B is a set of parameters to be determined. That is, $X[\beta_1 \ \cdots \ \beta_p] = [\mathbf{y}_1 \ \cdots \ \mathbf{y}_p]$. For more information, see the classic text by T. W. Anderson, *An Introduction to Multivariate Statistical Analysis*, John Wiley & Sons, New York, 1984 (and 1958). The preface of the text says, "A knowledge of matrix algebra is a prerequisite [for understanding the text]." Most modern multivariate statistics texts rely heavily on matrix notation and matrix algebra.

SOLUTIONS TO EXERCISES

1. Place the x-coordinates of the data in the second column of X and the y-coordinates in the vector **y**. So $X = \begin{bmatrix} 1 & 0 \\ 1 & 1 \\ 1 & 2 \\ 1 & 3 \end{bmatrix}$ and $\mathbf{y} = \begin{bmatrix} 1 \\ 1 \\ 2 \\ 2 \end{bmatrix}$. Compute

$$\underbrace{\begin{bmatrix} 1 & 1 & 1 & 1 \\ 0 & 1 & 2 & 3 \end{bmatrix}}_{X^T} \underbrace{\begin{bmatrix} 1 & 0 \\ 1 & 1 \\ 1 & 2 \\ 1 & 3 \end{bmatrix}}_{X} = \begin{bmatrix} 4 & 6 \\ 6 & 14 \end{bmatrix}, \qquad \underbrace{\begin{bmatrix} 1 & 1 & 1 & 1 \\ 0 & 1 & 2 & 3 \end{bmatrix}}_{X^T} \underbrace{\begin{bmatrix} 1 \\ 1 \\ 2 \\ 2 \end{bmatrix}}_{\mathbf{y}} = \begin{bmatrix} 6 \\ 11 \end{bmatrix}$$

The matrix normal equation and its solution are:

$$\begin{bmatrix} 4 & 6 \\ 6 & 14 \end{bmatrix} \begin{bmatrix} \beta_0 \\ \beta_1 \end{bmatrix} = \begin{bmatrix} 6 \\ 11 \end{bmatrix}$$

$$\begin{bmatrix} \beta_0 \\ \beta_1 \end{bmatrix} = \begin{bmatrix} 4 & 6 \\ 6 & 14 \end{bmatrix}^{-1} \begin{bmatrix} 6 \\ 11 \end{bmatrix} = \frac{1}{20}\begin{bmatrix} 14 & -6 \\ -6 & 4 \end{bmatrix}\begin{bmatrix} 6 \\ 11 \end{bmatrix} = \frac{1}{20}\begin{bmatrix} 18 \\ 8 \end{bmatrix} = \begin{bmatrix} .9 \\ .4 \end{bmatrix}$$

The least-squares line, $y = \beta_0 + \beta_1 x$, is $y = .9 + .4x$.

7. a. $y = X\beta + \varepsilon$, where $y = \begin{bmatrix} 1.8 \\ 2.7 \\ 3.4 \\ 3.8 \\ 3.9 \end{bmatrix}$, $X = \begin{bmatrix} 1 & 1 \\ 2 & 4 \\ 3 & 9 \\ 4 & 16 \\ 5 & 25 \end{bmatrix}$, $\beta = \begin{bmatrix} \beta_1 \\ \beta_2 \end{bmatrix}$, $\varepsilon = \begin{bmatrix} \varepsilon_1 \\ \varepsilon_2 \\ \varepsilon_3 \\ \varepsilon_4 \\ \varepsilon_5 \end{bmatrix}$.

b. In this problem, X^TX is invertible. You can use your matrix program to solve the normal equations without explicitly computing the entries in X^TX. For details, see the MATLAB box below or the corresponding box in the appendix for your matrix program. In any case,

$$\hat{\beta} = \begin{bmatrix} \hat{\beta}_1 \\ \hat{\beta}_2 \end{bmatrix} = \begin{bmatrix} 1.76 \\ -.20 \end{bmatrix} \quad \text{(to two decimal places)}$$

The desired least-squares equation is $y = 1.76x - .20x^2$.

13. Let **1** be the vector in $\mathbb{R}^3$ with 1 in each entry, let $\mathbf{t} = (0, \ldots, 12)$, and for $k = 2$ and 3, let $\mathbf{t}.\wedge k$ denote the vector whose entries are the kth powers of the entries in $\mathbf{t}$. (See the MATLAB box below.) Then the design matrix is:

$$X = [\mathbf{1} \quad \mathbf{t} \quad \mathbf{t}.\wedge 2 \quad \mathbf{t}.\wedge 3]$$

The observation vector **y** lists the measured positions of the plane.

a. Numerical solution of the normal equations yields

$$\beta = (-.8558, 4.7025, 5.5554, -.0274)$$

The least-squares polynomial (position of the plane at time t) is

$$y = -.8558 + 4.7025t + 5.5554t^2 - .0274t^3$$

b. The velocity is the derivative of the position function:

$$v(t) = 4.7025 + 11.1108t - .0822t^2$$

When $t = 4.5$ seconds, $v(4.5) = 53.0$ ft/sec.

15. From equation (1) on page 420,

$$X^TX = \begin{bmatrix} 1 & \cdots & 1 \\ x_1 & \cdots & x_n \end{bmatrix} \begin{bmatrix} 1 & x_1 \\ \vdots & \vdots \\ 1 & x_n \end{bmatrix} = \begin{bmatrix} n & \Sigma x \\ \Sigma x & \Sigma x^2 \end{bmatrix}$$

$$X^Ty = \begin{bmatrix} 1 & \cdots & 1 \\ x_1 & \cdots & x_n \end{bmatrix} \begin{bmatrix} y_1 \\ \vdots \\ y_n \end{bmatrix} = \begin{bmatrix} \Sigma y \\ \Sigma xy \end{bmatrix}$$

The equations (7) in the text follow immediately from the usual matrix normal equation $X^TX\beta = X^Ty$.

16. *Note:* The formulas you should derive are

$$\hat{\beta}_0 = \frac{(\Sigma x^2)(\Sigma y) - (\Sigma x)(\Sigma xy)}{n\Sigma x^2 - (\Sigma x)^2}, \quad \hat{\beta}_1 = \frac{n\Sigma xy - (\Sigma x)(\Sigma y)}{n\Sigma x^2 - (\Sigma x)^2}$$

Some statistics texts present other equivalent formulas for $\hat{\beta}_0$ and $\hat{\beta}_1$.

19. The equation to be proved is $\|\mathbf{y}\|^2 = \|X\hat{\beta}\|^2 + \|\mathbf{y} - X\hat{\beta}\|^2$. This follows from the Pythagorean Theorem (in Section 6.1) and the figure below.

Appendix: The Geometry of a Linear Model

The column space of the design matrix X is sometimes called the **design subspace**. If $\hat{\beta}$ is the least-squares solution of $\mathbf{y} = X\beta$, then the residual vector $\varepsilon = \mathbf{y} - X\hat{\beta}$ is orthogonal to the design subspace, and the equation $\mathbf{y} = X\hat{\beta} - \varepsilon$ is an orthogonal decomposition of the observed $\mathbf{y}$ into the sum of the least-squares predicted $\hat{\mathbf{y}}$ and the residual vector ε.

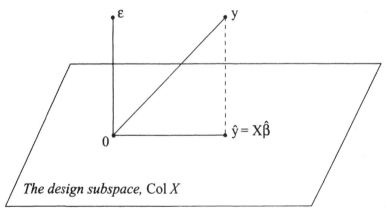

The design subspace, Col X

MATLAB   Least-squares Solutions and Functions of Vectors

Once you create the design matrix X and the observation vector $\mathbf{y}$, your computations for least-squares solutions here are the same as those described in the MATLAB box for Section 6.5. Here, A and $\mathbf{b}$ are replaced by X and $\mathbf{y}$, respectively. The MATLAB command

```
ref([X'*X   X'*y])
```

leads to the general description of all least-squares solutions. When X has linearly independent columns, the command `(X'*X)\(X'*y)` creates the least-squares solution. In subsequent courses, you may choose simply to use `X\y`, which also produces a least-squares solution, except when X is square and singular (or nearly singular).

To construct the design matrix for an exercise in this section, you may need MATLAB's ability to compute functions of vectors. If $\mathbf{x}$ is a vector and k is a positive integer, then $\mathbf{x.\char94 k}$ is a vector the same size as $\mathbf{x}$ whose entries are the kth powers of the entries in $\mathbf{x}$. The function $\mathbf{cos(x).\char94 k}$ was mentioned in the MATLAB box for Section 4.3. The exponential function, $\mathbf{exp(x)}$, and natural logarithm function, $\mathbf{log(x)}$, also act on each entry in $\mathbf{x}$. The entries in the vector $\mathbf{exp(-.02*x)}$, for example, are computed by applying the function $e^{-.02x}$ to the corresponding entries in $\mathbf{x}$.

6.7   INNER PRODUCT SPACES

Three examples of inner product spaces are described here, in Examples 1, 2, and 7. Corresponding applications appear in the next section. Material in Sections 6.7 and 6.8 will be useful for many careers, particularly science, engineering, and mathematics. If your course does not cover this now, the text and *Study Guide* can help you learn it later on your own.

KEY IDEAS

The concepts of length and orthogonality in $\mathbb{R}^n$ have analogues in a number of other vector spaces. The definition of an inner product identifies the basic properties needed for a theory that parallels the familiar theory for $\mathbb{R}^n$. Two useful facts, the Cauchy-Schwarz inequality and the triangle inequality, were not developed earlier, but they are important for applications both in $\mathbb{R}^n$ and in other inner product spaces. Every mathematics major will need to know these facts in other undergraduate courses.

The inner product in Example 1 is used in Section 6.8 to describe weighted least-squares problems. The inner product in Examples 2–6 provides a more sophisticated approach to the least-squares curve fitting discussed in Section 6.6. See the "trend analysis" in Section 6.8.

Be sure to read the paragraph preceding Example 6. The idea of "best approximation" to a function is of fundamental importance in mathematics. The most common applications of best approximation (such as Fourier series, introduced in Section 6.8) involve the inner product in Example 7.

SOLUTIONS TO EXERCISES

1. The inner product is $\langle \mathbf{x}, \mathbf{y} \rangle = 4x_1y_1 + 5x_2y_2$. Let $\mathbf{x} = (1, 1)$, $\mathbf{y} = (5, -1)$.

    a. $\|\mathbf{x}\|^2 = 4 \cdot 1 \cdot 1 + 5 \cdot 1 \cdot 1 = 9$, $\|\mathbf{x}\| = 3$

       $\|\mathbf{y}\|^2 = 4 \cdot 5 \cdot 5 + 5(-1)(-1) = 105$, $\|\mathbf{y}\| = \sqrt{105}$

       $\|\langle \mathbf{x}, \mathbf{y} \rangle\|^2 = |4 \cdot 1 \cdot 5 + 5 \cdot 1(-1)|^2 = |15|^2 = 225$

    b. A vector $\mathbf{z} = (z_1, z_2)$ is orthogonal to $\mathbf{y}$ if and only if $\langle \mathbf{z}, \mathbf{y} \rangle = 0$, that is,

$$4 \cdot z_1 \cdot 5 + 5 \cdot z_2 \cdot (-1) = 0, \quad 20z_1 - 5z_2 = 0, \quad \text{and} \quad z_2 = 4z_1$$

Thus (z_1, z_2) is orthogonal to $\mathbf{y}$ if and only if $z_2 = 4z_1$.

7. Given $p(t) = 4 + t$ and $q(t) = 5 - 4t^2$. The orthogonal projection $\hat{q}$ of q onto the subspace spanned by p is $[\langle q, p \rangle / \langle p, p \rangle]p$. The notation of Example 5 organizes the calculations nicely:

Polynomial: p q

Vector of values: $\begin{bmatrix} 3 \\ 4 \\ 5 \end{bmatrix}, \begin{bmatrix} 1 \\ 5 \\ 1 \end{bmatrix}$ ← value at -1 / ← value at 0 / ← value at 1

The inner product $\langle q, p \rangle$ equals the (standard) inner product of the two corresponding vectors in $\mathbb{R}^3$: $\langle q, p \rangle = 3 \cdot 1 + 4 \cdot 5 + 5 \cdot 1 = 28$. Similarly, $\langle p, p \rangle = 3^2 + 4^2 + 5^2 = 50$. Thus

$$\hat{q}(t) = \frac{28}{50}(4 + t) = \frac{56}{25} + \frac{14}{25}t$$

13. Suppose A is invertible and $\langle \mathbf{u}, \mathbf{v} \rangle = (A\mathbf{u}) \cdot (A\mathbf{v})$, for $\mathbf{u}, \mathbf{v}$ in $\mathbb{R}^n$. Note that $\langle \mathbf{u}, \mathbf{v} \rangle$ is in $\mathbb{R}$, and check each axiom in the definition on page 428:

i. $\langle \mathbf{u}, \mathbf{v} \rangle = (A\mathbf{u}) \cdot (A\mathbf{v}) = (A\mathbf{v}) \cdot (A\mathbf{u})$ Property of dot product
    $= \langle \mathbf{v}, \mathbf{u} \rangle$

ii. $\langle \mathbf{u} + \mathbf{v}, \mathbf{w} \rangle = [A(\mathbf{u} + \mathbf{v})] \cdot (A\mathbf{w}) = [A\mathbf{u} + A\mathbf{v}] \cdot (A\mathbf{w})$ Matrix multiplication
    $= (A\mathbf{u}) \cdot (A\mathbf{w}) + (A\mathbf{v}) \cdot (A\mathbf{w})$ Property of dot product
    $= \langle \mathbf{u}, \mathbf{w} \rangle + \langle \mathbf{v}, \mathbf{w} \rangle$

iii. $\langle c\mathbf{u}\ \mathbf{v} \rangle = [A(c\mathbf{u})] \cdot (A\mathbf{v}) = [c(A\mathbf{u})] \cdot (A\mathbf{v})$ Matrix multiplication
    $= c(A\mathbf{u}) \cdot (A\mathbf{v})$ Property of dot product
    $= c\langle \mathbf{u}, \mathbf{v} \rangle$

iv. $\langle \mathbf{u}, \mathbf{u} \rangle = (A\mathbf{u}) \cdot (A\mathbf{u}) = \|A\mathbf{u}\|^2 \geq 0$, and this quantity is zero if and only if the vector $A\mathbf{u}$ is **0**. But $A\mathbf{u} = \mathbf{0}$ if and only if $\mathbf{u} = \mathbf{0}$, because A is invertible.

Another method for verifying the axioms is to use properties of the transpose operation. The calculations are similar. However, for (i), you need to use the fact that the transpose of a scalar (which is a 1×1 matrix) is the scalar itself: $\langle \mathbf{u}, \mathbf{v} \rangle = \langle \mathbf{u}, \mathbf{v} \rangle^T = [(A\mathbf{u})^T(A\mathbf{v})]^T = (A\mathbf{v})^T(A\mathbf{u})^{TT} = (A\mathbf{v})^T(A\mathbf{u}) = \langle \mathbf{v}, \mathbf{u} \rangle$.

17. $\|\mathbf{u} + \mathbf{v}\|^2 = \langle \mathbf{u} + \mathbf{v}, \mathbf{u} + \mathbf{v} \rangle = \langle \mathbf{u}, \mathbf{u} + \mathbf{v} \rangle + \langle \mathbf{v}, \mathbf{u} + \mathbf{v} \rangle$ Axiom ii
$= \langle \mathbf{u} + \mathbf{v}, \mathbf{u} \rangle + \langle \mathbf{u} + \mathbf{v}, \mathbf{v} \rangle$ Axiom i
$= \langle \mathbf{u}, \mathbf{u} \rangle + \langle \mathbf{v}, \mathbf{u} \rangle + \langle \mathbf{u}, \mathbf{v} \rangle + \langle \mathbf{v}, \mathbf{v} \rangle$ Axiom ii
$= \langle \mathbf{u}, \mathbf{u} \rangle + 2\langle \mathbf{u}, \mathbf{v} \rangle + \langle \mathbf{v}, \mathbf{v} \rangle$ Axiom i

Next, replace $\mathbf{v}$ by $-\mathbf{v}$ and use the fact that $\mathbf{u} - \mathbf{v} = \mathbf{u} + (-1)\mathbf{v}$.

$$\|\mathbf{u} - \mathbf{v}\|^2 = \langle \mathbf{u}, \mathbf{u}\rangle + 2\langle \mathbf{u}, -\mathbf{v}\rangle + \langle -\mathbf{v}, -\mathbf{v}\rangle \qquad \text{Replacing } \mathbf{v} \text{ above by } -\mathbf{v}$$

$$= \langle \mathbf{u}, \mathbf{u}\rangle + 2\langle \mathbf{u}, \mathbf{v}\rangle + (-1)^2\langle \mathbf{v}, \mathbf{v}\rangle \qquad \text{Axiom iii and Exercise 15}$$

Subtracting, $\|\mathbf{u} + \mathbf{v}\|^2 - \|\mathbf{u} - \mathbf{v}\|^2 = 2\langle \mathbf{u}, \mathbf{v}\rangle - (-2\langle \mathbf{u}, \mathbf{v}\rangle) = 4\langle \mathbf{u}, \mathbf{v}\rangle$. Division by 4 gives the desired identity.

19. The full solution is in the text.

25. In the space $C[-1, 1]$ with the integral inner product, the polynomials t and 1 are orthogonal, because

$$\langle t, 1\rangle = \int_{-1}^{1} t \cdot 1 \, dt = \frac{1}{2}t^2\Big|_{-1}^{1} = \frac{1}{2}(1)^2 - \frac{1}{2}(-1)^2 = 0$$

So 1 and t can be in an orthogonal basis for $\text{Span}\{1, t, t^2\}$. Next, compute $\text{proj}_W\, t^2$, the orthogonal projection of the vector t^2 onto the subspace W spanned by 1 and t.

$$\langle t^2, 1\rangle = \int_{-1}^{1} t^2 \cdot 1 \, dt = \frac{1}{3}t^3\Big|_{-1}^{1} = \frac{1}{3}(1)^3 - \frac{1}{3}(-1)^3 = \frac{2}{3}$$

$$\langle 1, 1\rangle = \int_{-1}^{1} 1 \cdot 1\, dt = t\Big|_{-1}^{1} = 1 - (-1) = 2$$

$$\langle t^2, t\rangle = \int_{-1}^{1} t^2 \cdot t \, dt = \frac{1}{4}t^4\Big|_{-1}^{1} = \frac{1}{4}(1)^4 - \frac{1}{4}(-1)^4 = 0$$

There is no need to compute $\langle t, t\rangle$, because t^2 is orthogonal to t. Thus

$$\text{proj}_W\, t^2 = \frac{\langle t^2, 1\rangle}{\langle 1, 1\rangle}1 + \frac{\langle t^2, t\rangle}{\langle t, t\rangle}t = \frac{2/3}{2}1 + 0 = \frac{1}{3}$$

A polynomial orthogonal to W is $t^2 - \text{proj}_W\, t^2 = t^2 - \frac{1}{3}$. Another choice is this polynomial scaled by 3, namely, $3t^2 - 1$. Thus, the polynomials, 1, t^2, and $3t^2 - 1$ form an orthogonal basis for $\text{Span}\{1, t, t^2\}$.

Can you find the next Legendre polynomial, a cubic polynomial that is orthogonal to each of the first three Legendre polynomials?

6.8   APPLICATIONS OF INNER PRODUCT SPACES ———

Of the three applications in this section, the discussion of Fourier series is by far the most important. Such series have great practical value, particularly in mathematics, engineering, and the physical sciences. Calculations with Fourier series are simple because sine and cosine functions are orthogonal. This fact is often overlooked in undergraduate courses that do not assume a linear algebra background.

KEY IDEAS

The text gives the normal equations for the weighted least-squares solution of $A\mathbf{x} = \mathbf{y}$. When applied to a least-squares line problem, the most common situation, the normal equations are usually written as

$$(WX)^T WX \hat{\boldsymbol{\beta}} = (WX)^T \mathbf{y}$$

where W is the (diagonal) weighting matrix, X is the design matrix, $\hat{\boldsymbol{\beta}}$ is the least-squares parameter vector, and $\mathbf{y}$ is the observation vector.

Trend analysis is really a least-squares regression problem of the type described in Section 6.6, with data points $(x_1, y_1), \ldots, (x_n, y_n)$ fitted by a curve of the form

$$y = \beta_0 f_0(x) + \beta_1 f_1(x) + \cdots + \beta_k f_k(x)$$

where the functions $f_0, \ldots, f_k$ are polynomials that are orthogonal with respect to an inner product on $\mathbb{P}_{n-1}$ defined by

$$\langle p, q \rangle = p(x_1)q(x_1) + \cdots + p(x_n)q(x_n)$$

Usually, $x_1, \ldots, x_n$ are arranged to be evenly spaced and sum to zero, and the functions $f_1, \ldots, f_k$ are of degree 3 or 4 or less.

In $C[0, 2\pi]$ with the integral inner product, the set

$$\{1, \cos t, \cos 2t, \ldots, \cos nt, \sin t, \sin 2t, \ldots, \sin nt\} \tag{*}$$

is orthogonal. The nth order Fourier approximation to some f in $C[0, 2\pi]$ is simply the orthogonal projection of f onto the subspace W of trigonometric polynomials spanned by the functions in (*). The Fourier coefficients of f are the weights in the usual formula for the orthogonal projection of f onto W.

If an application involves an interval $[0, T]$ instead of $[0, 2\pi]$, then the inner product requires an integral over $[0, T]$, and the appropriate orthogonal set is obtained by replacing t in each function in (*) with $2\pi t/T$.

SOLUTIONS TO EXERCISES

1. For the data $(-2, 0)$, $(-1, 0)$, $(0, 2)$, $(1, 4)$, $(2, 4)$, construct

$$X = \begin{bmatrix} 1 & -2 \\ 1 & -1 \\ 1 & 0 \\ 1 & 1 \\ 1 & 2 \end{bmatrix} \text{ Design matrix,} \quad \boldsymbol{\beta} = \begin{bmatrix} \beta_0 \\ \beta_1 \end{bmatrix} \text{ Parameter vector,} \quad \mathbf{y} = \begin{bmatrix} 0 \\ 0 \\ 2 \\ 4 \\ 4 \end{bmatrix} \text{ Observation vector}$$

Since the first and last data points are about half as reliable as the other points, a suitable weighting matrix is

$$W = \begin{bmatrix} 1 & 0 & 0 & 0 & 0 \\ 0 & 2 & 0 & 0 & 0 \\ 0 & 0 & 2 & 0 & 0 \\ 0 & 0 & 0 & 2 & 0 \\ 0 & 0 & 0 & 0 & 1 \end{bmatrix}, \quad \text{so } WX = \begin{bmatrix} 1 & -2 \\ 2 & -2 \\ 2 & 0 \\ 2 & 2 \\ 1 & 2 \end{bmatrix}, \quad \text{and } W\mathbf{y} = \begin{bmatrix} 0 \\ 0 \\ 4 \\ 8 \\ 4 \end{bmatrix}$$

The remaining calculations are the same as in ordinary least squares, except that the *weighted* design matrix WX and the *weighted* observation vector $W\mathbf{y}$ appear in place of X and $\mathbf{y}$, respectively.

$$(WX)^T (WX) = \begin{bmatrix} 1 & 2 & 2 & 2 & 1 \\ -2 & -2 & 0 & 2 & 2 \end{bmatrix} \begin{bmatrix} 1 & -2 \\ 2 & -2 \\ 2 & 0 \\ 2 & 2 \\ 1 & 2 \end{bmatrix} = \begin{bmatrix} 14 & 0 \\ 0 & 16 \end{bmatrix}$$

$$(WX)^T (W\mathbf{y}) = \begin{bmatrix} 1 & 2 & 2 & 2 & 1 \\ -2 & -2 & 0 & 2 & 2 \end{bmatrix} \begin{bmatrix} 0 \\ 0 \\ 4 \\ 8 \\ 4 \end{bmatrix} = \begin{bmatrix} 28 \\ 24 \end{bmatrix}$$

The normal equations and solution are

$$\begin{bmatrix} 14 & 0 \\ 0 & 16 \end{bmatrix} \begin{bmatrix} \beta_0 \\ \beta_1 \end{bmatrix} = \begin{bmatrix} 28 \\ 24 \end{bmatrix}, \quad \begin{bmatrix} \beta_0 \\ \beta_1 \end{bmatrix} = \begin{bmatrix} 1/14 & 0 \\ 0 & 1/16 \end{bmatrix} \begin{bmatrix} 28 \\ 24 \end{bmatrix} = \begin{bmatrix} 2 \\ 3/2 \end{bmatrix}$$

The equation of the least-squares line is $y = 2 + (3/2)x$.

7. $\| \cos kt \|^2 = \int_0^{2\pi} \cos kt \cdot \cos kt \, dt = \int_0^{2\pi} \frac{1 + \cos 2kt}{2} \, dt$

$$= \left(\frac{1}{2}t + \frac{\sin 2kt}{4k} \right) \Big|_0^{2\pi} = (\frac{1}{2} \cdot 2\pi + 0) - 0 = \pi \quad \text{(if } k \neq 0\text{)}$$

$\| \sin kt \|^2 = \int_0^{2\pi} \sin kt \cdot \sin kt \, dt = \int_0^{2\pi} \frac{1 - \cos 2kt}{2} \, dt$

$$= \left(\frac{1}{2}t + \frac{\sin 2kt}{4k} \right) \Big|_0^{2\pi} = \left(\frac{1}{2} \cdot 2\pi + 0 \right) - 0 = \pi \quad \text{(if } k \neq 0\text{)}$$

9. $f(t) = 2\pi - t$. The definite integrals of $t \cos kt$ and $t \sin kt$, shown below, were computed in Example 4. The Fourier coefficients of f are:

$$\frac{a_0}{2} = \frac{1}{2} \cdot \frac{1}{\pi} \int_0^{2\pi} (2\pi - t)\, dt = \frac{1}{2\pi}(-\frac{1}{2})(2\pi - t)^2 \Big|_0^{2\pi} = 0 + \frac{1}{4\pi}(2\pi)^2 = \pi$$

and for $k > 0$,

$$a_k = \frac{1}{\pi} \int_0^{2\pi} (2\pi - t)\cos kt\, dt = \frac{1}{\pi}\int_0^{2\pi} 2\pi \cos kt\, dt - \frac{1}{\pi}\int_0^{2\pi} t\cos kt\, dt$$

$$= 0 - 0 = 0$$

$$b_k = \frac{1}{\pi} \int_0^{2\pi} (2\pi - t)\sin kt\, dt = \frac{1}{\pi}\int_0^{2\pi} 2\pi \sin kt\, dt - \frac{1}{\pi}\int_0^{2\pi} t\sin kt\, dt$$

$$= 0 - \left(-\frac{2}{k}\right) = \frac{2}{k}$$

The third-order Fourier approximation to f is

$$\pi + 2\sin t + \sin 2t + \frac{2}{3}\sin 3t$$

13. Take f and g in $C[0, 2\pi]$ and let m be a nonnegative integer. Then, the linearity of the inner product shows that

$$\langle (f + g), \cos mt \rangle = \langle f, \cos mt \rangle + \langle g, \cos mt \rangle$$

Dividing each term in this equality by $\langle \cos mt, \cos mt \rangle$, we conclude that the Fourier coefficient a_m of $f + g$ is the sum of the corresponding Fourier coefficients of f and of g. Similarly, the Fourier coefficient b_m of $f + g$ is the sum of the corresponding Fourier coefficients of f and of g.

Appendix: The Linearity of an Orthogonal Projection

The argument for Exercise 13 is a special case of a general principle. In any inner product space, the mapping $y \mapsto \dfrac{\langle y, u \rangle}{\langle u, u \rangle} u$ is linear, for any nonzero u. To verify this, take any x and y in the space and any scalar c. Then

$$\frac{\langle x + y, u \rangle}{\langle u, u \rangle} u = \frac{\langle x, u \rangle}{\langle u, u \rangle} u + \frac{\langle y, u \rangle}{\langle u, u \rangle} u,\ \text{and}\ \frac{\langle cx, u \rangle}{\langle u, u \rangle} u = \frac{c\langle x, u \rangle}{\langle u, u \rangle} u = c\frac{\langle x, u \rangle}{\langle u, u \rangle} u$$

Similarly, if $u_1, \ldots, u_p$ are any nonzero vectors, then the mapping

$$y \mapsto \frac{\langle y, u_1 \rangle}{\langle u_1, u_1 \rangle} u_1 + \cdots + \frac{\langle y, u_p \rangle}{\langle u_p, u_p \rangle} u_p$$

is a linear transformation. Thus, if $\{\mathbf{u}_1, \ldots, \mathbf{u}_p\}$ is an orthogonal basis for a subspace W, then the mapping $\mathbf{y} \mapsto \operatorname{proj}_W \mathbf{y}$ is a linear transformation.

In particular, if W is the vector space of trigonometric polynomials of order at most n, and if f and g are in $C[0, 2\pi]$, then

$$\operatorname{proj}_W(f + g) = \operatorname{proj}_W f + \operatorname{proj}_W g$$

That is, the nth order Fourier approximation to $f + g$ is the sum of the nth order Fourier approximations to f and to g. Can you use the linearity of the mapping $f \mapsto \operatorname{proj}_W f$ and the final result of Example 4 to produce (with practically no work) the answer to Exercise 9? [*Hint*: The nth order Fourier approximation to a constant function is the function itself.]

MATLAB   Graphing Functions

After you find f_4 and f_5 by hand computations, you can use **plot** to graph them. For instance, to plot $f(t) = \sin t + \sin 3t$, you can write

```
t = linspace(0,2*pi);
y = sin(t) + sin(3*t);
plot(t,y)
```

See the MATLAB box for Section 4.1 for more details.

CHAPTER 6   GLOSSARY CHECKLIST

Check your knowledge by attempting to write definitions of the terms below. Then compare your work with the definitions given in the text's Glossary. Ask your instructor which definitions, if any, might appear on a test.

angle (between nonzero vectors $\mathbf{u}$ and $\mathbf{v}$ in $\mathbb{R}^2$ or $\mathbb{R}^3$):   The angle ϑ between the . . . . Related to the scalar product by: $\mathbf{u} \cdot \mathbf{v} = \ldots$.

best approximation: The closest point . . . .

Cauchy-Schwarz inequality: . . . for all $\mathbf{u}$, $\mathbf{v}$.

component of y orthogonal to u (for $\mathbf{u} \neq \mathbf{0}$):   The vector . . . .

design matrix: The matrix X in the linear model . . . , where the columns of X are determined in some way by . . . .

distance between u and v: . . . , denoted by dist($\mathbf{u}$, $\mathbf{v}$).

Fourier approximation (of order n):   The closest point in . . . to . . . .

Fourier coefficients: The weights used to make . . . .

Fourier series: An infinite series that . . . in $C[0, 2\pi]$, with the inner product given by a definite integral.

fundamental subspaces (of A):   The . . . space and . . . space of A, and the . . . space and . . . space of A^T, with Col A^T commonly called the . . . .

general least-squares problem: Given an $m \times n$ matrix A and a vector **b** in $\mathbb{R}^m$, find . . . such that . . . for all . . . .

Gram-Schmidt process: An algorithm for producing . . . .

inner product: The scalar . . . , usually written as **u** • **v**, where **u** and **v** are vectors in $\mathbb{R}^n$ viewed as . . . . Also called the . . . of **u** and **v**. In general, a function on a vector space that assigns to each pair of vectors **u** and **v** a number . . . , subject to certain axioms.

inner product space: A vector space on which is defined . . . .

least-squares line: The line . . . that minimizes . . . in the equation . . . .

least-squares solution (of $A\mathbf{x} = \mathbf{b}$):   A vector . . . such that . . . .

length (of **v**):   The scalar $\|\mathbf{x}\| = $ . . . ; also called the . . . of **v**.

linear model (in statistics):   Any equation of the form . . . , where X and **y** are known and β is to be chosen to minimize . . . .

mean-deviation form (of a vector):   A vector whose entries . . . .

multiple regression: A linear model involving . . . variables and . . . .

normal equations: The system of equations represented by . . . , whose solution yields all . . . solutions of $A\mathbf{x} = \mathbf{b}$. In statistics, a common notation is . . . .

normalizing (a vector **v**):   The process of creating a . . . vector **u** that . . . .

observation vector: The vector . . . in the linear model $\mathbf{y} = X\beta + \varepsilon$, where the entries in . . . are the observed values of . . . .

orthogonal basis: A basis that . . . .

orthogonal complement (of W):   The set $W^\perp$ of . . . .

orthogonal matrix: A . . . matrix U such that . . . .

orthogonal projection of y onto u (or onto the line through **u** and the origin, for $\mathbf{u} \neq \mathbf{0}$):   The vector $\hat{\mathbf{y}}$ defined by . . . .

orthogonal projection of y onto W: The unique vector $\hat{\mathbf{y}}$ such that . . . .
Notation: $\hat{\mathbf{y}} = \text{proj}_W \mathbf{y}$.

orthogonal set: A set S of vectors such that . . . for . . . .

orthogonal to W: Orthogonal to every . . . .

orthonormal basis: A basis that is . . . .

orthonormal set: An . . . set of . . . .

parameter vector: The unknown vector . . . in the linear model . . . .

regression coefficients: The coefficients . . . in the . . . .

residual vector: The quantity . . . that appears in the general linear model: . . . , the difference between . . . and the . . . values (of y).

QR factorization: A factorization of an $m \times n$ matrix A with linearly independent columns, $A = QR$, where Q is an . . . matrix whose . . . , and R is an . . . matrix.

same direction (as a vector $\mathbf{v}$):   A vector that is . . . .

scale (a vector):   Multiply a vector by . . . .

trend analysis: The use of . . . to fit data, with the inner product . . . .

triangle inequality: . . . .

trigonometric polynomial: A linear combination of . . . and . . . functions such as . . . .

unit vector: A vector $\mathbf{v}$ such that . . . .

weighted least squares: Least-squares problems with a . . . inner product such as $\langle \mathbf{x}, \mathbf{y} \rangle = $ . . .

7 Symmetric Matrices and Quadratic Forms

7.1   DIAGONALIZATION OF SYMMETRIC MATRICES ⎯⎯⎯⎯⎯

To prepare for this section, review Section 5.3. Focus on the Diagonalization Theorem, Example 3, and Theorem 6. Also, review Example 3 in Section 6.2.

⎯⎯⎯⎯⎯

STUDY NOTES

If a symmetric matrix has distinct eigenvalues, as in Example 2, then the ordinary diagonalization process produces a matrix P with *orthogonal* columns, because eigenvectors from different eigenspaces are automatically orthogonal. However, the P you need here must have ortho**normal** columns. Forgetting to normalize the columns of P is the main error students make in this section.

The statements in Theorem 3 (The Spectral Theorem), together with the general approach used in Example 3, lead to the following outline for orthogonally diagonalizing any symmetric matrix.

Procedure for Orthogonally Diagonalizing a Symmetric Matrix A

1. Find the eigenvalues of A.

2. For each eigenvalue of A, construct an orthonormal basis for the eigenspace.

    a. If the eigenspace has a basis $\{\mathbf{v}\}$, normalize $\mathbf{v}$ to produce a unit vector $\mathbf{u}$.

    b. If the eigenspace has a basis $\{\mathbf{v}_1,\ \mathbf{v}_2\}$, first produce an orthogonal basis $\{\mathbf{v}_1,\ \mathbf{z}_2\}$, where

$$\mathbf{z}_2 = \mathbf{v}_2 - \frac{\mathbf{v}_2 \bullet \mathbf{v}_1}{\mathbf{v}_1 \bullet \mathbf{v}_1}\mathbf{v}_1$$

Then normalize $\mathbf{v}_1$ and $\mathbf{z}_2$ to produce an orthonormal basis $\{\mathbf{u}_1, \mathbf{u}_2\}$ for the eigenspace.

c. If the eigenspace has a basis $\{\mathbf{v}_1, ..., \mathbf{v}_k\}$, use the Gram-Schmidt process to construct an orthonormal basis $\{\mathbf{u}_1, ..., \mathbf{u}_k\}$ for the eigenspace.

3. The union of the bases for all the eigenspaces is an orthonormal basis for $\mathbb{R}^n$ (when A is $n \times n$). Use these vectors as the columns of an orthogonal matrix P.

4. Construct D from the eigenvalues, in an order corresponding to the columns of P. Repeat each eigenvalue according to the dimension of the eigenspace.

5. Finally, $A = PDP^{-1} = PDP^T$.

The exercises in this section have been constructed so that mastery of the Gram-Schmidt process is not needed, because some courses may omit Section 6.4. However, you do need to understand the calculations in Step 2(b) above.

SOLUTIONS TO EXERCISES

1. $A = \begin{bmatrix} 3 & 5 \\ 5 & -7 \end{bmatrix} = A^T$, because the (1,2) and (2,1) entries match. The entries on the main diagonal of A can have any values.

7. $P = \begin{bmatrix} .6 & .8 \\ .8 & -.6 \end{bmatrix} = [\mathbf{p}_1 \quad \mathbf{p}_2]$. To show that P is orthogonal by hand calculations, show that its columns are orthonormal: $\mathbf{p}_1 \cdot \mathbf{p}_2 = .48 - .48 = 0$, $\|\mathbf{p}_1\|^2 = (.6)^2 + (.8)^2 = 1$, and similarly, $\|\mathbf{p}_2\|^2 = 1$. Since P is square, P is an orthogonal matrix.

13. $A = \begin{bmatrix} 3 & 1 \\ 1 & 3 \end{bmatrix}$. Characteristic polynomial: $(3 - \lambda)^2 - 1 = \lambda^2 - 6\lambda + 8 = (\lambda - 4)(\lambda - 2)$. So the eigenvalues are 4 and 2.

For $\lambda = 4$: $[A - 4I \quad \mathbf{0}] = \begin{bmatrix} -1 & 1 & 0 \\ 1 & -1 & 0 \end{bmatrix} \sim \begin{bmatrix} ① & -1 & 0 \\ 0 & 0 & 0 \end{bmatrix}$, $\begin{array}{l} x_1 = x_2 \\ x_2 \text{ is free} \end{array}$

Take $x_2 = 1$ to get a basis for the eigenspace: $\begin{bmatrix} 1 \\ 1 \end{bmatrix}$. Then normalize to get a unit vector:

$\mathbf{u}_1 = \begin{bmatrix} 1/\sqrt{2} \\ 1/\sqrt{2} \end{bmatrix}$. (Don't forget this step.)

For $\lambda = 2$: $[A - 2I \quad \mathbf{0}] = \begin{bmatrix} 1 & 1 & 0 \\ 1 & 1 & 0 \end{bmatrix} \sim \begin{bmatrix} \boxed{1} & 1 & 0 \\ 0 & 0 & 0 \end{bmatrix}$, $\begin{array}{l} x_1 = -x_2 \\ x_2 \text{ is free} \end{array}$

Take $x_2 = 1$ to get a basis for the eigenspace: $\begin{bmatrix} -1 \\ 1 \end{bmatrix}$. Then normalize to get a unit vector:

$$\mathbf{u}_2 = \begin{bmatrix} -1/\sqrt{2} \\ 1/\sqrt{2} \end{bmatrix}.$$

Set $P = [\mathbf{u}_1 \quad \mathbf{u}_2] = \begin{bmatrix} 1/\sqrt{2} & -1/\sqrt{2} \\ 1/\sqrt{2} & 1/\sqrt{2} \end{bmatrix}$. The corresponding D is $\begin{bmatrix} 4 & 0 \\ 0 & 2 \end{bmatrix}$.

Study Tip: The fact that eigenvectors for distinct eigenvalues are orthogonal gives you a check on your work. After you find $\mathbf{u}_2$ in Exercise 13, verify that $\mathbf{u}_2 \cdot \mathbf{u}_1 = 0$. Actually, when $\mathbf{u}_1$ and $\mathbf{u}_2$ are in $\mathbb{R}^2$, you can easily guess what $\mathbf{u}_2$ must be, once you know $\mathbf{u}_1$. If you do this, you should compute $A\mathbf{u}_2$ to make sure that $\mathbf{u}_2$ is indeed an eigenvector. Ask your instructor how much work you should show on a test question similar to Exercise 13.

19. Be sure to *work* this problem before reading the solution. Use Exercises 12–24 to sharpen your skills. They are critical for the rest of the chapter. Here, $A = \begin{bmatrix} 3 & -2 & 4 \\ -2 & 6 & 2 \\ 4 & 2 & 3 \end{bmatrix}$, and the eigenvalues are given: 7 and –2.

For $\lambda = 7$: $[A - 7I \quad \mathbf{0}] = \begin{bmatrix} -4 & -2 & 4 & 0 \\ -2 & -1 & 2 & 0 \\ 4 & 2 & -4 & 0 \end{bmatrix} \sim \begin{bmatrix} \boxed{1} & .5 & -1 & 0 \\ 0 & 0 & 0 & 0 \\ 0 & 0 & 0 & 0 \end{bmatrix}$

Thus, $x_1 = -.5x_2 + x_3$, with x_2 and x_3 free. Instead of describing all vectors in the eigenspace, you can produce a basis quickly by choosing two linearly independent solutions of $(A - 7I)\mathbf{x} = \mathbf{0}$. The natural choices are the vector corresponding to $x_2 = 1$ and $x_3 = 0$ and the vector for $x_2 = 0$ and $x_3 = 1$. However, in this particular problem, the coefficient $-.5$ of x_2 suggests that a better choice for the first vector is to take $x_2 = 2$ and $x_3 = 0$. In this case, the two vectors are

$$\mathbf{v}_1 = \begin{bmatrix} -1 \\ 2 \\ 0 \end{bmatrix} \text{ and } \mathbf{v}_2 = \begin{bmatrix} 1 \\ 0 \\ 1 \end{bmatrix}$$

This basis for the eigenspace is not orthogonal. Keep $\mathbf{v}_1$ and subtract from $\mathbf{v}_2$ its orthogonal projection onto $\mathbf{v}_1$. The new vector, $\mathbf{z}_2$, is an eigenvector for the eigenvalue 7 because it is a linear combination of the vectors $\mathbf{v}_2$ and $\mathbf{v}_1$ in the eigenspace for $\lambda = 7$:

$$\mathbf{z}_2 = \mathbf{v}_2 - \frac{\mathbf{v}_2 \cdot \mathbf{v}_1}{\mathbf{v}_1 \cdot \mathbf{v}_1} \mathbf{v}_1 = \begin{bmatrix} 1 \\ 0 \\ 1 \end{bmatrix} - \frac{-1}{5} \begin{bmatrix} -1 \\ 2 \\ 0 \end{bmatrix} = \begin{bmatrix} 4/5 \\ 2/5 \\ 1 \end{bmatrix}$$

Instead of $\mathbf{z}_2$, use $\mathbf{z}_2' = \begin{bmatrix} 4 \\ 2 \\ 5 \end{bmatrix}$, which is easier to normalize. Check that $\mathbf{z}_2' \cdot \mathbf{v}_1 = 0$. An

orthonormal basis for the eigenspace is $\mathbf{u}_1 = \begin{bmatrix} -1/\sqrt{5} \\ 2/\sqrt{5} \\ 0 \end{bmatrix}$, $\mathbf{u}_2 = \begin{bmatrix} 4/\sqrt{45} \\ 2/\sqrt{45} \\ 5/\sqrt{45} \end{bmatrix}$.

For $\underline{\lambda = -2}$: $[A + 2I \quad \mathbf{0}] = \begin{bmatrix} 5 & -2 & 4 & 0 \\ -2 & 8 & 2 & 0 \\ 4 & 2 & 5 & 0 \end{bmatrix} \sim \begin{bmatrix} 1 & -4 & -1 & 0 \\ 5 & -2 & 4 & 0 \\ 4 & 2 & 5 & 0 \end{bmatrix} \sim \begin{bmatrix} 1 & -4 & -1 & 0 \\ 0 & 18 & 9 & 0 \\ 0 & 18 & 9 & 0 \end{bmatrix}$

$$= \begin{bmatrix} 1 & -4 & -1 & 0 \\ 0 & 1 & 1/2 & 0 \\ 0 & 0 & 0 & 0 \end{bmatrix} \sim \begin{bmatrix} ① & 0 & 1 & 0 \\ 0 & ① & 1/2 & 0 \\ 0 & 0 & 0 & 0 \end{bmatrix}, \quad \begin{array}{l} x_1 = -x_3 \\ x_2 = -(1/2)x_3 \\ x_3 \text{ is free} \end{array}$$

Take $x_3 = 2$ to get a basis for the eigenspace, $\begin{bmatrix} -2 \\ -1 \\ 2 \end{bmatrix}$, and normalize to obtain $\mathbf{u}_3 = \begin{bmatrix} -2/3 \\ -1/3 \\ 2/3 \end{bmatrix}$.

Finally, set $P = [\mathbf{u}_1 \quad \mathbf{u}_2 \quad \mathbf{u}_3] = \begin{bmatrix} -1/\sqrt{5} & 4/\sqrt{45} & -2/3 \\ 2/\sqrt{5} & 2/\sqrt{45} & -1/3 \\ 0 & 5/\sqrt{45} & 2/3 \end{bmatrix}$ and $D = \begin{bmatrix} 7 & 0 & 0 \\ 0 & 7 & 0 \\ 0 & 0 & -2 \end{bmatrix}$.

What other answers might someone produce? If the vectors $\mathbf{v}_1$ and $\mathbf{v}_2$ are interchanged, the

first two columns of P probably will be $\begin{bmatrix} 1/\sqrt{2} \\ 0 \\ 1/\sqrt{2} \end{bmatrix}$ and $\begin{bmatrix} -1/\sqrt{18} \\ 4/\sqrt{18} \\ 1/\sqrt{18} \end{bmatrix}$. If the entries in D are

rearranged, the columns of P must be rearranged to correspond to the new entries in D.

Study Tip: The matrix in Exercise 20 has only two distinct eigenvalues (according to the text's information), so one or both eigenspaces will be at least two-dimensional. You will have to construct an orthonormal basis for such an eigenspace. Exercises 23 and 24 are good models for exam questions because they give you information from which you can orthogonally diagonalize A without extensive computations.

25. a. See Theorem 2 and the paragraph preceding the theorem.

    b. See Theorem 1.

    c. See Example 3.

    d. See the paragraph following formula (2) and Exercise 35.

29. By hypothesis, $A = PDP^{-1}$, where P is orthogonal and D is diagonal. Since A is invertible, 0 is not an eigenvalue and D is invertible. Then

$$A^{-1} = (PDP^{-1})^{-1} = (P^{-1})^{-1}D^{-1}P^{-1} = PD^{-1}P^{-1}$$

Since D^{-1} is diagonal, A^{-1} is orthogonally diagonalizable.

    A second argument: By Theorem 2, A is symmetric. Since A is invertible, a property of transposes shows that $(A^{-1})^T = (A^T)^{-1} = A^{-1}$, so A^{-1} is symmetric. By Theorem 2, again, A^{-1} is orthogonally diagonalizable.

31. The solution is in the text.

35. a. The matrix $B = \mathbf{u}\mathbf{u}^T$ is an outer product, or a rank 1 matrix. Given $\mathbf{x}$ in $\mathbb{R}^n$, $B\mathbf{x} = (\mathbf{u}\mathbf{u}^T)\mathbf{x} = \mathbf{u}(\mathbf{u}^T\mathbf{x}) = (\mathbf{u}^T\mathbf{x})\mathbf{u}$, because $\mathbf{u}^T\mathbf{x}$ is a scalar. Using dot products, $B\mathbf{x} = (\mathbf{x} \cdot \mathbf{u})\mathbf{u}$. Since $\mathbf{u}$ is a unit vector, this is the orthogonal projection of $\mathbf{x}$ onto $\mathbf{u}$. See Section 6.2.

    b. B is symmetric, because $B^T = (\mathbf{u}\mathbf{u}^T)^T = \mathbf{u}^{TT}\mathbf{u}^T = \mathbf{u}\mathbf{u}^T = B$. Also, $B^2 = (\mathbf{u}\mathbf{u}^T)(\mathbf{u}\mathbf{u}^T) = \mathbf{u}(\mathbf{u}^T\mathbf{u})\mathbf{u}^T = \mathbf{u}\mathbf{u}^T = B$, because $\mathbf{u}^T\mathbf{u} = 1$.

MATLAB   Orthogonal Diagonalization

Use **eig(A)** for eigenvalues and **nulbasis** to obtain eigenvectors, as in Section 5.3. If you encounter a two-dimensional eigenspace with a basis $\{\mathbf{v}_1, \mathbf{v}_2\}$, use the command

```
v2 = v2 - (v2'*v1)/(v1'*v1)*v1
```

or

```
v2 = v2 - proj(v2,v1)
```

to make the *new* eigenvector $\mathbf{v}_2$ orthogonal to $\mathbf{v}_1$. See the MATLAB note for Section 6.4. After you normalize the eigenvectors and create P, check that $P^TP = I$, to verify that P is indeed an orthogonal matrix. For practice, you might use MATLAB to work Exercise 19.

7.2   QUADRATIC FORMS _____

Sections 7.1 and 7.2 provide the foundation for the rest of the chapter.

_____

KEY IDEAS

The main point here is to learn how a change of variable, $\mathbf{x} = P\mathbf{y}$, with P an orthogonal matrix, can transform a quadratic form into a new quadratic form with no cross-product terms.

The equation $\mathbf{x} = P\mathbf{y}$ expresses $\mathbf{x}$ as a linear combination of the columns of P, using the entries in $\mathbf{y}$ as weights. If the columns of P are used as a basis $\mathcal{B}$ for $\mathbb{R}^n$, then the entries in $\mathbf{y}$ are the coordinates of $\mathbf{x}$ relative to the basis $\mathcal{B}$. See Section 4.4 (or Section 2.9). As Section 7.2 shows, the columns of P are eigenvectors of the matrix A of the quadratic form $\mathbf{x}^T A\mathbf{x}$.

If you were fortunate enough to study Section 5.6 or 5.7, you saw the *same* equation $\mathbf{x} = P\mathbf{y}$, with P constructed from eigenvectors of A! The key difference here is that P must be an orthogonal matrix as well as a matrix that diagonalizes A. Nothing less will do. The change of variable $\mathbf{x} = P\mathbf{y}$ will work only if $P^T = P^{-1}$ (and $A = PDP^{-1}$).

If your course covers the various classes of quadratic forms (or, equivalently, classes of symmetric matrices), you should learn both the definitions and the characterizations (in Theorem 5) of these classes. Exercise 24 describes another useful way to characterize quadratic forms, often used in multivariable calculus courses. (The 2×2 case can be generalized to $n \times n$ matrices.)

_____

SOLUTIONS TO EXERCISES

1. a. $\mathbf{x}^T A\mathbf{x} = \begin{bmatrix} x_1 & x_2 \end{bmatrix} \begin{bmatrix} 5 & 1/3 \\ 1/3 & 1 \end{bmatrix} \begin{bmatrix} x_1 \\ x_2 \end{bmatrix} = \begin{bmatrix} x_1 & x_2 \end{bmatrix} \begin{bmatrix} 5x_1 + (1/3)x_2 \\ (1/3)x_1 + x_2 \end{bmatrix}$

$\qquad = 5x_1^2 + (2/3)x_1 x_2 + x_2^2$

b. When $\mathbf{x} = (6, 1)$, $\mathbf{x}^T A\mathbf{x} = 5(6)^2 + (2/3)(6)(1) + (1)^2 = 185$.

c. When $\mathbf{x} = (1, 3)$, $\mathbf{x}^T A\mathbf{x} = 5(1)^2 + (2/3)(1)(3) + (3)^2 = 16$.

7. The matrix of the quadratic form is $A = \begin{bmatrix} 1 & 5 \\ 5 & 1 \end{bmatrix}$. The characteristic polynomial is

$\lambda^2 - 2\lambda - 24 = (\lambda - 6)(\lambda + 4)$; eigenvalues are 6 and –4.

For $\underline{\lambda = 6}$: an eigenvector is $\begin{bmatrix} 1 \\ 1 \end{bmatrix}$, normalized: $\mathbf{u}_1 = \dfrac{1}{\sqrt{2}}\begin{bmatrix} 1 \\ 1 \end{bmatrix}$

For $\lambda = -4$:   an eigenvector is $\begin{bmatrix} -1 \\ 1 \end{bmatrix}$, normalized: $\mathbf{u}_2 = \dfrac{1}{\sqrt{2}}\begin{bmatrix} -1 \\ 1 \end{bmatrix}$

Thus $A = PDP^{-1}$ and $D = P^{-1}AP = P^TAP$, when $P = \dfrac{1}{\sqrt{2}}\begin{bmatrix} 1 & -1 \\ 1 & 1 \end{bmatrix}$ and $D = \begin{bmatrix} 6 & 0 \\ 0 & -4 \end{bmatrix}$.

The desired change of variable is $\mathbf{x} = P\mathbf{y}$, so that

$$\mathbf{x}^T A\mathbf{x} = (P\mathbf{y})^T A(P\mathbf{y}) = \mathbf{y}^T P^T A P\mathbf{y} = \mathbf{y}^T D\mathbf{y} \qquad (*)$$
$$= 6y_1^2 - 4y_2^2$$

Study Tip: To make the "change of variable" requested in Exercise 7, you should: (1) write the equation $\mathbf{x} = P\mathbf{y}$ and specify P; (2) show the matrix algebra in ($*$) that produces the new quadratic form; and (3) include the new quadratic form. Find out how much of this information you should supply if a problem like Exercise 7 were to appear on an exam.

13. The matrix of the quadratic form is $A = \begin{bmatrix} 1 & -3 \\ -3 & 9 \end{bmatrix}$. The characteristic polynomial is

$\lambda^2 - 10\lambda = \lambda(\lambda - 10)$; the eigenvalues are 10 and 0. Thus the quadratic form is positive semidefinite. To find the change of variable, proceed as in Exercise 7:

For $\lambda = 10$:   an eigenvector is $\begin{bmatrix} 1 \\ -3 \end{bmatrix}$, normalized: $\mathbf{u}_1 = \dfrac{1}{\sqrt{10}}\begin{bmatrix} 1 \\ -3 \end{bmatrix}$

For $\lambda = 0$:   an eigenvector is $\begin{bmatrix} 3 \\ 1 \end{bmatrix}$, normalized: $\mathbf{u}_2 = \dfrac{1}{\sqrt{10}}\begin{bmatrix} 3 \\ 1 \end{bmatrix}$

Take $P = \dfrac{1}{\sqrt{10}}\begin{bmatrix} 1 & 3 \\ -3 & 1 \end{bmatrix}$ and $D = \begin{bmatrix} 10 & 0 \\ 0 & 0 \end{bmatrix}$. Since P orthogonally diagonalizes A, the desired change of variable is $\mathbf{x} = P\mathbf{y}$, and

$$\mathbf{x}^T A\mathbf{x} = (P\mathbf{y})^T A(P\mathbf{y}) = \mathbf{y}^T P^T A P\mathbf{y} = \mathbf{y}^T D\mathbf{y} = 10y_1^2$$

The new quadratic form is $10y_1^2$.

19. Because 8 is larger than 5, you should make the x_2^2 term as large as possible. The constraint $x_1^2 + x_2^2 = 1$ keeps x_2 from exceeding 1. When $x_1 = 0$ and $x_2 = 1$, the value of the quadratic form is $5(0) + 8(1) = 8$.

21. a. See the definition before Example 1.
   b. See the paragraph following Example 3.
   c. See the Principal Axes Theorem and the Diagonalization Theorem (in Section 5.3).
   d. Check the definition.
   e. See Theorem 5.
   f. See the Numerical Note after Example 6.

25. The text's answer showed that $\mathbf{x}^T B^T B \mathbf{x} \geq 0$ for all $\mathbf{x}$. To show $B^T B$ is positive definite, suppose that $\mathbf{x}^T B^T B \mathbf{x} = 0$. Then $(B\mathbf{x})^T B \mathbf{x} = 0$, so that $\|B\mathbf{x}\|^2 = 0$ and $B\mathbf{x} = 0$. If B is invertible, then $\mathbf{x} = \mathbf{0}$, which shows that, in this case, the form $\mathbf{x}^T B^T B \mathbf{x}$ is positive definite.

26. The quadratic forms xTAx and xTBx are both positive definite, by Theorem 5, because all eigenvalues of A and B are positive. Then for any nonzero x, xT(A + B)x = xT(Ax + Bx) = xTAx + xTBx > 0, so the quadratic form xT(A + B)x is positive definite. Also, the matrix A + B is symmetric, because (A + B)T = AT + BT = A + B. By Theorem 5, A + B has positive eigenvalues.

Mastering Linear Algebra Concepts: Diagonalization and Quadratic Forms

Since the end of the course draws near, I recommend that you prepare a review sheet that contrasts the Diagonalization Theorem in Section 5.3 with Theorem 2 in Section 7.1. You might begin by copying the statements of these theorems, and then use the following questions to guide your review:

• What special properties does an orthogonal diagonalization have that are not present in all diagonalizations? Consider the eigenvalues, the eigenspaces, the eigenvectors, and the matrix P in PDP^{-1}.

• Suppose A is symmetric, and all eigenspaces are one-dimensional. What differences are there between a general diagonalization and an orthogonal diagonalization of A? What differences are there when A is symmetric and one eigenspace is two-dimensional?

• Why is an orthogonal diagonalization needed to simplify a quadratic form? Make the change of variable $\mathbf{x} = P\mathbf{y}$ in $\mathbf{x}^T A \mathbf{x}$ and show the algebra involved.

• If you studied Section 5.6 or 5.7, compare Figure 4 in Section 5.6 with Figure 3 in Section 5.7. How do the general shapes of the trajectories differ? Why do they differ? Could Figure 5 in Section 5.6 or Figure 5 in Section 5.7 be associated with a symmetric matrix?

7.3   CONSTRAINED OPTIMIZATION ———————

This section is important in its own right, since constrained optimization problems arise in many mathematical problems and applications. The main results of the section are also used in the following two sections.

KEY IDEAS

Theorem 6 gives the main idea. The maximum value of a quadratic form $\mathbf{x}^T A \mathbf{x}$ over the set of all unit vectors can be computed by finding the greatest eigenvalue of A; this maximum value is attained at a corresponding eigenvector. The key step in the proof is to diagonalize A by P, substitute $\mathbf{x} = P\mathbf{y}$, and use the fact that in this case, $\mathbf{x}$ and $\mathbf{y}$ have the same norm.

Example 6 presents a topic that is widely discussed in elementary economics texts.

SOLUTIONS TO EXERCISES

1. We are given an equality of two quadratic forms:

$$5x_1^2 + 6x_2^2 + 7x_3^2 + 4x_1x_2 - 4x_2x_3 = 9y_1^2 + 6y_2^2 + 3y_3^2$$

The matrix of the left quadratic form is $A = \begin{bmatrix} 5 & 2 & 0 \\ 2 & 6 & -2 \\ 0 & -2 & 7 \end{bmatrix}$.

The equality between the two quadratic forms indicates that the eigenvalues of A are 9, 6, 3. (Proof: The diagonal matrix D of the quadratic form $9y_1^2 + 6y_2^2 + 3y_3^2$ obviously has eigenvalues 9, 6, 3. Since A is similar to D, A has the same eigenvalues as D.) The standard calculations produce a unit eigenvector for each eigenvalue. Don't forget to normalize each eigenvector.

$$\lambda = 9: \mathbf{u}_1 = \begin{bmatrix} 1/3 \\ 2/3 \\ -2/3 \end{bmatrix}; \quad \lambda = 6: \mathbf{u}_2 = \begin{bmatrix} 2/3 \\ 1/3 \\ 2/3 \end{bmatrix}; \quad \lambda = 3: \mathbf{u}_3 = \begin{bmatrix} -2/3 \\ 2/3 \\ 1/3 \end{bmatrix}$$

These eigenvectors are mutually orthogonal because they correspond to distinct eigenvalues. So the desired change of variable is

$$\mathbf{x} = P\mathbf{y}, \text{ where } P = \begin{bmatrix} 1/3 & 2/3 & -2/3 \\ 2/3 & 1/3 & 2/3 \\ -2/3 & 2/3 & 1/3 \end{bmatrix}$$

Study Tip: Review the matrix algebra that leads from $\mathbf{x}^T A\mathbf{x}$ to $\mathbf{y}^T D\mathbf{y}$. Also, be sure you can show that $\|\mathbf{x}\| = \|\mathbf{y}\|$ when P is an orthogonal matrix.

7. The matrix of $Q(\mathbf{x}) = -2x_1^2 - x_2^2 + 4x_1x_2 + 4x_2x_3$ is $A = \begin{bmatrix} -2 & 2 & 0 \\ 2 & -1 & 2 \\ 0 & 2 & 0 \end{bmatrix}$.

The hint in the exercise lists 2, −1, and −4 as the eigenvalues. The greatest eigenvalue is 2, not −4, because "greatest" here refers to the eigenvalue that is farthest to the right on the real line. The maximum value of $Q(\mathbf{x})$ (for $\mathbf{x}$ a unit vector) is attained at a unit eigenvector for $\lambda = 2$. Standard calculations produce the eigenvector:

$$\mathbf{v}_1 = \begin{bmatrix} 1/2 \\ 1 \\ 1 \end{bmatrix}, \text{ scaled to } \begin{bmatrix} 1 \\ 2 \\ 2 \end{bmatrix}, \text{ and normalized to } \mathbf{u}_1 = \begin{bmatrix} 1/3 \\ 2/3 \\ 2/3 \end{bmatrix}.$$

Warning: Exercise 7 illustrates the potential error of selecting −4 instead of 2 as the greatest eigenvalue.

12. This exercise can be done by using a theorem, but try to do it by direct computation, using the hint in the text.

13. If $m = M$ and $\mathbf{x}$ is the unit eigenvector $\mathbf{u}_1$, then $\mathbf{x}^T A\mathbf{x} = \mathbf{u}_1^T A\mathbf{u}_1 = \mathbf{u}_1^T (m\mathbf{u}_1) = m$. Otherwise, set $\alpha = (t - m)/(M - m)$. The hint in the text shows that $0 \le \alpha \le 1$ when $m \le t \le M$. For such α, let $\mathbf{x} = \sqrt{1 - \alpha}\,\mathbf{u}_n + \sqrt{\alpha}\,\mathbf{u}_1$. Then the vectors $\sqrt{1 - \alpha}\,\mathbf{u}_n$ and $\sqrt{\alpha}\,\mathbf{u}_1$ are orthogonal because they are eigenvectors for different eigenvalues (or one of the vectors is $\mathbf{0}$). By the Pythagorean Theorem,

$$\mathbf{x}^T \mathbf{x} = \|\mathbf{x}\|^2 = \left\|\sqrt{1 - \alpha}\,\mathbf{u}_n\right\|^2 + \left\|\sqrt{\alpha}\,\mathbf{u}_1\right\|^2$$

$$= |1 - \alpha|\,\|\mathbf{u}_n\|^2 + |\alpha|\,\|\mathbf{u}_1\|^2$$

$$= (1 - \alpha) + \alpha = 1$$

because $\mathbf{u}_n$ and $\mathbf{u}_1$ are unit vectors and $0 \le \alpha \le 1$. Also, using the fact that $\mathbf{u}_n$ and $\mathbf{u}_1$ are orthogonal, compute

$$\mathbf{x}^T A\mathbf{x} = (\sqrt{1 - \alpha}\,\mathbf{u}_n + \sqrt{\alpha}\,\mathbf{u}_1)^T A(\sqrt{1 - \alpha}\,\mathbf{u}_n + \sqrt{\alpha}\,\mathbf{u}_1)$$

$$= (\sqrt{1 - \alpha}\,\mathbf{u}_n + \sqrt{\alpha}\,\mathbf{u}_1)^T (m\sqrt{1 - \alpha}\,\mathbf{u}_n + M\sqrt{\alpha}\,\mathbf{u}_1)$$

$$= |1 - \alpha|\, m\mathbf{u}_n^T\mathbf{u}_n + |\alpha|\, M\mathbf{u}_1^T\mathbf{u}_1 = (1 - \alpha)m + \alpha M = t$$

Thus the quadratic form $\mathbf{x}^T A\mathbf{x}$ assumes every value between m and M for a suitable unit vector $\mathbf{x}$.

7.4   THE SINGULAR VALUE DECOMPOSITION _____

This section is the capstone of the text. It completes the story of the linear transformation $\mathbf{x} \mapsto A\mathbf{x}$ for a general $m \times n$ matrix A and, in so doing, gives you an opportunity to review many basic concepts from Chapters 4–7. In addition, this section opens the door into the modern world of applied linear algebra. An understanding of the singular value decomposition is essential for advanced work in science and engineering that requires matrix computations.

_____

KEY IDEAS

The first singular value σ_1 of an $m \times n$ matrix A is the maximum of $\|A\mathbf{x}\|$ over all unit vectors. This maximum value is attained at a unit eigenvector $\mathbf{v}_1$ of $A^T A$ corresponding to the greatest eigenvalue λ_1 of $A^T A$. The second singular value is the maximum of $\|A\mathbf{x}\|$ over all unit vectors orthogonal to $\mathbf{v}_1$. The following algorithm produces the singular value decomposition for A. (As mentioned in the text, other more reliable methods are used in professional software.)

Procedure for Computing a Singular Value Decomposition

1. Find an orthonormal basis $\{v_1, \ldots, v_n\}$ for $\mathbb{R}^n$ consisting of eigenvectors of A^TA, arranged so that the associated eigenvalues satisfy $\lambda_1 \geq \cdots \geq \lambda_r > 0$ and $\lambda_{r+1} = \cdots = 0$, where $r = \text{rank } A$.

2. Construct the $n \times n$ orthogonal matrix $V = [v_1 \quad \cdots \quad v_n]$.

3. Let $\sigma_j = \sqrt{\lambda_j}$ $(1 \leq j \leq n)$, and construct the $m \times n$ diagonal matrix Σ whose (j, j)-entry is σ_j $(1 \leq j \leq n)$ and has zeros elsewhere.

4. The set $\{Av_1, \ldots, Av_r\}$ is orthogonal and $\sigma_j = \|Av_j\|$. Compute $u_j = (1/\sigma_j)Av_j$ for $1 \leq j \leq r$.

5. Extend $\{u_1, \ldots, u_r\}$ to an orthonormal basis $\{u_1, \ldots, u_m\}$ for $\mathbb{R}^m$. Write the $m \times m$ orthogonal matrix $U = [u_1 \quad \cdots \quad u_m]$.

6. $A = U\Sigma V^T$.

The diagram below illustrates how the SVD splits the action of A into first multiplication by V^T (which amounts to an orthogonal change of basis in $\mathbb{R}^n$), then a scaling by Σ in the directions of the standard basis vectors $e_1, \ldots, e_n$, and finally multiplication by U (an orthogonal change of basis in $\mathbb{R}^m$).

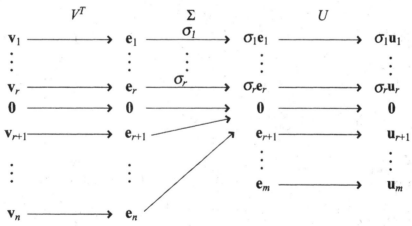

The Singular Value Decomposition: $A = U\Sigma V^T$

Note that the so-called *left* singular vectors of A are the columns of U (because U appears on the left in the factorization of A), even though $u_1, \ldots, u_n$ appear on the right side of the diagram above.

SOLUTIONS TO EXERCISES

1. $A = \begin{bmatrix} 1 & 0 \\ 0 & -3 \end{bmatrix}$, $A^T A = \begin{bmatrix} 1 & 0 \\ 0 & 9 \end{bmatrix}$. Eigenvalues and eigenvectors of $A^T A$ are:

$$\lambda_1 = 9: \mathbf{v}_1 = \begin{bmatrix} 0 \\ 1 \end{bmatrix}; \quad \lambda_2 = 1: \mathbf{v}_2 = \begin{bmatrix} 1 \\ 0 \end{bmatrix}$$

(Remember to arrange the eigenvalues in decreasing order.) Thus

$$V = \begin{bmatrix} 0 & 1 \\ 1 & 0 \end{bmatrix}$$

The singular values are $\sigma_1 = \sqrt{9} = 3$ and $\sigma_2 = 1$. The matrix Σ is the same shape as A, and

$$\Sigma = \begin{bmatrix} \sigma_1 & 0 \\ 0 & \sigma_2 \end{bmatrix} = \begin{bmatrix} 3 & 0 \\ 0 & 1 \end{bmatrix}$$

Next, compute

$$A\mathbf{v}_1 = \begin{bmatrix} 1 & 0 \\ 0 & -3 \end{bmatrix}\begin{bmatrix} 0 \\ 1 \end{bmatrix} = \begin{bmatrix} 0 \\ -3 \end{bmatrix}, \quad A\mathbf{v}_2 = \begin{bmatrix} 1 & 0 \\ 0 & -3 \end{bmatrix}\begin{bmatrix} 1 \\ 0 \end{bmatrix} = \begin{bmatrix} 1 \\ 0 \end{bmatrix}$$

and normalize:

$$\mathbf{u}_1 = \frac{1}{\sigma_1}A\mathbf{v}_1 = \frac{1}{3}\begin{bmatrix} 0 \\ -3 \end{bmatrix} = \begin{bmatrix} 0 \\ -1 \end{bmatrix}, \quad \mathbf{u}_2 = \frac{1}{\sigma_2}\begin{bmatrix} 1 \\ 0 \end{bmatrix} = \begin{bmatrix} 1 \\ 0 \end{bmatrix}$$

Finally, $\{\mathbf{u}_1, \mathbf{u}_2\}$ is already a basis for $\mathbb{R}^2$, so the basis for $\mathbb{R}^2$ is complete, and

This happens to equal V.

$$U = \begin{bmatrix} 0 & 1 \\ -1 & 0 \end{bmatrix}, \quad \text{and } A = U\Sigma V^T = \begin{bmatrix} 0 & 1 \\ -1 & 0 \end{bmatrix}\begin{bmatrix} 3 & 0 \\ 0 & 1 \end{bmatrix}\begin{bmatrix} 0 & 1 \\ 1 & 0 \end{bmatrix}$$

7. $A = \begin{bmatrix} 2 & -1 \\ 2 & 2 \end{bmatrix}$. $A^T A = \begin{bmatrix} 8 & 2 \\ 2 & 5 \end{bmatrix}$. Find the eigenvalues of $A^T A$ from the characteristic equation.

$$0 = \lambda^2 - 13\lambda + 36 = (\lambda - 9)(\lambda - 4); \quad \lambda_1 = 9, \quad \lambda_2 = 4$$

Corresponding unit eigenvectors for $A^T A$ (calculations omitted) are:

$$\lambda_1 = 9: \mathbf{v}_1 = \begin{bmatrix} 2/\sqrt{5} \\ 1/\sqrt{5} \end{bmatrix}; \quad \lambda_2 = 4: \mathbf{v}_2 = \begin{bmatrix} -1/\sqrt{5} \\ 2/\sqrt{5} \end{bmatrix}$$

Take

$$V = \begin{bmatrix} 2/\sqrt{5} & -1/\sqrt{5} \\ 1/\sqrt{5} & 2/\sqrt{5} \end{bmatrix}$$

The singular values are $\sigma_1 = \sqrt{9} = 3$ and $\sigma_2 = \sqrt{4} = 2$. The matrix Σ is the same shape as A

and $\Sigma = \begin{bmatrix} \sigma_1 & 0 \\ 0 & \sigma_2 \end{bmatrix} = \begin{bmatrix} 3 & 0 \\ 0 & 2 \end{bmatrix}$. Next, compute

$$A\mathbf{v}_1 = \begin{bmatrix} 2 & -1 \\ 2 & 2 \end{bmatrix}\begin{bmatrix} 2/\sqrt{5} \\ 1/\sqrt{5} \end{bmatrix} = \begin{bmatrix} 3/\sqrt{5} \\ 6/\sqrt{5} \end{bmatrix}, \quad A\mathbf{v}_2 = \begin{bmatrix} 2 & -1 \\ 2 & 2 \end{bmatrix}\begin{bmatrix} -1/\sqrt{5} \\ 2/\sqrt{5} \end{bmatrix} = \begin{bmatrix} -4/\sqrt{5} \\ 2/\sqrt{5} \end{bmatrix}$$

To check your work at this point, verify that $A\mathbf{v}_1$ and $A\mathbf{v}_2$ are orthogonal. (They are.) Then normalize:

$$\mathbf{u}_1 = \frac{1}{\sigma_1} A\mathbf{v}_1 = \frac{1}{3}\begin{bmatrix} 3/\sqrt{5} \\ 6/\sqrt{5} \end{bmatrix} = \begin{bmatrix} 1/\sqrt{5} \\ 2/\sqrt{5} \end{bmatrix}, \quad \mathbf{u}_2 = \frac{1}{\sigma_2} = \frac{1}{2}\begin{bmatrix} -4/\sqrt{5} \\ 2/\sqrt{5} \end{bmatrix} = \begin{bmatrix} -2/\sqrt{5} \\ 1/\sqrt{5} \end{bmatrix}$$

Since $\{\mathbf{u}_1, \mathbf{u}_2\}$ is a basis for $\mathbb{R}^2$, take $U = \begin{bmatrix} 1/\sqrt{5} & -2/\sqrt{5} \\ 2/\sqrt{5} & 1/\sqrt{5} \end{bmatrix}$. Thus

$$A = U\Sigma V^T = \begin{bmatrix} 1/\sqrt{5} & -2/\sqrt{5} \\ 2/\sqrt{5} & 1/\sqrt{5} \end{bmatrix}\begin{bmatrix} 3 & 0 \\ 0 & 2 \end{bmatrix}\begin{bmatrix} 2/\sqrt{5} & 1/\sqrt{5} \\ -1/\sqrt{5} & 2/\sqrt{5} \end{bmatrix} \quad \text{Use } V^T, \text{ not } V.$$

Study Tip: Your answer for a singular value decomposition may differ from that given in the text. To check your work, compute AV and $U\Sigma$. If $AV = U\Sigma$, then $A = U\Sigma V^T$ and your answer is correct (provided U and V truly are orthogonal matrices).

13. The matrix $A^T A$ is 3×3. Because the text has not given you practice computing and solving a cubic characteristic equation, the *Hint* suggests that you consider A^T instead of A. (You are free to work on A itself, if you prefer.) Using A^T, compute

$$(A^T)^T A^T = AA^T = \begin{bmatrix} 3 & 2 & 2 \\ 2 & 3 & -2 \end{bmatrix}\begin{bmatrix} 3 & 2 \\ 2 & 3 \\ 2 & -2 \end{bmatrix} = \begin{bmatrix} 17 & 8 \\ 8 & 17 \end{bmatrix}$$

The characteristic equation is

$$0 = \lambda^2 - 34\lambda + 225 = (\lambda - 25)(\lambda - 9); \quad \lambda_1 = 25, \ \lambda_2 = 9$$

The corresponding unit eigenvectors and the matrix V are

$$\lambda_1 = 25: \ \mathbf{v}_1 = \begin{bmatrix} 1/\sqrt{2} \\ 1/\sqrt{2} \end{bmatrix}; \quad \lambda_2 = 9: \ \mathbf{v}_2 = \begin{bmatrix} -1/\sqrt{2} \\ 1/\sqrt{2} \end{bmatrix}; \quad V = \begin{bmatrix} 1/\sqrt{2} & -1/\sqrt{2} \\ 1/\sqrt{2} & 1/\sqrt{2} \end{bmatrix}$$

The singular values are $\sigma_1 = 5$ and $\sigma_2 = 3$. Thus Σ is $\begin{bmatrix} 5 & 0 \\ 0 & 3 \\ 0 & 0 \end{bmatrix}$, the same size as A^T. To get $\mathbf{u}_1$

and $\mathbf{u}_2$, compute $A^T\mathbf{v}_1$ and $A^T\mathbf{v}_2$,

$$A^T[\mathbf{v}_1 \quad \mathbf{v}_2] = \begin{bmatrix} 3 & 2 \\ 2 & 3 \\ 2 & -2 \end{bmatrix} \begin{bmatrix} 1/\sqrt{2} & -1/\sqrt{2} \\ 1/\sqrt{2} & 1/\sqrt{2} \end{bmatrix} = \begin{bmatrix} 5/\sqrt{2} & -1/\sqrt{2} \\ 5/\sqrt{2} & 1/\sqrt{2} \\ 0 & -4/\sqrt{2} \end{bmatrix}$$

and normalize:

$$\mathbf{u}_1 = \begin{bmatrix} 1/\sqrt{2} \\ 1/\sqrt{2} \\ 0 \end{bmatrix}, \quad \mathbf{u}_2 = \begin{bmatrix} -1/\sqrt{18} \\ 1/\sqrt{18} \\ -4/\sqrt{18} \end{bmatrix}$$

We need one more vector, orthogonal to $\mathbf{u}_1$ and $\mathbf{u}_2$. So write the equations $\mathbf{u}_1^T\mathbf{x} = 0$ and $\mathbf{u}_2^T\mathbf{x} = 0$ and solve for $\mathbf{x}$. Simpler equations are

$$\begin{aligned} \sqrt{2}\,\mathbf{u}_1^T\mathbf{x} &= 0 \\ \sqrt{18}\,\mathbf{u}_2^T\mathbf{x} &= 0 \end{aligned} \quad \text{or} \quad \begin{aligned} x_1 + x_2 &= 0 \\ -x_1 + x_2 - 4x_3 &= 0 \end{aligned}$$

The solution is $x_1 = -2x_3$, $x_2 = 2x_3$, x_3 free. A suitable unit vector is

$$\mathbf{u}_3 = \begin{bmatrix} -2/3 \\ 2/3 \\ 1/3 \end{bmatrix}$$

Thus an SVD of A^T is

$$A^T = [\mathbf{u}_1 \quad \mathbf{u}_2 \quad \mathbf{u}_3] \begin{bmatrix} 5 & 0 \\ 0 & 3 \\ 0 & 0 \end{bmatrix} [\mathbf{v}_1 \quad \mathbf{v}_2]^T$$

So an SVD of A appears by taking transposes:

$$A = \begin{bmatrix} 1/\sqrt{2} & -1/\sqrt{2} \\ 1/\sqrt{2} & 1/\sqrt{2} \end{bmatrix} \begin{bmatrix} 5 & 0 & 0 \\ 0 & 3 & 0 \end{bmatrix} \begin{bmatrix} 1/\sqrt{2} & 1/\sqrt{2} & 0 \\ -1/\sqrt{18} & 1/\sqrt{18} & -4/\sqrt{18} \\ -2/\sqrt{3} & 2/3 & 1/3 \end{bmatrix}$$

This *is* an SVD because the outside matrices are orthogonal matrices, and the center matrix is a diagonal matrix of the proper type. Another way to find $\mathbf{u}_3$ is to realize that $\mathbf{u}_1$ and $\mathbf{u}_2$ form an orthonormal basis for Col A^T = Row A. The remaining $\mathbf{u}_3$ must be a basis for (Row A)$^\perp$ = Nul A.

Helpful Hint: The last remark in the solution of Exercise 13 applied to A^T because the main SVD construction was for A^T. In an SVD for A, the missing vectors $\mathbf{u}_{r+1}, \ldots, \mathbf{u}_n$ form an orthonormal basis for Nul A^T. (See Fig. 4 on page 479.) One way to obtain $\mathbf{u}_{r+1}, \ldots, \mathbf{u}_n$ is to construct a basis for the solution set of $A^T\mathbf{x} = \mathbf{0}$ and then perform the Gram-Schmidt process.

Study Tip: Exercises 15 and 16 make good exam questions.

19. Let $A = U\Sigma V^T$. Then

$$A^TA = (U\Sigma V^T)^T U\Sigma V^T = V\Sigma^T U^T U\Sigma V^T$$
$$= V(\Sigma^T\Sigma)V^{-1} \qquad \text{Because } U \text{ and } V \text{ are orthogonal}$$

If $\sigma_1, \ldots, \sigma_r$ are the nonzero diagonal entries in Σ, then $\Sigma^T\Sigma$ is diagonal, with diagonal entries $\sigma_1^2, \ldots, \sigma_r^2$ and possibly some zeros. Thus V diagonalizes A^TA. By the Diagonalization Theorem in Section 5.3, the columns of V are eigenvectors of A^TA, and $\sigma_1^2, \ldots, \sigma_r^2$ are the nonzero eigenvalues of A^TA. Hence $\sigma_1, \ldots, \sigma_r$ are the nonzero singular values of A. A similar calculation of AA^T shows that the columns of U are eigenvectors of AA^T.

23. From the proof of Theorem 10, $U\Sigma = [\sigma_1\mathbf{u}_1 \quad \cdots \quad \sigma_r\mathbf{u}_r \quad \mathbf{0} \quad \cdots \quad \mathbf{0}]$. The column-row expansion of a matrix product shows that

$$A = (U\Sigma)V^T = (U\Sigma)\begin{bmatrix} \mathbf{v}_1^T \\ \vdots \\ \mathbf{v}_n^T \end{bmatrix} = \sigma_1\mathbf{u}_1\mathbf{v}_1^T + \cdots + \sigma_r\mathbf{u}_r\mathbf{v}_r^T$$

This expansion generalizes the spectral decomposition in Section 7.1.

25. Consider the SVD for the standard matrix of T, say, $A = U\Sigma V^T$. Let $\mathcal{B} = \{\mathbf{v}_1, \ldots, \mathbf{v}_n\}$ and $\mathcal{C} = \{\mathbf{u}_1, \ldots, \mathbf{u}_m\}$ be bases constructed from the columns of V and U, respectively. Observe that, since the columns of V are orthonormal, $V^T\mathbf{v}_j = \mathbf{e}_j$, where $\mathbf{e}_j$ is the jth column of the $n \times n$ identity matrix. To find the matrix of T relative to $\mathcal{B}$ and $\mathcal{C}$, compute

$$T(\mathbf{v}_j) = A\mathbf{v}_j = U\Sigma V^T\mathbf{v}_j = U\Sigma\mathbf{e}_j = U\sigma_j\mathbf{e}_j = \sigma_j U\mathbf{e}_j = \sigma_j\mathbf{u}_j$$

So $[T(\mathbf{v}_j)]_\mathcal{C} = \sigma_j\mathbf{e}_j$. Formula (4) in the discussion at the beginning of Section 5.4 shows that the "diagonal" matrix Σ is the matrix of T relative to $\mathcal{B}$ and $\mathcal{C}$.

MATLAB   The Singular Value Decomposition

The command `[P  D]  =  eig(A'*A)` produces an orthogonal matrix P of eigenvectors and a diagonal matrix D of eigenvalues of A^TA, but the eigenvalues in D may not be in decreasing order. In such a case, you will have to rearrange things to form V and Σ (denoted below by S). For instance, if P is 3×3, the command

```
V  =  P(:,[1 3 2])
```

interchanges columns 2 and 3 of P to form V. The commands

```
S = zeros(size(A));   S(2,2) = sqrt(D(3,3))
```

produce a zero matrix for "Σ" the same size as A and place the square root of the (3,3)-entry of D into the (2,2)-entry of S. Other diagonal entries for S can be entered similarly. To form U for the SVD, first normalize the nonzero columns of $A*V$ and place them in a matrix U. If U is square, you are finished. If U is not square, the missing columns must form an orthonormal basis for Nul A^T. (See Fig. 4 on page 479.) The command `null(A')` produces this orthonormal basis. Thus, the square matrix U is given by

```
U = [U  null(A')]
```

This construction of the SVD helps you to think about properties of the factorization. In practical work, however, you should use the much faster and more numerically reliable command `[U S V] = svd(A)`.

7.5   APPLICATIONS TO IMAGE PROCESSING AND STATISTICS

If you find remote sensing or image processing interesting, or if you plan to use multivariate statistics later in your career, then you will want to study this section thoroughly. You may have difficulty finding an elementary explanation of this material elsewhere. The idea for the application to image processing came from a student in my linear algebra class—a geography major who was taking an undergraduate course in remote sensing. The book by Lillesand and Kiefer, referenced in the text, was one of the texts for her course.

KEY IDEAS

The **first principal component** of the data in the matrix of observations is a unit eigenvector $\mathbf{u}_1$ corresponding to the largest eigenvalue of the covariance matrix S. If $\mathbf{u}_1 = (c_1, \ldots, c_p)$, then the entries in $\mathbf{u}_1$ are weights in a linear combination of the original variables, $x_1, \ldots, x_p$, that creates a new variable y_1 (sometimes called a composite score or *index*):

$$y_1 = \mathbf{u}_1^T \mathbf{X} = c_1 x_1 + \cdots + c_p x_p$$

The variance of the values of this index is the largest possible among all indices whose coefficients $c_1, \ldots, c_p$ form a unit vector. (The variance of y_1 is the largest eigenvalue of S.) The **second principal component** is the unit eigenvector corresponding to the second largest eigenvalue of S. The entries in the second principal component determine the index with greatest variance among all possible indices (determined by a unit vector) that are uncorrelated (in a statistical sense) with y_1. Additional principal components are defined similarly.

Checkpoints: (1) If the variables x_1 and x_3 are uncorrelated, what can you say about the covariance matrix S? (2) What is the covariance matrix of the new variables $y_1, \ldots, y_p$ formed from the principal components of S?

SOLUTIONS TO EXERCISES

1. The matrix of observations is $X = \begin{bmatrix} 19 & 22 & 6 & 3 & 2 & 20 \\ 22 & 6 & 9 & 15 & 13 & 5 \end{bmatrix}$, and the sample mean **M** is

$\begin{bmatrix} 12 \\ 10 \end{bmatrix}$. Subtract **M** from each column of X to obtain

$$B = \begin{bmatrix} 7 & 10 & -6 & -9 & -10 & 8 \\ 2 & -4 & -1 & 5 & 3 & -5 \end{bmatrix}$$

The sample covariance matrix is

$$S = \frac{1}{N-1}BB^T = \frac{1}{5}\begin{bmatrix} 7 & 10 & -6 & -9 & -10 & 8 \\ 2 & -4 & -1 & 5 & 3 & -5 \end{bmatrix}\begin{bmatrix} 7 & 2 \\ 10 & -4 \\ -6 & -1 \\ -9 & 5 \\ -10 & 3 \\ 8 & -5 \end{bmatrix}$$

$$= \frac{1}{5}\begin{bmatrix} 430 & -135 \\ -135 & 80 \end{bmatrix} = \begin{bmatrix} 86 & -27 \\ -27 & 16 \end{bmatrix} \quad \text{Usually, } S \text{ contains decimals.}$$

Study Tip: Note that the formula for the sample mean involves division by N, but for statistical reasons, the covariance matrix formula involves division by $N-1$.

7. Let x_1, x_2 denote the variables for the two-dimensional data in Exercise 1. The characteristic equation of the covariance matrix S from Exercise 1 is $\lambda^2 - 102\lambda + 647 = 0$. By the quadratic formula, the roots of this equation are $\lambda_1 = 95.20$ and $\lambda_2 = 6.80$ (to two decimal places). The first principal component of the data is a unit eigenvector corresponding to λ_1, which turns out to be $(-.95, .32)$, or $(.95, -.32)$. The two possible choices for the new variable are $y_1 = -.95x_1 + .32x_2$ and $y_1 = .95x_1 - .32x_2$. The variance of y_1 is 95.20, while the total variance is $95.20 + 6.80 = 102$. Since $95.20/102 = .933$, the new variable y_1 explains about 93.3% of the variance in the data.

11. a. The solution in the text shows that the $\mathbf{Y}_k$ are in mean-deviation form, where $\mathbf{Y}_k = P\,\mathbf{X}_k$ for some $p \times p$ matrix P.

b. By part (a), the covariance matrix of $\mathbf{Y}_1, \ldots, \mathbf{Y}_N$ is

$$\frac{1}{N-1}[\mathbf{Y}_1 \quad \cdots \quad \mathbf{Y}_N][\mathbf{Y}_1 \quad \cdots \quad \mathbf{Y}_N]^T$$

$$= \frac{1}{N-1}P^T[\mathbf{X}_1 \quad \cdots \quad \mathbf{X}_N]\left(P^T[\mathbf{X}_1 \quad \cdots \quad \mathbf{X}_N]\right)^T$$

$$= P^T\left(\frac{1}{N-1}[\mathbf{X}_1 \quad \cdots \quad \mathbf{X}_N][\mathbf{X}_1 \quad \cdots \quad \mathbf{X}_N]^T\right)P$$

$$= P^T S P$$

because $\mathbf{X}_1, \ldots, \mathbf{X}_N$ are in mean-deviation form.

13. Let $\mathbf{M}$ be the sample mean of the data, and for $k = 1, \ldots, N$, write $\hat{\mathbf{X}}_k$ for $\mathbf{X}_k - \mathbf{M}$. Let $B = \left[\hat{\mathbf{X}}_1 \quad \cdots \quad \hat{\mathbf{X}}_N\right]$, the matrix of observations in mean deviation form. By the column-row expansion of BB^T, the sample covariance matrix is

$$S = \frac{1}{N-1}BB^T = \frac{1}{N-1}[\hat{\mathbf{X}}_1 \quad \cdots \quad \hat{\mathbf{X}}_N]\begin{bmatrix} \hat{\mathbf{X}}_1^T \\ \vdots \\ \hat{\mathbf{X}}_N^T \end{bmatrix}$$

$$= \frac{1}{N-1}\sum_1^N \hat{\mathbf{X}}_k \hat{\mathbf{X}}_k^T = \frac{1}{N-1}\sum_1^N (\mathbf{X}_k - \mathbf{M})(\mathbf{X}_k - \mathbf{M})^T$$

Answers to Checkpoints: (1) The (1, 3)-entry and (3, 1)-entry of S are zero. (2) The covariance matrix of $y_1, \ldots, y_p$ is the diagonal matrix formed from the eigenvalues of S. This matrix is diagonal because the new variables are pairwise uncorrelated.

MATLAB   Computing Principal Components

The command **mean(X')** produces a row vector whose jth entry lists the average of the jth row of X, and **diag(mean(X'))** creates a diagonal matrix whose diagonal entries are the row averages of X. (Be careful not to use **mean(X)**, which lists the averages of the *columns* of X.) Finally, the command **diag(mean(X'))*ones(size(X))** creates a matrix the size of X, whose columns are all the same, each one listing the row averages of X. To convert the data in X into mean-deviation form, use

```
B = X - diag(mean(X'))*ones(size(X))
```

The sample covariance matrix is produced by

```
S = B*B'/(N-1)
```

The principal component data is produced by

```
[U,D,V] = svd(B'/sqrt(N-1))
```

The columns of V are the principal components of the data, and the diagonal entries of `D^2` list the variances of the new variates.

CHAPTER 7   SUPPLEMENTARY EXERCISES _____

7. If $A = R^T R$, where R is invertible, then A is positive definite, by Exercise 25 in Section 7.2. Conversely, suppose that A is positive definite. Then by Exercise 26 in Section 7.2, $A = B^T B$ for some positive definite matrix B. Since the eigenvalues of B are positive, 0 is not an eigenvalue and so B is invertible. In particular, the columns of B are linearly independent. By Theorem 12 in Section 6.4, $B = QR$ for some $n \times n$ matrix Q with orthonormal columns and some upper triangular matrix R with positive elements on its diagonal. Since Q is square, $Q^T Q = I$. So

$$A = B^T B = (QR)^T (QR) = R^T Q^T QR = R^T R$$

and R has the required properties.

11. Start with an SVD decomposition, $A = U\Sigma V^T$. Since U is orthogonal, $U^T U = I$, and so $A = U\Sigma U^T UV^T = PQ$, where $P = U\Sigma U^T = U\Sigma U^{-1}$ and $Q = UV^T$. The matrix P is symmetric, because Σ is symmetric, and P has nonnegative eigenvalues because it is similar to Σ (which is diagonal with nonnegative entries). Thus P is positive semidefinite. The matrix Q is orthogonal because it is the product of orthogonal matrices.

CHAPTER 7   GLOSSARY CHECKLIST _____

Check your knowledge by attempting to write definitions of the terms below. Then compare your work with the definitions given in the text's Glossary. Ask your instructor which definitions, if any, might appear on a test.

condition number (of A):   The quotient σ_1/σ_n, where . . . .

covariance (of variables x_i and x_j, for $i \neq j$):   The entry in the covariance matrix S for a matrix of observations, where x_i and x_j vary over the . . . coordinates, respectively of the observation vectors.

covariance matrix (or **sample covariance matrix**):   The $p \times p$ matrix S defined by $S = \ldots$, where B is a $p \times N$ matrix of observations . . . .

indefinite matrix: A symmetric matrix A such that . . . .

indefinite quadratic form: A quadratic form Q such that $Q(\mathbf{x})$ . . . .

left singular vectors (of A):   The columns of . . . in the singular value decomposition $A = \ldots$.

matrix of observations: A $p \times N$ matrix whose columns are . . . , each column listing p measurements made on . . . .

mean-deviation form (of a matrix of observations):   A matrix whose . . . vectors are . . . .

Moore-Penrose inverse: *See* pseudoinverse.

negative definite matrix: A symmetric matrix A such that . . . .

negative definite quadratic form: A quadratic form Q such that $Q(\mathbf{x})$ . . . .

negative semidefinite matrix: A symmetric matrix A such that . . . .

negative semidefinite quadratic form: A quadratic form Q such that . . . .

orthogonally diagonalizable: A matrix A that admits a factorization, $A = PDP^{-1}$, with P . . . and D . . . .

positive definite matrix: A symmetric matrix A such that . . . .

positive definite quadratic form: A quadratic form Q such that $Q(\mathbf{x})$ . . . .

positive semidefinite matrix: A symmetric matrix A such that . . . .

positive semidefinite quadratic form: A quadratic form Q such that . . . .

principal axes (of a quadratic form $\mathbf{x}^T A\mathbf{x}$):   The orthonormal columns of an orthogonal matrix P such that . . . .

principal components (of the data in a matrix of observations B):   The . . . eigenvectors of a sample covariance matrix S for B, with the eigenvectors arranged so that the corresponding . . . .

projection matrix (or **orthogonal projection matrix**):   A symmetric matrix B such that . . . . A simple example is $B = \ldots$.

pseudoinverse (of A):   The matrix . . . , when UDV^T is a reduced singular value decomposition of A.

quadratic form: A function Q defined for $\mathbf{x}$ in $\mathbb{R}^n$ by $Q(\mathbf{x}) = \ldots$, where A is an $n \times n$ . . . matrix A (called the matrix of the quadratic form).

reduced singular value decomposition: A factorization $A = \ldots$, for an $m \times n$ matrix A of rank r, where U is ___ $\times$ ___ with orthonormal columns, D is ___ $\times$ ___ with . . . , and V is ___ $\times$ ___ with orthonormal columns.

right singular vectors (of A):   The columns of . . . in the singular value decomposition $A = \ldots$.

row sum: The sum of the entries . . . .

sample mean: The average M of a set of vectors, $\mathbf{X}_1, \ldots, \mathbf{X}_N$, given by $\mathbf{M} = \ldots$.

singular value decomposition (of an $m \times n$ matrix A): $A = \ldots$, where U is an __ $\times$ __ $\ldots$
  matrix, V is an __ $\times$ __ $\ldots$ matrix, and Σ is an __ $\times$ __ $\ldots$ matrix with $\ldots$.

singular values (of A):   The $\ldots$ of the eigenvalues of $\ldots$, arranged $\ldots$.

spectral decomposition (of A):   A representation $A = \ldots$, where $\ldots$.

symmetric matrix: A matrix A such that $\ldots$.

total variance: The $\ldots$ of the covariance matrix S of a matrix of $\ldots$.

uncorrelated variables: Any two variables x_i and x_j (with $i \neq j$) that range over the ith and jth
  coordinates of the observation vectors in an observation matrix, such that $\ldots$.

variance of a variable x_j: The diagonal entry $\ldots$ in the $\ldots$ matrix S for a matrix of
  observations, where x_j varies over the jth coordinates of the $\ldots$.

Technology Index of Procedures and Terms

The references here are to sections (not pages) that contain information about the use of technology with various procedures and objects considered in the text exercises. For MATLAB, consult the MATLAB boxes in the main Study Guide sections. For Maple, Mathematica, or the TI and HP calculators, consult the corresponding sections in the appropriate technology appendix.

Access exercise data, 1.1, 1.3
Augment a matrix, 1.3
Change-of-coordinates matrix, 4.7
Column of A, 2.1
Compare two vectors or matrices, 2.3
Compute a determinant, 3.2
Compute a product of numbers, 3.2
Condition number, 2.2
Construct a covariance matrix, 7.5
Construct an inverse, 2.2
Construct a matrix, 1.3, 2.4
Diagonalize a matrix, 5.3, 7.1
Display on screen, see format, 1.1, 5.8
Entry (in a matrix), 1.1, 2.1
Entrywise computation, 4.1, 5.6, 5.7, 6.6
Evaluate a polynomial at many points, 5.2
Exercise data, 1.1, 1.3
Extract diagonal entries, 3.2
Find a basis for a column space, 4.3
Find a basis for a null space, 4.3
Find a basis for an eigenspace, 5.1, 7.1
Find a characteristic polynomial, 5.2
Find eigenvalues and eigenvectors, 5.1
Find the zeros of a polynomial, 4.8
Function of a vector, 4.1, 6.6
Generate a sequence, 1.10, 5.6
Gram-Schmidt process, 6.4
Graph a polynomial, 5.2
Graph several functions, 4.1, 5.2, 6.8
Graph a trajectory, 5.6
Identity matrix, 2.1, 2.2
Inner product, 2.1, 6.1
Inverse power method, 5.8

LU factorization, 2.5
Matrix notation and operations, 2.1
Matrix power, 2.1
Matrix-vector multiplication, 1.4
Numbers in MATLAB,
  displayed, 1.1
  rational, 1.6, 5.1
Orthogonal projection, 6.2, 6.3, 6.4
Outer product, 2.1
Partitioned matrix, 2.4
Permuted LU factorization, 2.5
Power of a vector function, 4.1
Power method, 5.8
QR factorization, 6.4
Random matrix, 2.1, 2.3
Recall previous commands, 1.3, 1.10
Reduced echelon form, 2.8, 2.9, 4.3
Row of A, 2.1
Row operations, 1.1
Row reduction, 1.4
Scientific notation, 1.10
Select part of a matrix, 2.1
Sequence, generate, 1.10, 5.6
Singular value decomposition, 7.4
Solve $Ax = 0$, 1.5
Solve $Ax = b$, 1.4, 2.5
Special matrices, 2.1
  Hilbert matrix, 2.3
Trajectory, 5.6
Test for orthogonality, 6.2
Transpose, 2.1

Introduction to MATLAB

MATLAB stands for MATrix LABoratory. Originally written by Cleve Moler for college linear algebra courses, MATLAB has evolved into the premier software for linear algebra computations in science and industry, all over the world. Using MATLAB in this course will save you time on homework, help you learn linear algebra, and give you a glimpse of how linear algebra is applied in practical work.

At the ends of many *Study Guide* sections, MATLAB "boxes" will show you how to avoid routine arithmetic calculations, so you can focus on new concepts. Later in the course, you will run several state-of-the-art programs. For instance, given a 32×32 matrix A (which is small by today's standards), MATLAB's eigenvalue program provides important and accurate information about A in a fraction of a second, after performing about one million arithmetic operations.

GETTING STARTED WITH MATLAB

You need access to the MATLAB program, either on a personal computer (PC or Mac) or on a shared computer network system at your school. If the school network is prepared for your course, someone may have already installed special MATLAB programs written for this course, along with data for about 850 exercises in the text. Accessing a matrix or some vectors for a problem requires only a few keystrokes. (This avoids data entry error and saves you time.) And then MATLAB performs the computations for you. Once you see how much time you can save, you'll probably start using MATLAB for most of the numerical exercises!

If you plan to work at a computer away from school, you can purchase a Student Version of MATLAB at modest cost. You need to download the special programs and exercise data from the Web, at http://www.laylinalgebra.com. These programs and data are in a folder called the Laydata Toolbox. Download the folder into the main MATLAB folder and then add the Laydata Toolbox to the MATLAB PATH. A ReadMe file in the Toolbox has instructions. When that is done, MATLAB will have immediate access to everything in the Toolbox.

The next four paragraphs provide basic information about the MATLAB environment, including how to save and print a copy of your work. After that, the MATLAB boxes in this *Guide* will form your "lab manual" for the use of MATLAB during the course. (Begin with Section 1.1.) For your reference, an index of commands is included at the end of this appendix.

How to Start, Stop, And Run MATLAB

Once you start MATLAB (by clicking on a menu or icon), you will see the MATLAB logo as the program loads, followed by a prompt symbol ←, or EDU←, in what is called the *command window*. Recent versions of MATLAB have two other windows, which you may close or resize. The *Study Guide* focuses mainly on the command window. Type a MATLAB command at the prompt and press ⟨Enter⟩. MATLAB's response will follow. Here are some simple commands:

x=2*3	Sets up a variable *x* in your "workspace" and stores 6 in it. The symbols +, –, *, / , and ^ (for exponents) are used for arithmetic.
X = 5*x	Sets up a new variable with the value 30. MATLAB is case-sensitive. Extra spaces are ignored. They were inserted here for readability.
A = [1 2 3;4 –5 6]	Creates a matrix with 1, 2, 3 in the first row and 4, –5, 6 in the second row. One or more spaces separate entries in each row.
c1s1	Requests data for exercises in Chapter 1 Section 1, for example. See the MATLAB box at the end of *Study Guide* notes for Section 1.1.
who	Lists the variables in your workspace (a portion of computer memory that stores the variables created since you turned on the program).
clear	Removes all variables (and their contents) from the workspace.
help clear	Tells you about the **clear** command.
help	Displays a long list of topics for which help is available.
quit	Immediately clears the workspace and terminates MATLAB. **exit** does the same.

You can place several commands on one line, if you wish, separated by commas or semicolons. A semicolon after a command instructs MATLAB not to display the result of the command.

How To Format Matlab Output

Ordinarily, MATLAB displays up to 5 digits for each number and places a line feed between each object it displays. The following commands allow you to modify this standard display style.

format compact	Suppresses extra line-feeds, so you can see more output on one screen. Many students turn this on at the beginning of a work session.
format long	Displays all 15 digits that were used in the calculation.
format short	Displays only 5 digits.
format rat	Displays numbers as fractions, as accurately as possible.
format	Restores the standard display style.

Use **help format** to see other formatting options.

How To Save and Print a Copy of Your Work

You can save a record of every keystroke you make, along with MATLAB's responses, in a file you choose by turning on the "diary" feature. For instance, if you select the filename project1.txt, use the commands:

diary project1.txt	Writes output (following this command) to the file named. If the file already exists, the output is appended to the file.
diary off	Stops the recording
diary on	Starts writing output again to Project1.txt, named earlier. Places new output at the end of the file.

The file project1.txt will be located in the "current directory," which initially is the main directory that holds MATLAB and all of its subdirectories. You can change this directory, but the easiest way to control where your file goes is to include a path at the front of the name, for instance, **diary a:\project1.txt** . If you are working on a school network, the designation for the floppy disk drive might be something other than **a:** . MATLAB version 6 displays the current directory in a toolbar at the top of the screen; you can use that to change the current directory.

Once your output is in a text file, you can use any convenient text editor to add comments to your work, delete extraneous output or errors in typing, and then print the file. If you want to print while working on a network, you may need to find out what special network commands are required.

This concludes the main part of the MATLAB appendix. You should be ready to use MATLAB with your homework (as soon as the Laydata Toolbox is installed). Enjoy yourself!

SCRIPT M-FILES

When you have some experience using MATLAB, you may wish to experiment with script m-files to prepare homework assignments and projects. The diary process works reasonably well for short homework assignments. But, for longer projects, the diary file can be quite long and require extensive editing, particularly if you make multiple attempts to get the project working

A script M-file is simply a sequence of normal MATLAB commands stored in a text file that has ".m" as its extension. Instead of running these commands at the command prompt, you type the name of the script M-file at the prompt and press ⟨Enter⟩. For example, if you name your script file project1.m, and place it in the MATLAB working directory, then **project1** becomes a new MATLAB "command." Entering **project1** at the prompt causes the commands in the file to be executed, as if you had entered them one at a time.

The main advantage of using a script M-file instead of the diary to create a project report is that errors in the script can be corrected easily and the entire script run again with no effort. If you discover a mistake in the middle of a project that is being recorded by the diary, you have to repeat all the commands. (Later, you'll also have to edit out the first set of commands that did not work properly.) A second reason for using a script is that you can reuse it later if you need to solve a similar problem. Just make a few appropriate edits and you are ready to go.

How to Create a Script M-file

M-files are ordinary text files that you can create and modify with any text editor or word processor that can save a file as plain ASCII text. On Unix systems, this includes pico, vi, textedit, and emacs. On Windows and Macintosh systems, MATLAB comes with its own M-file Editor/Debugger, which you access from the toolbar or the `File` menu. The first line of a script M-file should be

`echo on`   Displays the commands along with results. Without this command, only the results will be displayed

Place the MATLAB commands next. The last line of the file should be

`echo off`   Turns off the previous echo command

Remember to place a `.m` extension on your file, and save it in the current working directory.

How to Prepare a Final Report

When the script M-file is running satisfactorily, you need to document with your work with comments, which are statements in the file that begin with a percent sign. Comments are used to outline the steps in your calculations and to interpret the results of various computations. For example:

`%The system is inconsistent, because 0 = 5/2 is not true.`

MATLAB ignores all such comments, but your instructor will appreciate them.

To create a printed copy of your work, use the diary command. If your file is project1.m, enter the following commands at the command prompt:

`diary project1.txt`   Records all commands and output that follow.

`problem1`   Executes the commands in project1 and displays all comments in the file on the screen. The diary captures these comments as well

`diary off`   Stops the recording.

Now you have a text file that shows all your work. Depending on the requirements of your course, you can either print the file for submission or attach the file to an email.

Index of MATLAB commands

Notes for the Maple
Computer Algebra System

GETTING STARTED WITH MAPLE

Using Maple with *Linear Algebra and Its Applications*

The Maple commands for most linear algebra operations, including the manipulation of vectors and matrices, can be found in the `linalg` package that comes with Maple. The `laylinalg` package contains Maple implementations of the special MATLAB commands presented in the text, data for many problems in the text, and data for the Maple Projects. The current version of the `laylinalg` package can be used in either Maple 7 or Maple 8 and can be downloaded from `http://www.laylinalgebra.com/`.

To make the `laylinalg` package accessible within Maple it is necessary to expand the downloaded archive and to instruct Maple where to find these files. First, unzip (or unstuff if you are using a Macintosh computer) the archive containing the `laylinalg` package into a new folder named `laylinalg`. When this has been successfully completed, the `laylinalg` folder will contain three files: `maple.hdb`, `maple.ind`, and `maple.lib`. (The fourth file in the archive, `MapleReadMe.txt`, contains a more detailed version of the information in this paragraph.) To instruct Maple where to find the `laylinalg` package, the full pathname of this new directory must be added to the list of directories in `libname`. For example, if the full pathname is `C:\\laylinalg` then the command

```
> libname := "C:\\laylinalg", libname:
```
must be executed. To ensure these steps are performed every time you use Maple, this command should be placed in a Maple initialization file. The `MapleReadMe.txt` file contains additional information about Maple initialization files; see also the Maple help page produced by the command `?mapleinit`. If you experience any difficulties with this installation process, please check with your instructor for specific instructions for your site.

To load a Maple "package" into a Maple session, use the `with` command. The command

```
> with( laylinalg );
```
loads the `laylinalg` package and displays all of the commands in the package. This particular command also loads Maple's own `linalg` package and modifies the display of Maple vectors so they appear as $n \times 1$ column matrices.[1]

Each Maple command must end with a semicolon or colon. The semicolon tells Maple to display the result of the command; a colon suppresses the output. Maple is case-sensitive but ignores all whitespaces (blanks) around numbers, names, keywords, and operators in com-

---

[1] To have vectors displayed horizontally, execute the command `pvac := false;`. The command `pvac` stands for print vectors as columns.

mands. For example, Maple will not recognize `with( LayLinAlg );` but will understand `with(laylinalg);`. The commands in these notes include extra spaces to increase readability.

Entering Matrices in Maple

There are many ways to create matrices in Maple. In general, a matrix is created with the `matrix` command. The first two arguments of this command are the number of rows and columns in the matrix, respectively. If there are only two arguments, then the elements of the matrix are unassigned (see matrix A, below). The optional third argument is used to specify the entries of the matrix. If the third argument is a list or vector, these quantities are used to fill the matrix row-by-row from left-to-right (see matrices B and C). The third argument can also be a Maple function of two variables. In this case each element of the matrix is assigned value by evaluating the function with the corresponding row and column indices (see matrix E). Diagonal matrices are easily defined using the `diag` command from the `linalg` package (see matrix F).

```
> A := matrix( 2, 3 );
```
$$A := \text{array}(1 \,..\, 2,\, 1 \,..\, 3,\, [\,])$$

```
> evalm( A );
```
$$\begin{bmatrix} A_{1,1} & A_{1,2} & A_{1,3} \\ A_{2,1} & A_{2,2} & A_{2,3} \end{bmatrix}$$

```
> B := matrix( 2, 3, [1,2,3,4,5,6] );
```
$$B := \begin{bmatrix} 1 & 2 & 3 \\ 4 & 5 & 6 \end{bmatrix}$$

```
> C := matrix( 2, 3, [1,2,3,4] );
```
$$C := \begin{bmatrix} 1 & 2 & 3 \\ 4 & C_{2,2} & C_{2,3} \end{bmatrix}$$

```
> E := matrix( 2, 3, (i,j) -> 1/(i+j) );
```
$$E := \begin{bmatrix} \frac{1}{2} & \frac{1}{3} & \frac{1}{4} \\ \frac{1}{3} & \frac{1}{4} & \frac{1}{5} \end{bmatrix}$$

```
> F := diag( 1, 2, 3 );
```
$$F := \begin{bmatrix} 1 & 0 & 0 \\ 0 & 2 & 0 \\ 0 & 0 & 3 \end{bmatrix}$$

The `entermatrix` command, from the `linalg` package, provides a somewhat more user-friendly interface for entering the entries of a matrix:

> G := matrix(2,3): # define the matrix G

> entermatrix(G);
enter element 1,1 > 1; # note that the semicolon is required
enter element 1,2 > 2;
enter element 1,3 > 3;
enter element 2,1 > 4;
enter element 2,2 > 5;
enter element 2,3 > 6;

A more graphical but seldom-used method for defining a matrix with between two and four rows and two and four columns is to use the Matrix Palette. You are strongly advised to not use the Matrix and Vector Palettes with the `laylinalg` package. The commands produced by the Matrix and Vector Palettes create Maple objects with type `Matrix` and `Vector`. These objects are designed for use with Maple's `LinearAlgebra` package, not the type `matrix` and `vector` objects used by the `linalg` and `laylinalg` packages.

Using Maple's Online Help

These notes are intended to be self-contained. If, however, you are interested in additional information, you should first consult Maple's online help. The command `help( laylinalg )`; or, equivalently, `?laylinalg` accesses the help worksheet with an overview the `laylinalg` package. Help worksheets exist for all Maple commands, including all commands in the `linalg` and `laylinalg` packages. For example, the command `?gauss` accesses the help worksheet for the `gauss` command in the `laylinalg` package.

The `Help` menu on the Maple user interface can also be used to search for help on a specific keyword. A `Full Text Search` lists all help worksheets containing keywords that you specify. A `Topic Search` interactively shows all help worksheets whose topic matches the search string. For example, there is a long list of help topics that begin with the string `la`. When the search string is expanded to `lay` the list of matching help topics reduces to all of the help worksheets for the `laylinalg` package.

STUDY GUIDE NOTES

At the beginning of each Maple session for this linear algebra course, enter the command:
```
> with( laylinalg );
```
Notice the use of a colon to suppress the output.

SECTION 1.1   Row Operations

The data for the exercises in this section are contained in the `laylinalg` package. The list of exercises within chapter 1 section 1 for which Maple data is available is displayed by executing the command: `c1s1();`. If Maple simply repeats your input, it is likely that the `with( laylinalg );` command has not been executed.

The data for a specific problem, such as Exercise 13 in Section 1.1, can be loaded with the command: `c1s1( 13 );` or `c1s1(13);`. (The spaces around **13** are optional.) The output from this command displays each assignment that Maple has made. In this section the data for each exercise are stored in a matrix M. To see the current contents of the matrix M, execute the command: `evalm( M );`.

Row operations on M can be performed using commands from either the `laylinalg` or the `linalg` package. Table 1 summarize the commands for the three elementary row operations.

laylinalg Command Syntax	Description
`replace( M, r, m, s );`	row $r \leftarrow$ row $r + m *$ row s
`swap( M, r, s );`	row $r \leftrightarrow$ row s
`scale( M, r, c );`	row $r \leftarrow c *$ row r of M

linalg Command Syntax	Description
`addrow( M, s, r, m );`	row $r \leftarrow$ row $r + m *$ row s
`swaprow( M, r, s );`	row $r \leftrightarrow$ row s
`mulrow( M, r, c );`	row $r \leftarrow c *$ row r of M

Table 1: Comparison of elementary row operations in the `laylinalg` and `linalg` packages.

Notes:

- The syntax and functionality of the commands in the `linalg` and `laylinalg` packages are almost identical. The names in the `linalg` package are used throughout Maple; those used in `laylinalg` more closely match those used elsewhere in the textbook. The `laylinalg` commands will be used throughout this appendix.

- The name of any matrix in your current Maple session can be inserted in place of `M` in the commands above.

- The letters r and s are positive integers corresponding to row numbers. The names m and c are scalar expressions (often numbers, but sometimes symbolic expressions) that you choose.

If you enter one of the row operation commands, say,
```
> swap( M, 1, 3 );
```
then the new matrix produced from M is not given a name. You can refer to this result using %, the history operator. If, instead, you type
```
> M1 := swap( M, 1, 3 );
```
then the answer is stored in the matrix M1. If the next operation is
```
> M2 := replace( M1, 2, 5, 1 );
```
then the result after performing this row operation on M1 is placed in M2, and so on. Note the use of := in *assignments*; the symbol = is used to create an *equation*.

One advantage of giving a new name to each new matrix is that you can easily go back a step if you do not like what you just did to a matrix. If you type
```
> M := replace( M, 2, 5, 1 );
```
then the result is placed back in M and the "old" M is lost. Of course, the "reverse" operation
```
> M := replace( M, 2, -5, 1 );
```
will recreate the original matrix M.

Notes:

- For the problems in this section and the next, the multiple m needed in the **replace** or **addrow** command will usually be a small integer or fraction that you can compute in your head. The next two paragraphs describe how to use Maple in situations where m is not easily computed mentally.

  The entry in row r and column c of a Maple matrix M is denoted by M[r,c]. If the number stored in M[r,c] is a floating-point number, that is, it is displayed with a decimal point, then the number may be accurate to only about eight (8) digits. It is generally more accurate — and simpler — to use M[r,c] instead of re-typing the displayed values in computations.

- For instance, to use the entry M[s,c] to change M[r,c] to 0, enter the commands:
  ```
  > m := -M[r,c] / M[s,c];         # mult of row s to add to row r
  > M1 := replace( M, r, m, s );   # adds m * row s to row r
  ```
  Or, you could use the single command:
  ```
  > M1 := replace( M, r, -M[r,c]/M[s,c], s );
  ```

- The environment variable **Digits** controls the number of significant digits Maple uses in floating-point computations. The default is 10. While this should be more than sufficient for the problems in this text, you can change this by assigning a new value to **Digits**. In general, you should expect a floating-point computation to be accurate to approximately **Digits**-2 digits.

SECTION 1.1   Symbolic Row Operations

A powerful feature of Maple is its ability to work with matrices and vectors that involve unspecified parameters. For example, Exercise 25 in Section 1.1 can be solved using Maple and
the `laylinalg` package in the following steps

```
> c1s1( 25 );
> evalm( M );                        # display the augmented matrix
> M1 := replace( M, 3, 2, 1 );   # add 2 × row 1 to row 3
> M2 := replace( M1, 3, 1, 2 );  # add row 2 to row 3
> M2[3,4] = 0;                       # eqn to make system consistent
```

Symbolic multipliers can be used in `replace` and `scale`, for example: `replace(A, 3, a,
1);`, but all row numbers must be explicit integers.

SECTION 1.3   Constructing a Matrix

To see the exercises with data for Section 1.3, execute the command:

```
> c1s3( );
```

The data for Exercise 25, for example, consists of the matrix A, the right-hand side vector $\mathbf{b}$,
and the columns of the matrix, $\mathbf{a}_1$, $\mathbf{a}_2$, $\mathbf{a}_3$. The corresponding Maple names, A, b, a1, a2, and
a3, can be assigned by executing the command:

```
> c1s1( 25 );
```

The command

```
> M := augment( a1, a2, a3, b );
```

creates a matrix with the given vectors as its columns. The same matrix could also be created
by

```
> M := augment( A, b );
```

Exercises 11–14, 24–28, and 31 can be solved using the `laylinalg` commands `replace`, `swap`,
and (occasionally) `scale` (or the similar `linalg` commands: `addrow`, `swaprow`, and `mulrow`),
described in the Maple Note for Section 1.1 on page MP-4.

SECTION 1.4   gauss and bgauss

To solve $A\mathbf{x} = \mathbf{b}$, row reduce the augmented matrix

```
> M := augment( A, b );
```

The command

```
> x := vector( [5,3,-7] );
```

creates a column vector, x, with entries 5, 3, and -7. Matrix-vector multiplication is

```
> evalm( A &* x );
```

The **laylinalg** package contains two commands, **gauss** and **bgauss**, that can be used to speed up row reduction of a matrix. Use the command **with(laylinalg);** to load the **laylinalg** package. A ZIP file containing the **laylinalg** package can be downloaded from the WWW.

To make row reduction of $M = [\ A \quad b\ ]$ more efficient, the command **gauss(M, r);** chooses the leading entry in row r of M as a pivot and uses row replacements to create zeros in the pivot column below this pivot entry. The result can be assigned to a name, otherwise the only way to access the result is with the history operator (%). In either case, the result is displayed to the screen.

For the backward phase of row reduction, use **bgauss(M, r);** which selects the leading entry in row r of M as the pivot, and creates zeros in the column above the pivot. Use **scale** to create leading 1's in the pivot positions.

Note:

- Exercise 19 of this section can be solved using symbolic row operations as discussed in the Maple Note for Section 1.1.

SECTION 1.5   Zero Matrices

The command **vector(m, 0);** creates the zero vector in R^m and **matrix(m, n, 0);** creates an $m \times n$ matrix of zeros. To solve an equation $Ax = 0$, use the command
```
> M := augment( A, vector( m, 0 ) );
```
to augment A with the column vector containing m zeros, then use **gauss, swap, bgauss,** and **scale** to complete the row-reduction of M.

SECTION 1.6   Rational and Floating-Point Format

Chemical equation-balance problems are studied best using exact or symbolic arithmetic. This is because the balance variables must be whole numbers (with no round-off allowed). When the chemical equations have integer coefficients — as they do here — Maple will perform exact computations. That is, when **laylinalg** row operations are applied to an integer- or rational-valued matrix, the resulting matrix will have integer or rational entries. No floating-point approximations will be used. Once you find a rational solution of a chemical equation-balance problem, you can multiply the entries in the solution vector by a suitable integer to produce a solution that involves only whole numbers.

Notes:

- To convert the entries of a Maple matrix, M, to floating-point numbers, use either **evalf(evalm(M));** or **map(convert, M, float);**.

- The command **map(convert, M, rational);** returns the matrix in which each entry of the matrix is replaced with an approximately equal rational number.

SECTION 1.10   Generating a Sequence

The data for Exercises 9–13 in Section 1.10 consists of the migration matrix, M, and the initial population vector $\mathbf{x}_0$. The following set of commands generates the first ten (10) terms in the sequence defined by $\mathbf{x}_{k+1} = M\mathbf{x}_k$:

```
> x||0 := evalm(x0);              # display initial population
> for k from 0 to 9 do           # execute loop 10 times
>    x||(k+1) := evalm( M &* x||k );  # compute x_{k+1} from x_k
> end do;
```

Note that the first command reassigns x0 to itself. This command can be omitted if you do not wish to include the initial population in the output.

Numbers are entered in Maple without commas. Maple uses scientific notation for numbers larger than 1,000,000 (10^6) or smaller than 10^{-6}. In Maple, numbers in scientific notation can be entered as `4.0 * 10^5` or `2.46 * 10^(-3)`. Note that the parentheses around a negative exponent are required. When the mantissa is entered as a floating point number, the number is displayed (and computed) with `Digits` digits. The default setting of `Digits`, 10, should be more than sufficient for all computations in this text.

SECTION 2.1   Matrix Notation and Operations

Several methods for defining a matrix in Maple have been described previously in this appendix. Refer to these discussions for information about creating your own matrices.

Recall that, in Maple, the $(i,\ j)$-entry of A is `A[i,j]`. One or more columns or rows of A can be obtained with the `col` or `row` command. For example, `col(A,3)` is column 3 of A and `row(A,2)` is row 2 of A. To extract the first two columns of A, use `col(A,1..2)`. (The expression `a..b` represents the integers from a to b.) Be aware that both `row` and `col` return a sequence of one or more (column) vectors.

Multiple vectors can be assembled as the columns of a new matrix with **augment**, as described in Section 1.4. For example, `augment( col(A,1..2) );`. To create a matrix from the rows of an existing matrix, either take the transpose of the result from **augment** or use the `stackmatrix` command.

The command `submatrix( A, a..b, c..d );` returns a matrix whose entries come from rows a through b and columns c through d of A. The number of rows in A is given by `rowdim(A)` and the number of columns by `coldim(A)`.

Maple uses + and − to denote matrix addition and subtraction, respectively. Multiplication of a matrix or vector by a scalar can be obtained with the commutative multiplication operator `*`; the non-commutative operator `&*` must be used for matrix multiplication, including matrix–vector products. If A is square and k is an integer, `A^k` denotes the kth power of A. The **evalm** command is needed to force Maple to display the result of commands that produce vectors or matrices.

The transpose of A is `transpose(A)`. The outer product of $\mathbf{u}$ and $\mathbf{v}$ can be obtained with the command `evalm( u &* transpose(v) );`. The command `innerprod(u,v);` gives the inner product of **u** and **v**, and this scalar is displayed without using `evalm`. (If the entries of **u** and/or **v** are complex, then you probably want to use `dotprod` instead of `innerprod`; see also the Maple Note for Section 6.1.)

Maple commands for the construction of many special matrices are given below:

```
> M := matrix( 5, 6, 0 );        # a 5 × 6 matrix of zeros
> M := matrix( 3, 5, 1 );        # a 3 × 5 matrix of ones
> M := diag( 3, 5, 7, 2, 4 );    # a 5 × 5 diagonal matrix
> M := diag( 1$6 );              # the 6 × 6 identity matrix
> M := randomint( 6, 4 );        # a 6 × 4 matrix (-9 ≤ entries ≤ 9)
```

SECTION 2.2   The Identity Matrix and A^{-1}

When A is 5×5, the command `M := augment( A, diag( 1$5 ) );` creates the augmented matrix $[A \quad I]$. Use `gauss`, `swap`, `bgauss`, and `scale` to reduce the augmented matrix to $[I \quad A^{-1}]$, if possible. See the Maple Note for Section 1.4 on page MP-6.

There are other Maple commands that row reduce matrices, invert matrices, and solve systems $A\mathbf{x} = \mathbf{b}$. They will be introduced later, after you have studied the concepts and algorithms of this section.

SECTION 2.3   `inverse`, `cond`, and `hilbert`

Determining whether a matrix is invertible is not always a simple matter. A fast and fairly reliable method is to use the command `inverse( A );`, which computes the inverse of A. The error message `Error, (in inverse) singular matrix` is displayed if the matrix is singular.

In Exercises 41–44 of this section, the Maple command `cond( A );` computes the condition number of any matrix A using quantities called the singular values of A (discussed in Section 7.4). To perform the experiment described in Exercise 42 you can use the following Maple commands:

```
> x := randvector( 4 );          # random 4-d vector
> b := evalm( A &* x );          # compute exact RHS
> x1 := evalm( inverse(A) &* b ); # solution using A^-1
> evalm( x - x1 );               # exact - computed
```

Because Maple uses exact arithmetic for integers, **x** and **x1** should be the same. To force Maple to use floating-point computations, replace the definition of **b** with `b := evalf( evalm( A &* x ) );`.

For Exercise 45, the command `hilbert( n );` produces the $n \times n$ Hilbert matrix. To obtain the $n \times n$ Hilbert matrix with floating-point entries use `evalf( hilbert( n ) );`.

SECTION 2.4   Partitioned Matrices

Partitioned matrices can be created in Maple with the **blockmatrix** command. For example, if A, B, C, E, F, and G are matrices of appropriate sizes, then the command
> M := blockmatrix(2, 3, [A,B,C, E,F,G]);

creates a larger matrix of the form

$$M = \begin{bmatrix} A & B & C \\ E & F & G \end{bmatrix}.$$

Once M is formed, there is no record of the partition that was used to create M. For instance, although B is the (1,2)-block used to create M, the value of M[1,2] is the same as the (1,2)-entry of A. You should be aware, however, that if A[1,2] is an unevaluated name at the time the **blockmatrix** command is executed and is subsequently assigned a value, this assignment to A[1,2] changes *all* instances of A[1,2] in the current Maple session. In this case, an assignment of a value to A[1,2] will also change the (1,2)-entry of M.

Other **linalg** commands that can be useful in the creation of new matrices from existing matrices include: **stackmatrix**, **augment**, **concat**, **delrows**, **delcols**, **diag**, **extend**, **minor**, and **submatrix**.

SECTION 2.5   LU Factorization and the **linsolve** Command

Row reduction of A using the command **gauss** will produce the intermediate matrices needed for an LU factorization of A. You can try this on the matrix in Example 2, stored as Exercise 33 of this section in the **laylinalg** package. The matrices in (5) for Example 2 (page 145 of the text) are produced by the commands:
> c2s5(33);   # load matrix A
> U = gauss(A,1); # U has 0's below the first pivot
> U = guass(U,2); # now U has 0's below pivots 1 and 2
> U = gauss(U,3); # the echelon form

You can copy the information from the screen onto your paper, and divide by the pivot entries to produce L as in the text.

To construct a permuted LU factorization, use the Maple command
> U := gauss(U, r, v);

where r is the row index of the pivot and v is a row vector that lists the rows to be changed by replacement operations. For example, if A has 5 rows and the first pivot is in row 4, use U := gauss(A, 4, [1,2,3,5]);. If the next pivot is in row 2, use U := gauss(U, 2, [1,3,5]);. To build the permuted matrix L, use full columns from A or the partially reduced U, divided by the corresponding pivot. Then change entries to zero if they are in a row already selected as a "pivot row".

The Maple command U := LUdecomp(A, L='L', P='P'); produces a permuted LU factorization, $A = PLU$, for any matrix A.

When A is invertible, the simplest way to solve $A\mathbf{x} = \mathbf{b}$ with Maple is to use the `linsolve` command: `x := linsolve(A,b);`. The command `x := inverse(A) &* b;` is less efficient and can be less accurate.

SECTION 2.8   rref

The Maple command `rref(A);` produces the reduced echelon form of the matrix A. From this matrix you can write a basis for Col A or write the homogeneous equations that describe Nul A. (Do not forget that A is a coefficient matrix, not an augmented matrix.)

SECTION 2.9   rank

You can use `rref( A );` to check the rank of a matrix A. For a matrix with floating-point entries, roundoff error or an extremely small pivot can produce an incorrect echelon form. A more reliable command for computing the rank of a matrix A is `rank( A );`.

SECTION 3.2   Computing Determinants

To compute det A, define `U:=A;` and then repeatedly use the commands `U:=gauss(U,r);` and `U:=swap(U,r,s);` as needed to reduce A to an echelon form U (see the Maple Note for Section 1.4 on page MP-6). Then, the determinant of A is given by the command
```
> detA := (-1)^k * mul( U[i,i], i=1..n );
```
where A is an $n \times n$ matrix and k is the number of row swaps made while reducing A to U.

The command `det( A );` can be used to check your work, but the longer sequence of commands is preferred — at this time — because it emphasizes the *process* of computing determinants.

SECTION 4.1   Graphing Functions

The following command can be used to graph the function f in Exercise 37:
```
> plot( f, t=0..2*Pi );
```
In Maple's `plot` command, the first argument is the function (or list of functions) to be plotted and the second argument contains the name of the independent variable (here, t) and an interval on which the plot is to be created. Note that `Pi` is Maple's name for the constant π. Consult Maple's online help for the `plot` command and `plots` package for optional arguments and additional examples.

SECTION 4.3   rref and genmatrix

The Maple command `rref(A)` produces the reduced row echelon form of the matrix A. From this matrix you can write a basis for Col A or write the homogeneous equations that describe Nul A. (Do not forget that A is a coefficient matrix, not an augmented matrix.)

For Exercises such as 37 and 38 the `genmatrix` command can be used to generate the co-efficient matrix A from a list or set of linear equations involving a set of unknowns, say `c[1]`, `c[2]`, ... For example, the coefficient matrix for Exercise 38 can be obtained by the following commands:

```
> EQN := add( c[j] * cos(t)^j, j=0..6 ) = 0;    # genl form of eqn
> eq[0] := eval( EQN, t=0 );                     # eqn w/ t = 0
  ⋮
> eq[6] := eval( EQN, t=Pi );                    # eqn w/ t = π
> A := genmatrix( [ eq[k] $ k=0..6 ],            # coef matrix
>                  [ c[k] $ k=0..6 ] );
```

SECTION 4.4   `linsolve`

The Maple command `linsolve( A, b )`; produces the solution of $Ax = \mathbf{b}$ when A is invertible. The `linsolve` command has additional capabilities that will be introduced later, after you have the appropriate background.

SECTION 4.6   `rref`, `rank` and `randomint`

In this course, you can use either `rref( A )`; or `rank( A )`; to check the rank of a matrix A. In practical work, particularly with floating-point matrices, you should use the more reliable `rank( A )`;.

The `laylinalg` command `randomint( m, n )`; creates an $m \times n$ matrix of random integers between -9 and 9.

Note:

- In earlier editions the `randomint` command was called `randint`. The name change was necessary because MATLAB now has a `randint` command in its Communications Tool-box.

SECTION 4.7   Change-of-Coordinates Matrix

The `laylinalg` package includes data for Exercises 7–10 and 17–19. The Maple command `rref` can be used to row reduce a matrix such as

$$[\; \mathbf{c}_1 \quad \mathbf{c}_2 \quad \mathbf{b}_1 \quad \mathbf{b}_2 \;]$$

to the desired form. The command `augment`, introduced in the Maple Note for Section 1.3 on page MP-6 can be used to create a matrix from a collection of column vectors.

SECTION 4.8   `solve` and `polyroots`

In Exercises 7–16 and 25–28, the coefficients of the polynomial in the auxiliary equation are stored in a vector $\mathbf{p}$. The coefficients are stored in descending order. For example, the polynomial $x^2 + 6x + 9$ would be represented with `p := vector(3,[1,6,9])`;.

The `polyroots` command in the `laylinalg` package computes floating-point approximations to all (real and complex) roots of the polynomial whose coefficients are stored in the vector
p. The syntax is `polyroots( p );`.

Given the coefficient vector **p** the corresponding polynomial can be created with the following commands:

```
> n := vectdim(p);                      # degree of polynomial
> P := add( p[k]*x^(n-k), k=1..n );  # polynomial w/ coefs p
```

The exact roots of the polynomial can now be found using

```
> solve( P=0, x );
```

Note, however, that it might not be possible to find all roots for a polynomial of degree 5 or
higher.

SECTION 4.9

The Maple box for Section 1.10 contains information that is useful for the Exercises in this
section.

SECTION 5.1   Finding Eigenvectors

The `laylinalg` package contains a command that will simplify your homework by automatically
producing a basis for an eigenspace when you know the eigenvalue. For example, if A is a 3×3
matrix with eigenvalue $\lambda = 7$, then the command `nulbasis( A-7*diag(1$3) );` returns a
matrix whose columns form a basis for the corresponding eigenspace. The command `nulbasis(
C);` returns a matrix whose columns form a basis for Nul C (the same basis you would get if
you started with `rref(C);` and computed the basis by hand).

Additional Maple commands for computing eigenvalues and eigenvectors will be discussed
later. You should use `nulbasis` now, to reinforce the basic concepts of this chapter.

For Exercises 37–40 use the `eigenvalues` command to obtain the eigenvalues of a matrix.
For example, the commands

```
> lambda := eigenvalues( A );            # eigenvalues of A
> nulbasis( A - lambda[2]*diag(1$5) ); # basis for e-space of λ₂
```

compute a basis for the eigenspace corresponding to the second eigenvalue of a 5×5 matrix
A. To avoid typing or numerical errors, you are encouraged to avoid looking at the list of
eigenvalues and explicitly typing the eigenvalue in the argument of the `nulbasis` command. It
is much safer — and faster — to reference the eigenvalues as in the commands above.

SECTION 5.2   charpoly

The Maple command `charpoly` can be used to check your answers in Exercises 9–14. Note
that if A is $n \times n$, `charpoly(A,lambda)` is $\det(\lambda I - A)$. If n is even, this is the characteristic
polynomial of A. When n is odd, this is the negative of the characteristic polynomial of A (as
defined in the text).

For Exercises 28 and 29, create a 4×4 matrix with random integer entries with the command
`randomint( 4 );`. For Exercise 29, use `gauss` and perhaps `swap` to create the echelon form

without row scaling. See also the Maple Note for Section 1.4 on page MP-6.

   The following commands create the graph of the characteristic polynomial of the matrix A in Exercise 30 for all values of the parameter a in a list a_vals:

```
> p := charpoly( A, lambda );  # char poly in terms of parameter a
> P := [seq( p, a=a_vals )];    # char poly for each value of a
> plot( P, lambda=0..3 );       # graph all char polys on same axes
```

SECTION 5.3   Diagonalization and `eigenvalues`

To practice the diagonalization procedure in this section, you should use the `nulbasis` command to produce eigenvectors. For Exercises 33–36, the command `ev := eigenvalues( A );` can be used to provide the eigenvalues. See also the Maple Note for Section 5.1 on page MP-13.

SECTION 5.5   Diagonalization and `eigenvectors`

Once you become familiar with the diagonalization process described in the Maple Note for Section 5.3, you can use other Maple commands to simplify the process of determining the eigenvectors for a matrix.

   The command `eigenvectors(A);` returns, for each eigenvalue of A, a list containing three pieces of information: the eigenvalue, the (algebraic) multiplicity of the eigenvalue, and a basis for the corresponding eigenspace. The data structure is a little complicated, but the following sequence of commands illustrates how to find the matrices P and D such that $AP = PD$:

```
> EV := [ eigenvectors( A ) ];           # e-values and e-spaces
> lambda := seq( op(1,e)$op(2,e), e=EV ); # e-values (w/repetition)
> espace := seq( op(op(3,e)), e=EV );     # e-space basis
> DD := diag( lambda );                   # diag e-value matrix
> P := augment( espace );                 # e-vector matrix
```

Note that the matrix P may not be the same as the one you find by hand. A quick check that your answer is correct is: `iszero( A&*P - P&*DD );`. The result will be *true* if the argument is a zero matrix. When the matrices involve floating-point entries, Maple may return *false* when one entry is very small (but not exactly zero). In any case, a visual inspection should be performed on the difference $AP - PD$.

SECTION 5.5   Complex Eigenvalues, `Re`, and `Im`

If A is a 2×2 real matrix and if `eigenvalues( A );` returns a pair of complex conjugate eigenvalues, you can use the `nulbasis` command to find a complex eigenvector $\mathbf{v}$ corresponding to one of these eigenvalues. Then, build a 2×2 real matrix P from the real and imaginary parts of $\mathbf{v}$ and compute $C = P^{-1}AP$. This can be done with the following commands:

```
> V := nulbasis( A - lambda[1]*diag(1$2) );   # complex e-vector
> Vr := evalm( Re( V ) );                       # real part
> Vi := evalm( Im( V ) );                       # imaginary part
> P := augment( Vr, Vi );
> C := evalm( inverse(P) &* A &* P );
```
In general, the real and imaginary parts of a complex-valued vector **v** are obtained with the commands evalm(Re(v)); and evalm(Im(v));, respectively.

SECTION 5.6   Plotting Discrete Trajectories

The following sequence of commands creates the sequence of vectors $\mathbf{x}$, $A\mathbf{x}$, $A^2\mathbf{x}$, ..., $A^{15}\mathbf{x}$ for a given initial vector **x**:
```
> x := vector( [1,0] );            # initial vector (change as needed)
> T := convert( x, list ):         # initialize list of points
> for j from 1 to 15 do            # change 15 to number of iterations
>    x := evalm( A&*x );           # compute the "next" point
>    T := T, convert( x, list );   # add new point to list
> end do;
```
These points can be plotted with the command
```
> plot( [T], style=POINT );
```
Omitting the second argument of the plot command will "connect the dots" between the points. Other optional arguments can be used to further customize the plot. The display command — in the plots package — can be used to combine multiple trajectories into a single plot. (Do not forget to execute the command with(plots); to load the plots package.)

For Exercise 17, the following commands create a graph of the first component in each data point in the list T constructed above
```
> T1 := [seq( [i-1,T[i,1]], i=1..nops([T]) )];
> plot( T1 );
```
To graph the sum of the first two components of each data point in T replace T1 with T2 defined by
```
> T2 := [seq( [i-1,T[i,1]+T[i,2]], i=1..nops([T]) )];
```
Similarly, to plot the quotient of the first two components in each data point in T replace T1 with T2 defined by
```
> T3 := [seq( [i-1,T[i,1]/T[i,2]], i=1..nops([T]) )];
```

SECTION 5.7   Plotting Trajectories for Differential Equations

The following template of Maple commands can be used to create a plot of the direction field and a solution curve for the 2×2 system of differential equations in Example 1.

```
> with( DEtools );                                      # load DEtools pkg
> A := matrix( 2, 2,
>                 [ [-1.5, 0.5], [1, -1] ] );           # enter coef matrix
> SYS := [ diff(y1(t),t)                                # DO NOT CHANGE
>             = A[1,1]*y1(t) + A[1,2]*y2(t),            # DO NOT CHANGE
>           diff(y2(t),t)                               # DO NOT CHANGE
>             = A[2,1]*y1(t) + A[2,2]*y2(t) ];          # DO NOT CHANGE
> IC := [ y1(0)=5, y2(0)=4 ];                           # enter init cond
> RANGE := t = 0..10;                                   # enter time interval
> WINDOW := y1 = -5..5, y2 = -4..4;                     # enter plot window
> DEplot( SYS, [y1(t),y2(t)],                           # DO NOT CHANGE
>           RANGE, [IC], WINDOW );                       # DO NOT CHANGE
```

To create plots for other systems, simply modify the coefficient matrix and the initial condition then modify the time interval and window to create the desired plot.

Note:

- The with(DEtools); command needs to be entered only once — anytime before the first use of DEplot.

SECTION 5.8   Power Method and Inverse Power Method

A Maple implementation of the Power Method (page 363 of the text) assumes that A is an $n \times n$ matrix with a strictly dominant eigenvalue and an initial vector $\mathbf{x}_0$:

```
> x := evalm( x0 );            # initial vector
> for k from 1 to 15 do        # change 15 to number of iterations
>   mu := norm( x );           # largest element in x
>   x := evalm( A &* x/mu );   # next iterate in power method
> end do;
```

The value of μ should approach the dominant eigenvalue and the vector $\mathbf{x}$ should approach a corresponding eigenvector.

One possible implementation of the Inverse Power Method is:

```
> I3 := diag(1$3);                     # 3 × 3 identity matrix
> alpha := 3.3;                        # initial guess at e-value
> x := vector( [1,1,1] );              # initial guess at e-vector
> for k from 1 to 15 do                # change bounds as needed
>   y := linsolve( A-alpha*I3, x );    # solve (A − αI)y_k = x_k
>   mu := norm( y );                   # largest element in abs value
>   nu := alpha + 1/mu;                # updated approx to e-value
>   x := evalm( y/mu );                # updated approx to e-vector
> end do;
```

Notes:

- The initial value of α should be reasonably close to the desired eigenvalue and $\mathbf{x}$ is a vector whose largest element (in absolute value) is 1.

- Recall that `norm(x)` returns the maximum of the absolute values of the entries in $\mathbf{x}$.

SECTION 6.1   `innerprod` and `norm`

The Maple command `innerprod(u,v)` returns the inner product of two real-valued column vectors $\mathbf{u}$ and $\mathbf{v}$. The associated length of a vector $\mathbf{v}$ is `norm(v,2)`. (See also the Maple Note for Section 2.1 on page MP-8.)

SECTION 6.2   Orthogonality

In Exercises 1–9 and 17–22, a fast way to test the orthogonality of a set such as $\{\mathbf{u}_1, \mathbf{u}_2, \mathbf{u}_3\}$ is to create a matrix
```
> U := augment( u1, u2, u3 );
```
whose columns are the vectors from the set, and test whether
```
> evalm( transpose(U) &* U);
```
yields a diagonal matrix.

Notes:

- To understand how this works, see Theorem 6 on page 390.

- The `orthog` command could also be used, but this does not emphasize the essential properties of orthogonality.

SECTION 6.3   Orthogonal Projections

The orthogonal projection of a vector onto a single vector was described in the Maple Note for Section 6.2. The orthogonal projection onto the set spanned by an orthogonal set of vectors is the sum of the one-dimensional projections. Another way to construct this projection is to normalize the orthogonal vectors, place these orthonormal vectors in the columns of a matrix U, and apply Theorem 10. For instance, if $\{\mathbf{y}_1, \mathbf{y}_2, \mathbf{y}_3\}$ is an orthogonal set of nonzero vectors, then the matrix U defined by the Maple command:
```
> U := augment( y1/norm(y1,2),
>                y2/norm(y2,2),
>                y3/norm(y3,2) );
```
has orthonormal columns. The orthogonal projection of $\mathbf{y}$ onto Span$\{\mathbf{y}_1, \mathbf{y}_2, \mathbf{y}_3\}$ is obtained with the command:
```
> evalm( U &* (transpose(U) &* y) );
```

SECTION 6.4   The Gram-Schmidt Process, `proj`, and `gs`

If A has only two linearly independent columns, then the Gram-Schmidt process is:

```
> v1 := col(A,1);
> v2 := evalm( col(A,2)
>              - innerprod(col(A,2),v1)/innerprod(v1,v1) * v1 );
```

If A has three columns, add the command:

```
> v3 := evalm( col(A,3)
>              - innerprod(col(A,3),v1)/innerprod(v1,v1) * v1
>              - innerprod(col(A,3),v2)/innerprod(v2,v2) * v2 );
```

The generalization to larger matrices should now be apparent.

Notice that Maple commands in the form

```
> evalm( innerprod(y,u)/innerprod(u,u) * u );
```

provide the orthogonal projection of a vector **y** onto a vector **u**. A simpler, but less illustrative, alternative is to use the `proj` command from the `laylinalg` package. The syntax is:

```
> proj( y, convert(u,matrix) );
```

After you are familiar with the Gram-Schmidt process, use the command `proj( x, V );` to compute the projection of a vector **x** onto the subspace spanned by the columns of a matrix V. For example:

```
> v2 := evalm( col(A,2) - proj( col(A,2), augment(v1) ) );
> v3 := evalm( col(A,3) - proj( col(A,3), augment(v1,v2) ) );
```

Although you should construct the QR decomposition of a matrix A using the approach in the text, you can check your construction of the matrix Q with the command `Q := gs(A);`. The `gs` command from the `laylinalg` package provides an even faster method of applying the Gram-Schmidt process. The syntax is simply: `gs(A);`. After you are very familiar with the Gram-Schmidt process the use of `gs` saves time.

SECTION 6.5   `linsolve` and `leastsqrs`

When A has linearly dependent columns, you can write the general description of all least-squares solutions on paper after you row reduce the augmented matrix for the normal equations:

```
> rref( augment( transpose(A)&*A, transpose(A)&*b ) );
```

When A has linearly independent columns, a Maple command to solve the normal equations is:

```
> linsolve( transpose(A)&*A, transpose(A)&*b );
```

Other approaches to solving the normal equations include the use of `rref` as described above and the explicit use of $(A^T A)^{-1}$. For Exercises 15 and 16, see the Numerical Note on page 415 in the text and use `linsolve( R, transpose(Q)&*b );` to solve $Rx = Q^T b$.

When A is not square but has linearly independent columns, a common algorithm for solving the overdetermined system $Ax = b$ is to find Q and R and then solve $Rx = Q^T b$. You should use the normal equations or the QR factorization. This will give you a solid conceptual background for applying least-squares techniques later in your career.

For Exercise 26, the command `A := stackmatrix( A1, A2 );` creates the (partitioned) matrix whose top block is A_1 and bottom block is A_2. Of course, each block must have the same number of columns. (See also the Maple Note for Section 2.4 on page MP-10 for more information on partitioned matrices.)

SECTION 6.6   Computing Regression Coefficients

Maple can be used to construct the design and observation vectors from a list of data points. For example, for Exercise 1:
```
> c6s6( 1 );                        # load data from laylinalg
> n := vectdim( x );                # number of data points
> X := augment( vector(n,1), x ); # design matrix
```
The least squares solution to $X\beta = \mathbf{y}$ as described in the Maple Note for Section 6.5.

SECTION 6.6   leastsqrs, Functions of Vectors, and map

Once you create the design matrix X and the observation vector $\mathbf{y}$, your computation for least-squares solutions here are the same as those described in the Maple Note for Section 6.5. Here, A and $\mathbf{b}$ are replaced by X and $\mathbf{y}$, respectively. The Maple command
```
> rref( augment( transpose(X)&*X, transpose(X)&*y ) );
```
leads to the general description of all least-squares solutions. When X has linearly independent columns, the command
```
> linsolve( transpose(X)&*X, transpose(X)&*y );
```
creates the least-squares solution. In subsequent courses, you may choose simply to use leastsqrs(X, y);, which also produces all least-squares solutions.

To construct the design matrix for an exercise in this section, you may need Maple's map command to apply a function to each element of a vector. If x is a vector and f is a Maple function, then map(f, x); is a vector the same size as x whose entries are the result of applying f to the entries in x. For example, in Exercise 7 the x vector can be created as above and the vector of the square of each element is x2 := map(u->u^2, x);. The design matrix is now easily constructed with the use of augment. When f is a built-in Maple function, say the exponential function exp, the syntax is somewhat simpler: map(exp, x);.

SECTION 6.7   int and Inner Products on $C[0, 2\pi]$

The Maple command to evaluate the definite integral $\int_a^b f\ dt$ is int(f, t=a..b);. To solve Exercise 28, begin by creating a Maple procedure to compute the inner product of two functions in $C[0, 2\pi]$:
```
> ip := (f,g) -> int(f*g,t=0..2*Pi);
```
The inner product procedure, ip, and the four functions p_0, p_1, p_2, and p_3 can be defined by executing c6s7(28);. The Gram–Schmidt process described in the Maple Note for Section 6.4 on page MP-17 is applicable here. For example, the first two steps are:
```
> q0 := p0;
> q1 := p1 - ip(p1,q0)/ip(q0,q0) * q0;
```
Observe that the only difference between these commands and the ones introduced in Section 6.4 is that **innerprod** is replaced with the newly-defined inner product **ip**.

SECTION 6.8 Graphing Functions

The `plot` command can be used to plot one or more functions. For example, the two plots in Figure 3 for Example 4 on page 442 in the text can be created as follows:

```
> f := t;                              # define function f
> F3 := Pi - 2*sin(t) - sin(2*t)
>           - 2/3*sin(3*t);            # 3rd-order Fourier approx
> plot( [f,F3], t=0..2*Pi );          # Figure 3 (a)
> F4 := F3 - 2/4*sin(4*t);            # 4th-order Fourier approx
> plot( [f,F4], t=0..2*Pi );          # Figure 3 (b)
```

The Maple Note for Section 4.1 on page MP-11 contains additional information about Maple's `plot` command.

SECTION 7.1 Orthogonal Diagonalization

Use the `eigenvalues` and `nulbasis` commands to find the eigenvalues and corresponding eigenvectors as in Section 5.3 on page MP-14. If you encounter a two-dimensional eigenspace, with a basis $\{v_1, v_2\}$, use the command

```
> v2 := evalm( v2 - innerprod(v2,v1)/innerprod(v1,v1) * v1 );
```

or

```
> v2 := evalm( v2 - proj(v2,augment(v1)) );
```

to make the two eigenvectors orthogonal. See the Maple Note for Section 6.4 on page MP-17. After you normalize the vectors and create P, check whether $P^T P = I$ to verify that P is indeed an orthogonal matrix.

SECTION 7.4 The Singular Value Decomposition

The singular value decomposition of a matrix A is constructed from the orthogonal diagonalization of $A^T A$. The orthogonal diagonalization of a matrix is discussed in the Maple Note for Section 7.1 on page MP-20. Note that the eigenvalues of $A^T A$ may not be sorted in decreasing order. You may need to reorder some of the information from the orthogonal diagonalization when creating the diagonal matrix D and matrix of right singular vectors V. For example, if $A^T A$ is a 3×3 matrix and `lambda` is a list of the eigenvalues of $A^T A$ with, say, `lambda[3] > lambda[1] > lambda[2]`, then the command

```
> Sigma := diag( seq( sqrt(lambda[i]), i=[3,1,2] ) );
```

creates the diagonal matrix of singular values of A. Note that Σ must have the same dimensions as A; the `extend` command can be used to add one or more rows or columns of zeros to the diagonal matrix created above. For example,

```
> Sigma := extend( Sigma, 1, 0, 0 );
```

adds one row of zeros to the bottom of the diagonal matrix. The matrix V would then be created by:

```
> V := augment( seq( col(P,i), i=[3,1,2] ) );
```

The matrix of left singular vectors, U, can be formed by normalizing the nonzero columns of the product AV. If U is not square, the missing columns must be chosen to form an orthonormal

basis for Nul A^T. (See Figure 4 on page 479 in the text.) The command
```
> GramSchmidt( nullspace( transpose(A) ), normalized );
```
produces this orthonormal basis. Thus, the square matrix U defined by
```
> U := augment( U, op(GramSchmidt( nullspace( transpose(A) ),
>                                   normalized )) );
```

This construction helps you to think about the fundamental properties of the singular value decomposition. There are two additional Maple commands that can be used to directly obtain information about the singular values of a matrix. The **singularvals** command returns the singular values of A as a list but does not provide any information about the singular vectors. The **Svd** command computes floating-point approximations to the singular values and vectors of a numeric matrix A. The command
```
> SV := evalf( Svd( A, U, V ) );
```
returns the singular values in the table SV. The matrices U and V contain the right and left singular vectors, respectively. To create the diagonal matrix Σ from the table of singular values, use the command:
```
> Sigma := diag( SV[i] $ i=1..vectdim(SV) );
```

SECTION 7.5   Computing Principal Components

The p-dimensional vector containing the mean of each row of a $p \times N$ matrix X can be created with the command:
```
> Rmean := vector([ seq( add( x, x=convert(row(X,i),list) )
>                         / coldim(X), i=1..rowdim(X) ) ]);
```
The $p \times N$ matrix containing N copies of the row averages is obtained with:
```
> augment( Rmean $ coldim(X) );
```
Thus, to convert the data in X into mean-deviation form, use
```
> B := evalm( X - augment( Rmean $ coldim(X) ) );
```
The sample covariance matrix is produced by:
```
> S := evalm( B &* transpose(B) / (coldim(X)-1) );
```
The principal component analysis is completed by performing a singular value decomposition on the matrix $A = \frac{1}{\sqrt{N-1}} B^T$. (See the Numerical Note at the end of Section 7.5 of the text.)

INDEX OF MAPLE COMMANDS

Notes for the Mathematica
Computer Algebra System

GETTING STARTED WITH MATHEMATICA

Using Mathematica with Linear Algebra and Its Applications

To use Mathematica for your homework in this course, you will need additional Mathematica files which were created specifically for your text. At some schools, the campus-wide version of Mathematica already has these files available on some or all computers. Ask your instructor. If you have your own copy of Mathematica at home, you will need to download the additional Mathematica files off the Web. To do this, go to www.laylinalgebra.com and follow the instructions there for downloading Mathematica files. One of the extracted files, LayFunctions.m, contains additional Mathematica commands needed in this course. We will refer to this file as the LayFunctions package. The other downloaded files contain selected exercise data for the text. In order for the LayFunctions package to work properly, you must create a new folder called LayData (be sure to spell this exactly as it appears here with a capital L and a capital D) inside the Mathematica folder called ExtraPackages and then place LayFunctions.m inside this folder. For Mathematica 3.0, the path to ExtraPackages is

Wolfram Research / Mathematica / 3.0 / AddOns / ExtraPackages

(If you have a later version of Mathematica, the path to ExtraPackages is the same except 3.0 is replaced by the version number of your copy of Mathematica.)

Mathematica Notebooks

The electronic data downloaded from the web correspond to selected exercises in *Linear Algebra and Its Applications*. These files are interactive documents called *notebooks* and each of these notebooks has a name of the form C(chapter#)S(section#).nb or C(chapter#)Ssuppl.nb. For example, the notebook C2S3.nb contains electronic data for selected exercises in Section 3 of Chapter 2 and C2Ssuppl.nb contains electronic data corresponding to the Supplementary Exercises at the end of Chapter 2.

Each notebook is divided up into a sequence of individual units called *cells*. Each cell, except for the cell containing the title, has a corresponding *cell bracket* appearing in the right margin of the notebook. The first cell contains the title of the notebook. The second cell in the notebook contains instructions for working with the *palette* appearing in the third cell. A palette is similar to a set of calculator buttons, providing you with shortcuts to typing in commands from the keyboard. To activate the palette, follow the instructions appearing in the second cell.

The remaining cells in the notebook contain selected data for the homework exercises followed by a final cell group containing an *initialization cell*. The initialization cell has been deliberately placed out of the way at the end of the notebook because it will automatically be evaluated for you when you begin working with the notebook. You do not need to do anything with this cell.

For the purposes of documentation, the contents of the initialization cell are briefly described here, but it is not necessary than you understand this information and therefore feel free to skip the remainder of this paragraph if you like. The initialization cell contains the following five commands (separated by semicolons):

```
Off[General::spell1];
Needs["LayData`LayFunctions`"];
Needs["LinearAlgebra`GaussianElimination`"];
Needs["LinearAlgebra`MatrixManipulation`"];
$Post := If[MatrixQ[#], MatrixForm[#], #] &;
```

The first command in the initialization cell instructs Mathematica to ignore displaying particular types of warning messages. The next three lines in the initialization cell are Needs commands that instruct Mathematica to load additional commands into memory. The first Needs instruction loads all of the commands found in the LayFunctions package. This package was created specifically for *Linear Algebra and Its Applications*. The remaining two Needs commands in the initialization cell instruct Mathematica to load all commands found in the GaussianElimination and MatrixManipulation packages. We will refer to the commands in these three packages as *standard add-on commands* since they are part of the standard Mathematica software package. Since Mathematica does not display matrices in the standard matrix form you find in your textbook, a final command was added to the initialization cell instructing Mathematica to output all matrices in matrix form. (This command involves the advanced concept of *pure functions* and it is not necessary to understand how this command works.)

Mathematica Representation of a Matrix

Most of the data in the electronic notebooks for *Linear Algebra and Its Applications* consists of matrices. The Mathematica representation of the matrix

$$\begin{pmatrix} a_{11} & a_{12} & \cdots & a_{1n} \\ a_{21} & a_{22} & \cdots & a_{1n} \\ \vdots & \vdots & \ddots & \vdots \\ a_{m1} & a_{m2} & \cdots & a_{mn} \end{pmatrix}$$

is $\{\ \{\ a_{11}, a_{12}, \ldots, a_{1n}\ \},\ \{\ a_{21}, a_{22}, \ldots, a_{2n}\ \},\ \ldots,\ \{\ a_{m1}, a_{m2}, \ldots, a_{mn}\ \}\ \}$. Each set of objects contained in brackets $\{\ldots\}$ is called a *list* and therefore a matrix is just a list of lists.

Using Notebook Data

To illustrate how to work with the electronic data found in the notebooks, start the Mathematica program and then open the notebook C2S3.nb. After the palette cell, you will see a cell containing the text Exercise 1. This cell is called a *closed* cell because its contents are hidden from view. To see the contents of the cell (i.e. to *open* the cell), double click on right cell bracket associated with this cell. After the cell is opened, you will see a *text* cell with the phrase Think before you compute. This is just some general advice telling you to study the problem before solving it with Mathematica. The next cell contains the assignment statement A = { { 5, 7 } , { -3, -6 } }. To execute this assignment statement, click anywhere inside the cell containing the assignment statement and then press Shift+Enter (on some computer platforms, pressing Return or Enter will also work). When you do this, a window with the title Ask Init will appear asking you the question Do you want to automatically evaluate the initialization cells in the notebook "S2C3.nb"?. *Always choose* **Yes** so that Mathematica automatically executes the commands in the initialization cell. A label In[6]:= will then appear in the assignment statement and a output cell beginning with Out[6]:= will appear displaying the matrix used in the assignment statement.

In[6]:=A = { { 5, 7 } , { -3, -6, } }

$$\text{Out[6]//MatrixForm=} \begin{pmatrix} 5 & 7 \\ -3 & -6 \end{pmatrix}$$

The labels In[6]:= and Out[6]= indicate that five commands (contained in the initialization cell) were automatically executed prior to this input statement. Throughout these notes, dialogs with Mathematica are shown with the In[n]:= and Out[n]= labels. The first input statement in each section of these notes will usually be In[1]:= for the sake of simplicity, even though the first input number that you see will be larger.

Typically you will only be working with a small number of exercises in a given notebook. You can work problems within a notebook in any order, skipping any homework data cells you want. If you wish to print out the notebook after you complete your work, first delete all of the cells you skipped to save paper. To do this, click on all the brackets corresponding to exercises you skipped while holding down the Ctrl key and then press Delete.

Creating Matrices with Mathematica

When working with one of the notebooks, you may on occasion want to enter your own data into the notebook. For instance, to create your own matrix, you first need to form a new input cell. To do this, use your mouse and click between the two cells where you want to put your matrix and a new cell will appear as soon as you begin entering information in from the keyboard or palette. Suppose you want to define A to be the matrix $\begin{pmatrix} 1 & 2 & 3 & 4 \\ 5 & 6 & 7 & 8 \\ 9 & 10 & 11 & 12 \end{pmatrix}$. One way to do this is to enter the following assignment statement.

In[1]:=A = { { 1, 2, 3, 4 } , { 5, 6, 7, 8 } , { 9, 10, 11, 12 } }

$$\text{Out[1]//MatrixForm} = \begin{pmatrix} 1 & 2 & 3 & 4 \\ 5 & 6 & 7 & 8 \\ 9 & 10 & 11 & 12 \end{pmatrix}$$

A semicolon (;) can be placed at the end of an input statement to prevent output from being displayed.

In[1]:=A = { { 1, 2, 3, 4 } , { 5, 6, 7, 8 } , { 9, 10, 11, 12 } };

The entry in the ith row and jth column of matrix A is accessed by entering A[[i, j]]. (Comments appearing in (*...*) are used to help you better understand the input and output statements and they ignored by Mathematica.)

In[2]:=A[[2, 3]]   (* Display the entry in row 2 and column 3 of A *)
Out[2]:=7   (* The corresponding entry in A is displayed *)

A more natural way to enter a matrix is to use the palette. Activate the notebook palette and then select a location in the notebook where you wish to begin a new cell. Type in A= and then click on the button in the palette containing $\begin{pmatrix} \Box & \Box \\ \Box & \Box \end{pmatrix}$. To create another row in the matrix, press Ctrl+Enter. To create another column, press Ctrl+, (Ctrl+comma). Suppose you have created a matrix with 3 rows and 4 columns. To enter numbers, click on the box in row 1 and column 1, enter the first number, and use the Tab key to move from one entry to the next. When finished, press Shift+Enter. Your notebook might contain lines such as

$$\text{In[3]}:=A = \begin{pmatrix} 1 & 2 & 3 & 4 \\ 5 & 6 & 7 & 8 \\ 9 & 10 & 11 & 12 \end{pmatrix} \qquad \text{Out[3]//MatrixForm} = \begin{pmatrix} 1 & 2 & 3 & 4 \\ 5 & 6 & 7 & 8 \\ 9 & 10 & 11 & 12 \end{pmatrix}$$

Mathematica commands are reserved words that cannot be used in assignment statements. For example, executing the following input statement results in an error message stating that N is a protected symbol.

$$\text{In[4]}:=N = \begin{pmatrix} 1 & 2 \\ 3 & 4 \end{pmatrix}$$

Set::wrsym: Symbol N is Protected.

$$\text{Out[4]//MatrixForm} = \begin{pmatrix} 1 & 2 \\ 3 & 4 \end{pmatrix}$$

One way to avoid this problem is to use a lower case letter n instead of a capital N in the assignment statement. This is due to the fact that all Mathematica commands and functions begin with a capital letter and therefore a word beginning with a lower case letter will never result in a protected symbol error. Another way around this problem is to use an entire word (starting with a lower case letter) for the name of a matrix in an assignment statement. For example, we let matN represent the matrix in the following assignment statement (the output is not displayed here). Any such word created by the user may end with a digit (0−9), but cannot begin with a digit and no spaces can occur in the word.

$$\text{In[5]:=matN} = \begin{pmatrix} 1 & 2 \\ 3 & 4 \end{pmatrix}$$

It is not always necessary to assign a name to a matrix since Mathematica automatically attaches the name Out[n] to the matrix entered immediately after the In[n]:= prompt. For example, the name Out[6] is assigned to the matrix in the output of the following assignment statement.

$$\text{In[6]:=} \begin{pmatrix} 1 & 2 \\ 3 & 4 \end{pmatrix} \qquad \text{Out[6]//MatrixForm} = \begin{pmatrix} 1 & 2 \\ 3 & 4 \end{pmatrix}$$

The matrix is retrieved simply by typing and executing %6, which is short for Out[6].

$$\text{In[7]:=%6} \qquad \text{Out[7]//MatrixForm} = \begin{pmatrix} 1 & 2 \\ 3 & 4 \end{pmatrix}$$

BasicInput Palette

On some homework exercises, you will find it useful to use the BasicInput palette which contains additional shortcuts for computing powers, square roots and so forth. To open this palette, choose File from the menu, then Palettes followed by BasicInput. Drag BasicInput to the side of the notebook window and change the size of your notebook window if necessary so that the palette and the notebook appear separately in two nonoverlapping windows.

Online Help

The question mark (?) is used to obtain a precise definition of a given Mathematica command. For example, by executing ?Pi, a brief description of the Pi command will be displayed. For more information about Mathematica commands, consult the Help menu.

STUDY GUIDE NOTES

SECTION 1.1   Row Operations

The LayFunctions package contains the following commands that perform elementary row operations on a matrix, denoted here by M:

```
ReplaceRow[ M, r, m, s ]      (* Replaces row r of M by row r plus m times row s. *)
Swap[ M, r, s ]               (* Interchanges rows r and s of the matrix M. *)
Scale[ M, r, c ]              (* Multiplies row r of M by a scalar c. *)
```

The letters r and s, are integers while m and c can be any scalar expressions. Any of these variables can be a symbolic expression as well.

To illustrate how these commands are used, open C1S1.nb and activate the notebook palette. Then open the cell for Exercise 11 and execute the assignment statement (be sure to choose **Yes** in the Ask Init window). You will then see input and output statements.

In[6]:= M = { { 0, 1, 4, -5 }, { 1, 3, 5, -2 }, { 3, 7, 7, 6} }

$$\text{Out[6]//MatrixForm} = \begin{pmatrix} 0 & 1 & 4 & -5 \\ 1 & 3 & 5 & -2 \\ 3 & 7 & 7 & 6 \end{pmatrix}$$

Suppose you want to begin the row reduction process by switching rows 1 and 2 of M. Then execute the command Swap[M, 1, 2] or Swap[%6, 1, 2]. Then the matrix, whose rows have been switched, appears in Out[7]. Next, ReplaceRow[%7, 3, -3 , 1] can be used in the next step of row reduction and so on.

Note: For simple problems, the multiple m you need in ReplaceRow[%n, r, m, s] is usually a small integer or fraction that can be computed in your head. In general, m may not be so easy to compute mentally. The next two paragraphs describe how to handle such cases.

For a given matrix appearing in Out[n], %n[[s, c]] is the entry in row s and column c of this matrix. Now suppose you want to use the entry in row s and column c of the matrix appearing in Out[8] to change $\%8[[r, c]]$ to 0. Enter the commands (where / is the division key)

In[9]:=m = −%8[[r, c]] / %8[[s, c]];   (* The multiple of row s to be added to row r *)
In[10]:=ReplaceRow[%8, r, m, s]   (* Add m times row s to row r *)
Or, just use one command: M=ReplaceRow[%8, r, −%8[[r, c]]/ %8[[s, c]], s].

Exact and Approximate Calculations

Mathematica will perform exact calculations as long as all the numbers used in your work are integers or rational numbers not containing any decimal points. For example, $4/9$ is an exact number, but $4.0/9$ or $4./9$ are approximate numbers which Mathematica assumes are only accurate to a fixed number of decimal places.

If you perform a calculation involving both an approximate number and an exact number, then the result will be an approximate number. For example, Mathematica interprets the sum $2 + 1.$ to equal the approximate number $3.$ (containing a decimal point).

At first it may seem that you would always want to perform exact calculations in order to obtain exact answers. But it takes Mathematica more computer time to perform exact calculations and when performing computations with large matrices, results may be slow in coming if you use exact numbers. In other instances, Mathematica will not be able to perform exact calculations. Therefore at times it will be necessary to enter approximate numbers to obtain approximate results.

Symbolic Elementary Row Operations

Mathematica works with variables and unspecified parameters as in the following example.

$$\text{In[11]}:=\begin{pmatrix} a & b \\ c & d \end{pmatrix} \qquad \text{Out[11]}//\text{MatrixForm}=\begin{pmatrix} a & b \\ c & d \end{pmatrix}$$

$$\text{In[12]}:=\text{ReplaceRow}[\ \%11,\ 2,\ -c/a,\ 1\ ] \qquad \text{Out[12]}//\text{MatrixForm}=\begin{pmatrix} a & b \\ 0 & \frac{-bc}{a}+d \end{pmatrix}$$

SECTION 1.3   Transpose

The *transpose* of a given matrix A is a matrix whose ith column is the ith row of A. The command Transpose[A] creates this matrix. For example, suppose $A = \begin{pmatrix} 1 & 3 & 5 \\ -7 & 0 & 8 \end{pmatrix}$. Then executing

Transpose[A] results in an output of $\begin{pmatrix} 1 & -7 \\ 3 & 0 \\ 5 & 8 \end{pmatrix}$.

Representation of a Vector

A vector **v** which is either represented as $(\ v_1,\ v_2,\ \ldots,\ v_n\ )$ or as $\begin{pmatrix} v_1 \\ v_2 \\ \vdots \\ v_n \end{pmatrix}$ in the textbook, can either be represented by a list $\{\ v_1,\ v_2,\ \ldots,\ v_n\ \}$ or as the matrix $\{\ \{\ v_1\},\ \{\ v_2\},\ \ldots,\ \{\ v_n\ \}\ \}$ in Mathematica. We also refer to $\{\ v_1,\ v_2,\ \ldots,\ v_n\ \}$ as being in *row form*. Executing the command Transpose[$\{$ v $\}$] turns $v = \{\ v_1,\ v_2,\ \ldots,\ v_n\ \}$ into $\{\{\ v_1\},\ \{\ v_2\},\ \ldots,\ \{\ v_n\ \}\ \}$.

Constructing a Matrix

For matrix A and vector **b**, each with the same number of rows, Augment[A, **b**] forms the augmented matrix $[A\ b]$. Consider the following example. (The semicolon appearing in In[1] allows you to enter multiple commands into a single input cell and the output corresponding to the command preceding each semicolon is not displayed.)

In[1]:=A = { { a, b, c }, { d, e, f } }; b={ { g }, { h } } ; Augment[A, b]

$$\text{Out[1]}//\text{MatrixForm}=\begin{pmatrix} a & b & c & g \\ d & e & f & h \end{pmatrix}$$

The command TakeColumns[A, {i}]] forms column i of A. The data for Exercise 25, for example, consists of a matrix A, the vector **b** and then three TakeColumns commands which are used to create a1, a2, a3, namely columns 1, 2, and 3 of A, respectively. Executing the command

Augment[a1, a2, a3, b] creates the matrix $[\, a_1 \;\; a_2 \;\; a_3 \;\; b\,]$. The same matrix can be created using Augment[A, b].

Exercises 11-14 and 25-28, and 31 can be solved using the commands ReplaceRow, Swap, and (occasionally) Scale.

SECTION 1.4   Gauss and BGauss

To solve $Ax = b$, row reduce the matrix $M = [A \;\; b]$. For example, the assignment statement b={ {0}, {-5}, {7} }   creates a column vector b with entries 0, -5 and 7 and the command Augment[A b] can be used to form M provided that A contains exactly three rows. The commands Gauss, BGauss and Scale speed up the row reduction process. Gauss[A, r] uses the leading entry in row r of A as a pivot, and uses row replacements to create zeros in the pivot column below this pivot entry. In the backward phase of row reduction, use BGauss[A, r], which selects the leading entry in row r of A as a pivot, and creates zeros in the pivot column *above* the pivot. Use Scale to create leading 1's in the pivot positions.

Multiplying a Matrix and a Vector

A period (.) is used to multiply a matrix and a vector.

$$\text{In[1]:=}A \;=\; \begin{pmatrix} 0 & 1 & 2 & 3 \\ 5 & 5 & 8 & 7 \end{pmatrix}; \; x \;=\; \begin{pmatrix} -1 \\ 6 \\ 0 \\ 10 \end{pmatrix}; \; \text{A.x} \qquad \text{Out[1]=//MatrixForm=} \begin{pmatrix} 36 \\ 95 \end{pmatrix}$$

Mathematica also allows multiplication of a matrix and a vector when the vector is in row form .

$$\text{In[2]:=}A \;=\; \begin{pmatrix} 0 & 1 & 2 & 3 \\ 5 & 5 & 8 & 7 \end{pmatrix}; \; x \;=\; \{\, -1, 6, 0, 10 \,\}; \; \text{A.x} \qquad \text{Out[2]} = \{\, 36, \;\; 95 \,\}$$

SECTION 1.5   The Zero Matrix

The command ZeroMatrix[m, n] creates a $m \times n$ matrix of zeros. The following example demonstrates how to set up an augmented matrix M corresponding to $Ax = 0$ using the Augment command.

$$\text{In[1]:=}A = \begin{pmatrix} 1 & 2 \\ 3 & 4 \end{pmatrix}; \; M = \text{Augment[A, ZeroMatrix[2, 1]]};$$

Then use Gauss, Swap, BGauss, and Scale to row reduce M completely.

SECTION 1.6 Rational Format

Chemical equation-balance problems are studied best using exact or symbolic arithmetic, because the balance variables must be whole numbers (with no round-off allowed). Mathematica will display the exact (rational) value of every entry during row reduction as long as each number you enter is an integer or rational number not containing a decimal point (.).

Once you find a rational solution of a chemical equation-balance problem, you can multiply the entries in the solution by a suitable integer to produce a solution that involves only whole numbers.

SECTION 1.10 Generating a Sequence

Recall that we can identify a geometric point (a, b) with the column vector $\begin{pmatrix} a \\ b \end{pmatrix}$. With Mathematica, the ordered pair (a, b) is represented by $\{ a,\ b \}$.

The data for Exercise 13, Section 1.10, is used here to demonstrate how to generate a sequence. Open C1S10.nb and then open the cell corresponding to Exercise 13. Execute the input cell.

In[1]:= M = {{ .95, .03 }, { .05, .97 }}; x0 = { 600000, 400000 };
       y0 = { 350000, 650000 };

Suppose you want to create the first 20 terms of the sequence $\{x_1, x_2, x_3, \ldots\}$ using the linear difference equation $x_{k+1} = Mx_k$ for $k = 0, 1, 2, \ldots$, where $x_0 = x0$. You can do this by first creating a function x satisfying $x[k] = x_k$ for $k = 0, 1, 2, \ldots$. When defining a new function with Mathematica, it is good practice to first clear out any previously defined values of the function using the Clear command as demonstrated in the first part of In[2] below. Next, the assignment statement x[0]:=x0 defines the value of $x[0]$ to be x0 and then $x[1], x[2], x[3], \ldots$ are recursively defined by letting x[k_]:=M.x[k-1]. The underscore immediately following k on the left hand side of the assignment statement is necessary in order to define k to be a variable and the colon-equals sign(:=), which is called a *delayed equals* sign, must be used instead of an equals sign (=), when defining a function recursively

In[2]:=Clear[x]; x[0] := x0; x[k_] := M.x[k−1];

Now that the function x is defined, the value of x_n is retrieved by simply executing $x[n]$.

In[3]:=x[100]   (* For example, determine the value of x_{100} *)
Out[3]={ 375054., 624946. }   (* The value of x_{100} is displayed in row form *)

Notice that Mathematica displays 6 significant digits for each entry in Out[3]. To see n digits of precision, replace x[100] with N[x[100], n] in In[3].

Executing Table[x[k], {k, 1, n}]//TableForm generates and displays the values of $x[1], x[2], \ldots, x[n]$ in a table where $x[k]$ is displayed horizontally on the kth line of the table (the output is not displayed here).

In[4]:=Table[x[k], { k, 1, 20 }]//TableForm   (* Generates the values of $x_1, x_2, \ldots, x_{20}$ *)

Note that Mathematica notation for evaluating the function x at k is x[k] where square brackets ([...]) are used instead of the more natural notation of $x(k)$ that is typically seen in mathematics. Be sure to use square brackets because Mathematica uses round brackets to group terms, but not in defining or using functions.

SECTION 2.1   Matrix Operations

Matrix addition and subtraction are performed using the plus (+) and minus (−) signs, respectively. Matrix multiplication is performed using a period (.). To compute A^k for an $n \times n$ matrix A, as defined on page 114 in the textbook, execute x1 = MatrixPower[A, k]. Computing A^k returns a matrix whose entries are the kth powers of the corresponding entries in A. Multiplying a scalar c and matrix A is performed by entering $c\,A$, leaving a space (or asterisk (*)) between the c and the A. This product can also be produced using an asterisk (*) or parenthesis by executing $c * A$, $c(A)$ or $(c)A$. To compute $A^T F$, enter Transpose[A].F . Here are a few other examples (the output is not displayed):

$$\text{In[1]}{:}{=}A = \begin{pmatrix} 1 & 2 & 3 \\ 0 & -1 & 4 \end{pmatrix}; \; B = \begin{pmatrix} 5 & 0 & -3 \\ 8 & 7 & 5 \end{pmatrix}; \; F = \begin{pmatrix} 2 & 3 \\ 1 & 0 \end{pmatrix};$$

In[2]:=A+B   In[3]:=A−B   In[4]:=F.A   In[5]:=5 A   In[6]:=MatrixPower[F, 25]

If $\mathbf{u}$ and $\mathbf{v}$ are vectors are in row form and have the same size, then $\mathbf{u.v}$ represents their inner product, and the outer product $\mathbf{u} \cdot \mathbf{v}^{\mathbf{T}}$ is computed with Transpose[{ u }].{ v } .

Special Matrices

Mathematica has commands that construct many special matrices. For example,

In[7]:=ZeroMatrix[5, 6]	(* A 5 × 6 matrix of zeros *)
In[8]:=IdentityMatrix[6]	(* The 6 × 6 identity matrix *)
In[9]:=DiagonalMatrix[{3, 5, 7, 3}]	(* A 4 × 4 diagonal matrix *)
In[10]:=RandInt[6]	(* A 6 × 6 matrix with random integer entries *)
In[11]:=RandInt[6, 4]	(* A 6 × 4 matrix consisting of random integer entries *)

Place the question mark (?) in front of any command to learn all the features of the command. For example, execute ?RandInt to learn more about the RandInt command.

SECTION 2.2   IdentityMatrix

The $n \times n$ identity matrix is denoted by IdentityMatrix[n]. If A is 5×5, then the command M=Augment[A, IdentityMatrix[n]] creates the augmented matrix $[A \;\; I]$. Use Gauss, BGauss and Scale (introduced in Section 1.4) to reduce $[A \;\; I]$. See page 1-17 in the Study Guide for more details on finding an inverse.

Mathematica has other commands that row reduce matrices, invert matrices, and solve equations $Ax = b$. They will be introduced later, after you have studied the concepts and algorithms in this section.

SECTION 2.3   Inverse

Determining whether a matrix is invertible is not always a simple matter. A fast and fairly reliable method is to use the command Inverse[A], which computes the inverse of A. If all the entries in A are in exact form (i.e., none of the entries in A contain decimal points) and if A is singular (noninvertible), then a message is displayed stating that A is singular. If at least one of the entries in A contains a decimal point, a warning message is given if A is singular or almost singular. In the latter case, Mathematica might not be able to find the inverse. But if you first replace each entry in A with its exact, rational equivalent, then the inverse is computed and displayed, even if it is almost singular.

Note that for large matrices, computing the inverse of A with Mathematica will be very slow if all the entries in A are exact, rational numbers. If you have already created a matrix A whose entries are exact integers or exact rational numbers, then the assignment statement A=N[A] or A=A//N will convert all entries in A to approximate real numbers. Mathematica will then be able to compute the approximate value the inverse of A much more quickly.

For Exercises 41-44, the command ConditionNumber[A] computes the condition number of a square matrix A, using what are called the singular values of A (discussed in Section 7.4). All of the entries in A must be approximate real numbers containing a decimal point. To perform the experiment described in Exercise 42, be sure the A contains approximate real numbers and then execute the following Mathematica instructions

   In[1]:=x=RandInt[4,1]; b=A.x; x1=Inverse[A].b; x-x1

Since A consists of approximate real numbers, the remaining computations will also result in approximate numbers since computations involving exact and approximate numbers always result in approximate numbers. Displaying the value of $x - x1$ is the best way to compare x and $x1$. You can execute the assignment statement again to repeat the experiment.

SECTION 2.4   Partitioned Matrices

The command A[[$\{i_i, i_2, \ldots, i_r\}, \{j_1, j_2, \ldots, j_s\}$]] creates a $r \times s$ submatrix of A whose entries consist of the numbers contained in rows $i_i, i_2, \ldots, i_r$ and in columns $j_1, j_2, \ldots, j_s$.

The BlockMatrix command creates partitioned matrices. For instance, if A, B, F, G, H, and J are matrices having appropriate sizes, the command BlockMatrix[{{A,B,F},{G,H,J}}] creates the larger matrix $\begin{pmatrix} A & B & F \\ G & H & J \end{pmatrix}$.

SECTION 2.5   LU Factorization

Row reduction of A using the command Gauss will produce the intermediate matrices needed for an LU factorization of A. You can try this on the matrix in Example 2, stored as Exercise 33 in C2S5.nb. The matrices in (5) on page 145 in the text are produced by the following commands (assuming we have created an assignment statement for matrix A in statement In[1]).

   In[2]:=Gauss[A, 1]   In[3]:=Gauss[%2, 2]   In[4]:=Gauss[%3, 3]

You can copy the information from the screen onto your paper, and divide by the pivot entries to produce L as in the text. For most text exercises, the pivots are integers and so are displayed accurately.

   To construct a permuted LU factorization, use Gauss[U,r,v], where r is the row index of the pivot and v is a row vector that lists the rows to be changed by replacement operations. For example, if A has 5 rows and the first pivot is in row 4, use

   In[5]:= Gauss[A, 4, {1, 2, 3, 5}]

If the next pivot is in row 2, use

   In[6]:=Gauss[%5, 2, {1, 3, 5}]

To build the permuted matrix L, use full columns from A or the partially reduced %6, divided by the pivots. Then change entries to zero if they are in a row already selected as a "pivot row".

   The add-on package GaussianElimination in the LinearAlgebra folder contains a command LU-Factor that produces a permuted LU factorization of any square matrix A. Another command in the package, LUSolve, uses LUFactor to solve $Ax = b$. These commands become useful when you need to solve $Ax = b$ repeatedly with a different b each time. For details, consult the Mathematica documentation.

   When A is invertible, an efficient way to solve $Ax = b$ with Mathematica is to use the command LinearSolve[A, b]. Mathematica proceeds to compute a permuted LU factorization of A and then uses L and U to compute x. The alternative command x = Inverse[A].b is less efficient and can be less accurate.

SECTION 2.8   RowReduce **and Rank**

The command RowReduce[A] produces the reduced row echelon form of A. From this matrix, a basis for the column space of A is obtained and the homogeneous equations describing the null space of A are formed by appending an extra column of zeros to the matrix.

   Mathematica does not contain a Rank command. If the entries of A are all exact, rational numbers, RowReduce[A] reveals the rank of A, since Mathematica performs exact arithmetic in finding the reduced row echelon form of A. If at least one of the entries of A contains a decimal point, then roundoff error or an extremely small pivot entry can produce an incorrect echelon form.

SECTION 3.2   Computing Determinants

To compute x^n, execute x^n or use $\square^\square$ in the BasicInput palette to enter x^n in a more natural way.

To compute det A, define A and then repeatedly use Gauss[%n, r] and Swap[%n, r, s] (where n is chosen so that %n refers to the output of the previous step of row reduction) as needed to reduce A to an echelon form.

Now suppose the echelon form is obtained on line Out[10]. On line In[11], let $U = \%10$. The command Product[f[i], {i, 1, n}] computes the product $f[1]f[2]\cdots f[n]$. Thus, if k represents the number of row swaps used in the reduction of A to U, then the determinant of A is given by

In[12]:=(−1)^k Product[U[[i, i]], { i, 1, Length[U]}]

A space between $(-1)^k$ and Product indicates scalar multiplication in Mathematica. An asterisk (∗) or parenthesis can also be used for scalar multiplication.

In[12]:=(−1)^k ∗ Product[U[[i, i]], { i, 1, Length[U]}]

In[12]:=(−1)^k(Product[U[[i, i]], { i, 1, Length[U]}])

You can use the command Det[A] to check your work, but the longer sequence of commands is preferred - at this time - because it emphasizes the process of computing determinants.

SECTION 4.1   Graphing Functions

You can use the following commands to define and plot the function f given in Exercise 37.

In[1]:=f[t_]:=1 - 8 Cos[t]^2 + 8 Cos[t]^4;
In[2]:=Plot[f[t],{t, 0, 2π}]

SECTION 4.3   RowReduce and LinearEquationsToMatrices

The command RowReduce[A] produces the reduced row echelon form of A. From that you can write a basis for Col A or write the homogeneous equations that describe Nul A. (Don't forget that A is a coefficient matrix, not an augmented matrix.)

Two consecutive equal signs, == , represent an equal sign appearing in an equation. For Exercise 37 in Section 4.3, an assignment statement is used to let eq represent equation (5), where c1, c2, c3, and c4 are constants representing $c_1, c_2, c_3,$ and c_4, respectively.

In[1]:=eq = c1 t + c2 Sin[t] + c3 Cos[2 t] + c4 Sin[t] Cos[t] == 0

To generate a system of four equations and 4 unknowns, use the Table command to replace equation (5) with the values of $t = 0, 1, 2,$ and 3 (Out[2] is not displayed).

In[2]:=Table[eq, { t, 0, 3 }]

Suppose eqlist is a list of linear equations containing unknowns $x_1, x_2, \ldots, x_n$. The command LinearEquationsToMatrices[eqlist, { x1, x2, ..., xn }] returns the list { A, b }, where A and

b form the matrix equation $Ax = b$ equivalent to eqlist. The coefficient matrix A and vector b, corresponding to set of equations given in Out[2], are found using In[3] (Out[3] is not displayed).

   In[3]:={ A, b } = LinearEquationsToMatrices[%2, { c1, c2, c3, c4 }]

For Exercise 34: If **t** is a vector and k is a positive integer, then Cos[t]^k is a vector of the same size as **t**, formed entry wise from **t** using the function $\cos^k(\ )$.

SECTION 4.4   LinearSolve

The command LinearSolve[A, b] solves $Ax = b$ where A is $m \times n$. If $Ax = b$ is inconsistent, then a message is returned stating that a solution does not exist. If more than one solution exists, one possible solution is returned. LinearSolve has additional capabilities that will be introduced later, after you have the appropriate background.

SECTION 4.6   Rank and Random Matrices

Mathematica does not contain a Rank command. If the entries of A are in exact, rational form, RowReduce[A] reveals the rank of A, since Mathematica performs exact arithmetic in finding the reduced row echelon form of A. If at least one of the entries of A contains a decimal point, then roundoff error or an extremely small pivot entry can produce an incorrect echelon form.

The command RandInt[m, n, r, k] can be used to create an $m \times n$ matrix of rank r containing random integers between $-k$ and k and RandInt[m, n, r] creates a $m \times n$ matrix of rank r with random integers between -9 and 9.

SECTION 4.7   Change-of-Coordinates Matrix

The notebook C4S7.nb has data for Exercises 7-10 and 17-19 and the vectors are provided in matrix form. Given vectors c_1, c_2, b_1, and b_2 in $\mathbf{R}^2$, the command

   In[1]:=M = Augment[c1, c2, b1, b2]

produces a 2×4 matrix whose columns are c_1, c_2, b_1, and b_2, respectively. (See the note for Section 1.3 for more information about Augment.) The command RowReduce[M] row reduces M to the desired form.

SECTION 4.8   Solve

In Exercises 7–16 and 25–28, the coefficients of the polynomial in the auxiliary equation are stored in row vector form, with coefficients in ascending order. For instance, the polynomial $r^2 - 25$ from Exercise 15 is represented by p = {-25, 0, 1}.

The Mathematica command PolyRoots[p] produces a row vector whose entries are the roots of the polynomial described by p.

SECTION 4.9   RandomStochastic

The command RandomStochastic[n] creates a random $n \times n$ stochastic matrix P with exact, rational entries. In Exercise 22, execute the assignment statement P=RandomStochastic[n]//N so that the entries in P are approximate numbers. You can then compute P^k with the MatrixPower command. (The option //N is added because otherwise Mathematica will compute P^k using exact arithmetic which could be very slow.)

Mathematica

See Section 1.10 for information on how to use Mathematica in these exercises.

SECTION 5.1   Finding Eigenvectors

The LayFunctions package contains a command that simplifies your homework by automatically producing a basis for an eigenspace when you know an eigenvalue. For example, if A is a 3×3 matrix with eigenvalue 7, the command

    In[1]:=NulBasis[A − 7 IdentityMatrix[3]]

produces a matrix whose columns form a basis for the eigenspace corresponding to $\lambda = 7$. Note the space before the identity matrix, which indicates multiplication by the number 7. (An asterisk (∗) can also be used for scalar multiplication.) In general, IdentityMatrix[k] is the $k \times k$ identity matrix, and NulBasis[C] is a matrix whose columns from a basis for Nul C (the same basis you would get if you started with the reduced echelon form of A and computed the basis by hand).

    For Exercises 37-40, you need the command Eigenvalues[A], which lists the eigenvalues of A. For example, enter

    In[2]:=ev=Eigenvalues[A];
    In[3]:=NulBasis[A − ev[[2]] IdentityMatrix[3]]

to compute a basis for the eigenspace corresponding to the second eigenvalue listed in ev, represented by ev[[2]] (in general, ev[[k]] is the kth number in the list ev).

SECTION 5.2   The Characteristic Polynomial and Plot

To obtain the characteristic polynomial of an $n \times n$ matrix A, execute Det[A-λ IdentityMatrix[n]], leaving a space between λ and IdentityMatrix[n] for scalar multiplication (an asterisk (∗) or parenthesis can also be used for scalar multiplication). You can use this to check your answers in Exercises 9-14. For example,

    In[1]:=Det[A-λ IdentityMatrix[5]]

returns the corresponding characteristic polynomial of a 5×5 matrix A. To plot the characteristic polynomial represented by %1 (the output to In[1]), execute

    In[2]:=Plot[%1, { λ, a, b }]

and the graph will be displayed for values of λ ranging from a to b. (The greek letter λ can be found in the palette.) The option PlotStyle→RGBColor[r, g, b] will add color to your plot. For instance, RGBColor[1,0,0], RGBColor[0,1,0], RGBColor[0,0,1], RGBColor[1,1,0], and RGBColor[1,.5,0] represent the colors red, green, blue, yellow and orange respectively. For instance, to plot the curve in blue, execute Plot[%1, { λ, a, b }, PlotStyle→RGBColor[1,0,0]].

For Exercises 28 and 29, use RandInt[4] to create a 4×4 matrix with random integer entries. For Exercise 29, use Gauss and perhaps Swap to create the echelon form without scaling. See the Mathematica notes for Section 1.4.

To produce graphs of the characteristic polynomials for the matrix A in Exercise 30, first execute the assignment statement containing the data for Exercise 30 in the notebook C5S2.nb.

In[3]:= a = 32; A = {{ -6, 28, 21 }, { 4, -15, -12 }, { -8, a, -25 }};

You can then form the characteristic polynomial for A, assign it a name such as p and then plot the polynomial. The polynomial is assigned the name plot1.

In[4]:=p=Det[A-λ IdentityMatrix[3]]; plot1 = Plot[p, { λ, 0, 3 }]

Feel free to add color to the plot using the option PlotStyle→RGBColor[r, g, b]. To create the characteristic polynomial for A when $a = 31.9$, change the assignment statement in In[3] so that a equals 31.9 and then reexecute In[3]. Then change the name plot1 to plot2 in In[4] and reexecute the command to obtain a new plot. In a similar manner, obtain plots plot3, plot4 and plot5 corresponding to the remaining values of a. Show[plot1, plot2, plot3, plot4, plot5] will then display all the graphs on one coordinate system.

SECTION 5.3   Diagonalization and Eigensystem

To practice the diagonalization procedure in this section, you should use NulBasis to produce eigenvectors. For Exercises 33-36, enter the command ev=Eigenvalues[A] to provide the eigenvalues. See the Mathematica notes for Section 5.1.

In later work, you may automate the diagonalization process. The command Eigensystem[A] finds both eigenvalues and eigenvectors, returning a list of eigenvalues followed by a list of corresponding eigenvectors (displayed in row form). Assuming that A has been defined in In[1], the assignment statement in In[2] below assigns the names evals and evecs to the list of eigenvalues and eigenvectors, respectively. In this way evecs[[i]] is an eigenvector corresponding to the eigenvalue evals[[i]] for $i = 1, 2, \ldots, n$. Line In[3] creates the diagonal matrix D (we use the name matD since D is a reserved Mathematica symbol). Since evecs lists the eigenvectors in row form, the transpose of this list produces P in In[4]. The command In[5] verifies that the diagonalization is correct. The //Simplify command is needed to simplify the products. A warning message will occur if matrix P in not invertible. If A is 4×4 or larger and the entries of A are rational numbers, then Mathematica may not be able to quickly compute the eigenvalues and eigenvectors of A. In this case, the entries of A should be entered in decimal form.

In[2]:={evals, evecs} = Eigensystem[A];
In[3]:= matD = DiagonalMatrix[evals];
In[4]:=P = Transpose[evecs];
In[5]:=A−P.matD.Inverse[P] //Simplify

SECTION 5.5   Complex Eigenvalues, Re and Im

The letter I is a protected Mathematica symbol representing the imaginary number i. If A is a 2×2 real matrix and if Eigenvalues[A] contains a complex eigenvalue, then you can use NulBasis to find a corresponding complex eigenvector $\mathbf{v}$. To build a 2×2 real matrix P from the real and imaginary parts of $\mathbf{v}$, use the Mathematica expressions Re[v] and Im[v], respectively. Then compute $C = P^{-1}AP$.

SECTION 5.6   Plotting Discrete Trajectories

For a given 2×2 matrix A, if $\mathbf{x}_{k+1} = A\mathbf{x}_k$ then the following commands are used to create $\mathbf{x}_k$ for $k = 1, 2, 3, \ldots$. As an example, the matrix A and initial vector x0 defined in In[1] are found in Example 6 in your textbook. The purpose of statement In[2] is to define a function x where x[k] represents $\mathbf{x}_k$. Statement In[3] uses the Table command to define a list called pts representing x[0] through x[50]. Then the command ListPlot[pts] points is used plot the points in the xy-plane and the trajectory is assigned the name traj1. The optional PlotStyle command is added to increase the size of the displayed points on the screen (without this optional command, the plotted points are small and barely visible).

In[1]:=A = $\begin{pmatrix} .8 & .5 \\ -.1 & 1.0 \end{pmatrix}$; x0={0, 2.5};
In[2]:=Clear[x]; x[0] = x0; x[i_] := A.x[i − 1];
In[3]:=pts = Table[x[i], { i, 0, 50 }]; traj1 = ListPlot[pts, PlotStyle→PointSize[.02]]

To create another trajectory, repeat the assignment statement on line In[1] with a different value of x0 and then repeat In[2] and In[3], replacing traj1 with traj2 to represent the new trajectory. If necessary, repeat this process to obtain traj3, traj4, ..., trajn and then execute Show[traj1, traj2,...,trajn] to display all the trajectories on a single plot.

For Exercise 17, create x[0] through x[8] representing $x_0, \ldots, x_8$. Then x[0][[1]] through x[8][[1]] represent the number of juveniles in years 0 through 8, respectively and x[0][[2]] through x[8][[2]] represent the number of adults in years 0 through 8, respectively. The following commands will produce a graph of the juvenile population.

In[4]:=juv = Table[{i, x[i][[1]]}, { i, 0, 8 }];
In[5]:=ListPlot[juv, PlotStyle→PointSize[.02]]

To graph the sum of the two entries in each of x[0] through x[8], execute the following two commands.

```
In[6]:=sum = Table[ {i, x[i][[1]]+x[i][[2]]}, { i, 0, 8 } ];
In[7]:=ListPlot[ sum, PlotStyle→PointSize[.02] ]
```

To graph the ratio of juveniles to adults, execute the following two commands, where / is the division key.

```
In[8]:=ratio = Table[ {i, x[i][[1]]/ x[i][[2]]}, { i, 0, 8 } ];
In[9]:=ListPlot[ sum, PlotStyle→PointSize[.02] ]
```

SECTION 5.7   Solutions to Differential Equations

To find the eigenvalues of A, use evals=Eigenvalues[A]. The eigenvectors shown in the text's answers were produced using commands such as evecs=NulBasis[A−evals[[1]] IdentityMatrix[3] . If the eigenvalue evals[[1]] is complex, the corresponding eigenvector v will be complex. The real and imaginary parts of evecs are Re[evecs] and Im[evecs], respectively.

If you use the command { evals, evecs } = EigenSystem[A], your eigenvectors will be in row form and they should be multiples of those in the text's answers (when the eigenspaces are one-dimensional). To test whether a vector v is a multiple of a vector w, compute v/w. This divides each entry in v by the corresponding entry in w. If v is a multiple of w, the result of v/w should be a vector whose entries are all equal.

SECTION 5.8   The Power Method

The steps for estimating the dominant eigenvalue appear on page 365 in your textbook. The function MaxMag determines the number in a list having the largest absolute value. In[1] below is an application of the power method algorithm to Example 2, beginning on the bottom of page 359, in your textbook. The goal of this string of commands is to create a list called lis which contains the sequence of numbers and vectors converging to the eigenvalue and a corresponding eigenvector, respectively. The command For[k=0, k<=5, k++, *body*] evaluates the *body* of commands for $k = 0, 1, \ldots, 5$. The value of y stores the current value of Ax_k, μ stores the current value of μ_k and then Append[lis, {μ, x}] adds the values of μ and x to lis. Finally x is overwritten with $(1/\mu)$ y, representing x_{k+1} in the power method algorithm. After the For loop, the lis statement causes the list to appear in an output statement (which is not displayed here). The greek letter μ can be found in the BasicInput palette.

```
In[1]:=A = ( 6 5
            1 2 ); lis = { }; x = { 0., 1. };
      For[ k=0, k<=5, k++,
        y = A.x;
        μ = MaxMag[ y ];
        lis = Append[ lis, { μ, x } ];
        x = (1/μ) y; ];
      lis
```

The Inverse Power Method

The inverse power method is a modification of the power method. The following code corresponds to Example 3 starting on page 367 in the textbook. The greek letters α and μ can be found in the BasicInput palette.

$$\text{In[1]}{:=}A \ = \ \begin{pmatrix} 10 & -8 & -4 \\ -8 & 13 & 4 \\ -4 & 5 & 4 \end{pmatrix}; \text{ lis} = \{\ \}; \text{x} = \{\ 1., 1., 1.\ \}; \alpha \ = \ 1.9;$$

```
For[k=0, k<=5, k++,
  y = LinearSolve[ A− α IdentityMatrix[3], x ];
  μ = MaxMag[ y ];
  v = α + 1/μ;
  lis = Append[ lis,{ μ, x } ];
  x = (1/μ) y ];
lis
```

SECTION 6.1   Inner product and Norm

If **u** and **v** are vectors in row form, then u.v is their inner product. The length or *norm* of **v** is Sqrt[v.v]. Using the $\sqrt{\square}$ button on the palette, you can also enter $\sqrt{v.v}$ to compute the norm. See the Mathematica notes for Section 2.1.

SECTION 6.2   Orthogonality

In Exercises 1 - 9 and 17 - 22, a quick way in Mathematica to test a set $\{\mathbf{u}_1, \mathbf{u}_2, \mathbf{u}_3\}$ for orthogonality is to create a matrix U = Transpose[{ u1, u2, u3}], where u1, u2, and u3 represent the vectors from the set in row form, and to test whether Transpose[U].U is a diagonal matrix. *Note:* U is constructed using the Transpose command in order to produce a matrix whose *columns* are $\mathbf{u}_1, \mathbf{u}_2, \ldots, \mathbf{u}_n$.

For row vectors **y** and **u**, the orthogonal projection of **y** onto **u** is

In[1]:=(y.u)/(u.u) u

SECTION 6.3   Orthogonal Projections

The orthogonal projection of a single vector onto a single vector is described in the Mathematica note for Section 6.2. The orthogonal projection onto the set spanned by an orthogonal set of vectors is the sum of the one-dimensional projections. Another way to construct this projection is to normalize the orthogonal vectors, place them in the columns of a matrix U, and use Theorem 10. For instance, if $\{\mathbf{y}_1, \mathbf{y}_2, \mathbf{y}_3\}$ is an orthogonal set of nonzero vectors (in row form), then the command

In[1]:=U=Transpose[{ y1/$\sqrt{\text{y1.y1}}$, y2/$\sqrt{\text{y2.y2}}$, y3/$\sqrt{\text{y3.y3}}$ }]

creates U, containing orthonormal columns, and

> In[2]:=U.(Transpose[U].y)

produces the orthogonal projection of y onto Span$\{y_1, y_2, y_3\}$. (The parentheses around the command Transpose[U].y speeds up the computation by avoiding a matrix-matrix product.)

SECTION 6.4   The Gram-Schmidt Process

Let A be a matrix with n linearly independent columns. In Mathematica, the columns of A (in row form) are Transpose[A][[k]] for $k = 1, \ldots, n$. If $n = 2$, the Gram-Schmidt process is

> In[1]:= x1 = Transpose[A][[1]]; x2 = Transpose[A][[2]];
> In[2]:=v1 = x1
> In[3]:=v2 = x2 − (x2.v1) / (v1.v1) v1

If A has three columns, add x3 = Transpose[A][[3]]; to the end of In[1] and then compute

> In[4]:=v3 = x3 − (x3.v1) / (v1.v1) v1 − (x3.v2) / (v2.v2) v2

You should use these commands for a while, to learn the general procedure. After that, you can use the command Proj[x, V], which computes the projection of the vector x onto the subspace spanned by the columns of a matrix V. It is important to note that V must be a matrix in floating point form, i.e., a matrix whose entries are numbers containing decimal points. The command N[V] converts the entries in V to floating point form. In the following, Transpose[{ v1 }] and converts row vector v1 into a matrix and Transpose[{ v1, v2 }] forms the matrix [v1 v2] whose columns are v1 and v2.

> In[3]:=v2 = x2 − Proj[x2 , N[Transpose[{ v1 }]]]
> In[4]:=v3 = x3 − Proj[x3 , N[Transpose[{ v1, v2 }]]]

To check your work, use the command GS[A], which uses the Gram-Schmidt process on the columns of A to produce a matrix whose columns are orthogonal. When the columns are normalized, they form the columns of Q in the QR factorization of A.

SECTION 6.5   Finding a Least Squares Solution

When A has linearly dependent columns, you can write the general description of all least-squares solutions on paper after you row reduce the augmented matrix for the normal equations:

> In[1]:= RowReduce[Augment[Transpose[A].A, Transpose[A].Transpose[{ b }]]]

When A has linearly independent columns, use the LinearSolve command to find a solution to the normal equations $A^T Ax = A^T b$.

> In[2]:=LinearSolve[Transpose[A].A, Transpose[A].b]

You can also enter (Inverse[Transpose[A].A).(Transpose[A].b) or use RowReduce as shown in In[1] above. For Exercises 15 and 16, see the Numerical Note on page 415 in the text and use the command LinearSolve[R, Transpose[Q].b] to solve $Rx = Q^T b$.

The command A=BlockMatrix[{ {A1}, {A2} }] can be used in Exercise 26 to create a partitioned matrix whose top block is $A1$ and whose bottom block is $A2$.

SECTION 6.6   Least-squares Solutions and Functions of Vectors

Once you create the design matrix X and the observation vector **y**, your computations for the least-squares solutions are the same as those described in the MATLAB box for Section 6.5. Here, A and **b** are replaced by X and **y**, respectively. The Mathematica command

In[1]:= RowReduce[Augment[Transpose[X].X, Transpose[X].Transpose[{ y }]]]

leads to the general description of all least-squares solutions. When X has linearly independent columns, the following command creates the least-squares solution.

In[2]:= LinearSolve[Transpose[X].X, Transpose[X]. y]

To construct the design matrix for an exercise in this section, you may need Mathematica's ability to compute functions of vectors. If **x** is a vector and k is a positive integer, then x^k is a vector the same size as **x** whose entries are the kth powers of the entries in **x**. The function Cos[x]^k was mentioned in the Mathematica notes in Section 4.3. The exponential function, Exp[x], and the natural logarithm function, Log[x], also act on each entry in **x**. The entries in the vector Exp[−.02 x], for example, are computed by applying the function $e^{-.02x}$ to the corresponding entries in **x**.

SECTION 6.7   Integrate **Command**

The Mathematica command Integrate[f[t], {t, a, b}] will attempt to compute $\int_a^b f(t)dt$ when f is integrable on $[a, b]$. Using the palette, you can enter the integral in a more natural way: $\int_a^b$f[t] dt.

To solve Exercise 28 with the help of Mathematica, begin by creating an inner product function where $\langle f, g \rangle = \int_a^b f(t)g(t)dt$.

In[1]:= Clear[ip]; ip[f_, g_] := $\int_0^{2\pi}$ (f g) dx

Now define p1 through p4:

In[2]:= Clear[p1,p2,p3,p4]; p1 = 1; p2 = Cos[t]; p3 = Cos[t]^2; p4 = Cos[t]^3;

Then perform the Gram-Schmidt process in a similar fashion as illustrated in the Mathematica notes for Section 6.4, but using the newly defined inner product. The first two steps are shown here:

In[3]:= Clear[q1,q2,q3,q4]; q1 = p1;

In[4]:= q2 = p2 - (ip[p2, q1] / ip[q1, q1]) q1;

SECTION 6.8   Graphing Functions

In Exercise 16, after you find f_4 and f_5 by hand computations, you can use Plot to graph them. For instance, to plot $f(t) = \sin t + \sin 3t$, you can execute

```
In[1]:= Clear[f]; f[t_]:=Sin[t]+Sin[3t]
In[2]:= Plot[ f[t], { t, 0, 2π } ]
```

See the Mathematica notes for Sections 4.1 and 5.2 for more details.

SECTION 7.1   Orthogonal Diagonalization

Use Eigenvalues[A] for eigenvalues and NulBasis to obtain eigenvectors, as in the Mathematica notes for Section 5.3. If you encounter a two-dimensional eigenspace, with a basis $\{v_1, v_2\}$, use the command

```
In[1]:= v2 = v2 − (v2.v1)/(v1.v1) v1;
```

to make a *new* eigenvector v_2 orthogonal to v_1. See the Mathematica note for Section 6.4. After you normalize the vectors and create P, verify that $P^T P - I$ equals zero to verify that P is indeed an orthogonal matrix.

SECTION 7.4   The Singular Value Decomposition

The command { evals, evecs } = Eigensystem[Transpose[A].A] produces a list of eigenvalues evals, in decreasing order, and a matrix evecs whose rows are the eigenvectors corresponding to the eigenvalues in evals. Therefore, assigning Σ to DiagonalMatrix[$\sqrt{\text{evals}}$] produces Σ and letting V equal Transpose[evecs] produces the matrix V. To form U for the SVD, normalize the nonzero columns of $A.V$. If U needs more columns, use the method of Example 4.

This construction helps you think about properties of the factorization. In practical work, however, you should use the command { Ut, σ, V }=SingularValues[Transpose[A].A]] which produces a list of singular values σ along with U and V where U=Transpose[Ut].

SECTION 7.5   Computing Principal Components

The command rmean=RowMean[X] produces a row vector whose jth entry lists the mean average of the jth row of X and DiagonalMatrix[rmean] creates a diagonal matrix whose diagonal entries are the row averages of X. Finally, the command rmean.Table[1, {Length[X]},{Length[X]}] creates a matrix the size of X, whose columns are the same, each one listing the row averages of X. To compute the data in X into mean-deviation form, use

```
In[1]:=B = X−rmean.Table[ 1, {Length[X]}, {Length[X]} ]
```

The sample covariance matrix is produced by

In[2]:=S = B.Transpose[B]/(Length[X]−1)

and the principle component data is produced by

In[3]:= { Ut , σ, V } = SingularValues[Transpose[B] / $\sqrt{\text{Length}[X] - 1}$]

where U=Transpose[Ut]. See the Numerical Note at the end of Section 7.5 of the text.

INDEX OF MATHEMATICA COMMANDS

Notes for the TI-83+/86/89

GETTING STARTED WITH A TI-83+ CALCULATOR

Entering Matrices: There are several ways to enter matrices in the TI-83+. One method consists of entering the matrix directly as an expression, while another method is through the use of the matrix editor found in the 2nd MATRX [EDIT] menu. Below are examples that illustrate the creation of matrices and vectors using these methods. On the TI-83+, a (column) vector is simply an n×1 column matrix. For additional information, consult the TI-83+ Guidebook.

Direct Entry: Begin a matrix with a square bracket, and each row of the matrix with a matching pair of square brackets; separate the entries in the rows with commas. Figure 1 below shows how you can enter the matrix $\begin{bmatrix} 1 & 2 & 3 \\ 4 & 5 & 6 \end{bmatrix}$ directly and assign to it the name [A] from the 2nd MATRX [EDIT] menu. The symbol → appears after pressing the STO▸ key.

```
[[1,2,3][4,5,6]]
→[A]
              [[1 2 3]
               [4 5 6]]
```

Figure 1: [A] entered directly.

The Matrix Editor: Press 2nd MATRX [EDIT]. The editor allows you to create a matrix by specifying a variable name, a size—row size and column size—and the entries of the matrix. After choosing [EDIT], a screen such as Figure 2 will appear. Select a name by pressing ENTER on a particular matrix, input rows, columns, and entries, your screen should be similar to Figure 3. To view the matrix from the home screen, press 2nd [QUIT] to return home, and then choose 2nd MATRX and select the matrix. You can create vectors in a similar fashion. Simply enter vectors in as n×1 column matrices.

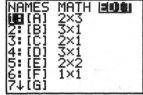

Figure 2: Matrix, edit menu

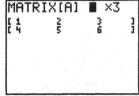

Figure 3: Matrix editor

GETTING STARTED WITH A TI-86 CALCULATOR

Entering Matrices: There are several ways to enter matrices and vectors into the TI-86. One method consists of entering the matrix directly as an expression, another method is through the use of the matrix (vector, respectively) editor found in the [MATRX] ([VECTR], respectively) menu. Below are examples that illustrate the creation of matrices and vectors using these methods. The entries in a matrix or vector can be either real or complex (pairs of real numbers). For additional information, consult the TI-86 Guidebook.

Direct Entry: Begin a matrix with a square bracket, and each row of the matrix with a matching pair of square brackets; separate the entries in the rows with commas. Figure 4 below shows how you can enter the matrix $\begin{bmatrix} 1 & 2 & 3 \\ 4 & 5 & 6 \\ 7 & 8 & 9 \end{bmatrix}$ directly and assign to it the name MAT1 (case sensitive)—

the symbol → appears after pressing the [STO▸] key—press [ENTER] to display the matrix MAT1 on the screen. For complex numbers, follow the same procedure, only enter the entries as pairs of real numbers (i.e. (real, imaginary)).

Figure 4: MAT1 entered directly.

The Matrix Editor: Press [2nd] (yellow key), [MATRX] ([7] key), to see [EDIT] in the menu. The editor allows you to create a matrix by specifying a variable name, a size—row size and column size—and the entries of the matrix. The editor, Figure 5, is accessed by pressing the [F2] menu key. Select from the matrix names shown or enter a new name for the matrix. For example to enter the matrix $Q = \begin{bmatrix} -1 & 0 & 4 \\ 5 & 1 & 3 \end{bmatrix}$, press Q and [ENTER]; your screen should be similar to Figure 6.

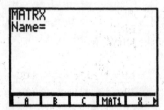

Figure 5: The matrix editor

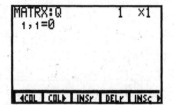

Figure 6: Matrix editor, ready
to receive the matrix Q.

The cursor is blinking at the top right, ready for you to enter the number of rows, press [2] and [ENTER]; for the number of columns, press [3] and [ENTER]. Type in the entries as prompted by rows. With the TI-86 use the cursor arrows to move up or down in a column, or left and right in a row. Press the [EXIT] key to exit the editor. To view the matrix Q on the screen, press [ALPHA] (blue key), [Q] ([4] key) and [ENTER] .

You can create vectors in a similar fashion. If using direct entry, a vector is entered as a row vector: [1, 2, 3, 4]. If using the vector editor, the procedure is similar to the matrix editor described above.

WARNING: Vectors and matrices are different objects on the TI-86. Operations in the [MATRX] menu are only for matrices, and operations in the [VECTR] menu are only for vectors. Vectors can also be entered as n×1 matrices; it will be impossible to use the operations in the [VECTR] menu on these. In the following notes the term vector is never used for an n×1 matrix. The programs [MtoV] and [VtoM] convert an n×1 matrix into a vector, or a vector into a matrix respectively.

GETTING STARTED WITH A TI-89 CALCULATOR

Entering Matrices: There are several ways to enter matrices into the TI-89. One method consists of entering the matrix directly as an expression, another method is through the use of the matrix editor found in the Data/Matrix Editor in the [APPS] menu. Below are examples that illustrate the creation of matrices using these methods. On the TI-89, a (column) vector is simply an n×1 column matrix. For additional information, consult the TI-89 Guidebook.

Direct Entry: Begin a matrix with a square bracket, and each row of the matrix with a matching pair of square brackets; separate the entries in the rows with commas. Figure 7 below shows how

you can enter the matrix $\begin{bmatrix} 1 & 2 & 3 \\ 4 & 5 & 6 \end{bmatrix}$ directly and assign to it the name aa—the symbol → appears

after pressing the [STO▸] key, press [ENTER] to display the matrix aa on the screen. Matrices can be denoted by nearly any combination of letters and numbers, but double lower-case letters will be used to denote matrices in this text. Entries can either be real or complex. To input complex entries, simply input them in with the form a+bi.

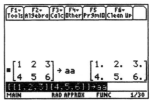

Figure 7: aa entered directly.

The Matrix Editor: Press [APPS], scroll down to the Data/Matrix Editor and press [ENTER]. See Figure 8. You will be asked whether you want to work on the current data item, open a data item, or make a new data item. Choose [New] to construct a new matrix. See Figure 9.

Figure 8: Choose data/matrix editor   Figure 9: Choose new

Use the right arrow to select either Data, Matrix, or List. Choose [Matrix] by pressing [ENTER]. Scroll down to Variable, type in the name of the matrix, input number of rows and columns in the appropriate places, and press [ENTER] twice. See Figure 10. If the variable name is not being used for anything else, the TI-89 will allow you to start inputing entries directly into the matrix. See Figure 11.

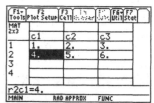

Figure 10:Choose name, rows and columns   Figure 11:Input the matrix

Once the matrix has been input, go home by pressing the [HOME] button. To view the matrix, type in the name that you assigned to the matrix and press [ENTER]. You can create vectors in a similar fashion. Simply enter vectors in as n×1 column matrices. If you want to change a matrix, select [Open] instead of [New], and open an existing matrix.

STUDY GUIDE NOTES

Data and Program Files

To use your TI calculator with the text, you will need to download to a computer the appropriate TI-data from the website www.laylinalgebra.com. Also available on this site is a selection of supplemental programs that will be beneficial in your study of linear algebra. The files are compressed and must be decompressed by a ZIP or Stuffit program (available from the website) into a single new folder on the computer. Then, transfer the files to your calculator. (Consult your TI guidebook for details.)

TI-83+ The data for each chapter of the text are stored in separate programs with names such as CHAPTER1 or CHAPTER4. The program that includes the supplemental programs is called ALINEAR. The files for each chapter are very large, so the chapters that are not in use must be either archived or deleted from memory. If other large files are stored into the calculator's RAM, they may have to be archived as well. To put data files in archive memory, press [2nd] [MEM] [5] and then choose the file you want to archive. Pressing [ENTER] will send the file to archive memory where it will no longer be accessible. To unarchive a file, follow the same steps with the UnArchive function.

When downloaded to the calculator, these programs will appear in the [PRGM] menu. To run a program for a particular chapter, go to the [PRGM] menu, scroll down to the chapter in question and press [ENTER]. The program will ask you for a section number and then an exercise number. It will then display where the data for the exercise is located; in matrices ranging from [A] to [F]. To run the program [ALINEAR], follow the same procedures and select the desired program and follow the directions for what to input.

TI-86   The data for each section of the text are stored in separate programs with names such as c1s2 or c2s3 for chapter 1 section 2, or chapter 2 section 3 respectively. The supplemental programs have names such as gauss, mdiav, or char. The files for each chapter are very large, and if all are downloaded at the same time, your calculator may not have enough memory to support them. Therefore you may want to download each chapter separately during the course. However, there is space for every file on a calculator with average memory use. You can clear unwanted things from your calculator's memory by accessing the [MEM] [DELET] menu.

When downloaded to the calculator, these programs will appear in the [PRGM] menu. To run a program for a particular chapter, go to the [PRGM] menu, select the chapter and section in question and press [ENTER]. The program will ask you for an exercise number. It will then display where the data for the exercise is located. To run any of the supplemental programs, simply select the desired program from the Program menu and follow the directions for what to input.

TI-89   The data for each chapter of the text are stored in separate programs with names such as chapter1 or chapter4. The program that includes the supplemental programs is called alinear. The files for each chapter are very large, so the chapters that are not in use may have to be either archived or deleted from memory. If other large files are stored into the calculator's RAM, they may have to be archived as well. However, there is space for every file on a calculator with average memory use. To put data files in archive memory, press [2nd] [VAR-LINK], choose the file you want to archive, and then press [F1] [8]. Pressing [ENTER] will send the file to archive memory where it will no longer be accessible. To unarchive a file, follow the same steps with the UnArchive function.

When downloaded to the calculator, these programs will appear in the [VAR-LINK] menu. To run a program for a particular chapter, go to the [VAR-LINK] menu, scroll down to the chapter in question and press [ENTER]. The program will ask you for a section number and then an exercise number. It will then display where the data for the exercise is located; in matrices ranging from aa to ff. To run the program alinear, follow the same procedures and select the desired program and follow the directions for what to input.

SECTION 1.1 Row Operations

TI-83+ In this exercise set, the data for each exercise are stored in a matrix [A]. Row operations on [A] are performed by the following commands:

rowswap([A],s,r)   Interchanges row r and s of [A]

*row(m,[A],s)   Multiples row s of [A] by a nonzero scalar m

row+(m,[A],s,r)   Replaces row r of [A] by (row r) + m(row s)

These operations are found by pressing [2nd] [MATRX] [MATH] and scrolling down to the bottom, Figures 12 and 13.

Figure 12: Matrix, math menu   Figure 13: Matrix, math menu extended

If you enter one of these commands, say, rowswap([A],1,3), then the new matrix, produced from [A], is stored in the "ANS" (for "answer"). If instead you enter, rowswap([A],1,3)→[B], the answer is stored in a new matrix [B].

The advantage of giving a new name to each new matrix is that you can easily go back a step if you don't like what you just did to a matrix. If, instead you enter rowswap([A],1,3)→[A], then the result is placed back in [A] and the "old" [A] is lost. Of course the "reverse" operation rowswap([A],1,3)→[A], will bring back the old [A].

Note: For the simple problems in this section and the next, the multiple m you need in the command *row+(m,[A],s,r) will usually be a small integer or fraction that you can compute in your head. In general, m may not be so easy to compute mentally. The next two paragraphs describe how to handle such a case.

The entry in row r and column c of a matrix [A] is denoted by [A](r, c). If the number stored in [A](r, c) is displayed with a decimal point, then the displayed values may be accurate to only about five digits. In this case, use the symbol [A](r, c) instead of the displayed value in calculations.

For instance, if you want to use the entry (pivot) [A](s, c) to change [A](r, c) to 0, enter the commands

    -[A](r,c)/[A](s,c)→m   Multiple of row s to be added to row r

    *row+(m,[A],s,r)→[A]   Adds m times row s to row r

or you can just use the command

    *row+(-[A](r,c)/[A](s,c),[A],s,r)→[A] .

TI-86   In this exercise set, the data for each exercise are stored in a matrix M. Row operations on M are performed by the following commands:

    rSwap(M,r,s)   Interchanges rows r and s of M
    multR(m,M,r)   Multiplies row r of M by a nonzero scalar m
    mRAdd(m,M,s,r)   Replaces row r of M by (row r) + m × (row s)

These operations are found on the second page of the [MATRX] [OPS] menu. (Press the [2nd] key followed by [7] to bring up the [MATRX] menu, Figure 14. The row operations are found when you press the [F4] menu key followed by the [MORE] key, Figure 15.)

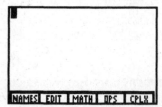
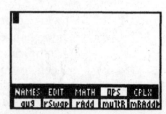

Figure 14: The matrix menu   Figure 15: Matrix menu: row operations

To use the row operations efficiently it is convenient to store matrices in memory. The name of any matrix in your calculator memory can be inserted in place of M; the letters r, m, and s stand for numbers you choose. Press ENTER after each command.

If you enter one of these commands, say, rSwap(M,1,3), then the new matrix, produced from M, is stored in the matrix "ANS" (for "answer"). If instead you enter rSwap(M,1,3)→M1, then the answer is stored in a new matrix M1. If the new operation is mRAdd(5,M1,1,2)→M2, then the result of changing M1 is placed in M2, and so on.

The advantage of giving a new name to each new matrix is that you can easily go back a step if you don't like what you just did to a matrix. If, instead you enter mRAdd(5,M,1,2)→M, then the result is placed back in M and the "old" M is lost. Of course the "reverse" operation mRAdd(-5,M,1,2)→M, will bring back the old M.

Note: For the simple problems in this section and the next, the multiple m you need in the command mRAdd(m,M,s,r) will usually be a small integer or fraction that you can compute in your head. In general, m may not be so easy to compute mentally. The next two paragraphs describe how to handle such a case.

The entry in row r and column c of a matrix M is denoted by M(r, c). If the number stored in M(r, c) is displayed with a decimal point, then the displayed values may be accurate to only about five digits. In this case, use the symbol M(r, c) instead of the displayed value in calculations.

For instance, if you want to use the pivot entry M(s, c) to change M(r, c) to 0, enter the commands

    -M(r,c)/M(s,c))→m   Multiple of row s to be added to row r
    mRAdd(m,M,s,r)→M   Adds m times row s to row r

or you can just use the command
    mRAdd(-M(r,c)/M(s,c),M,s,r)→M.

TI-89 In this exercise set, the data for each exercise are stored in a matrix aa. Row operations on aa are performed by the following commands:

    mRowAdd(m,aa,s,r)   Replaces row r of aa by (row r) + m × (row s)
    rowSwap(aa,r,s)   Interchanges rows r and s of aa
    mRow(m,aa,r)   Multiplies row r of aa by a nonzero scalar m

These operations are found by pressing 2nd [MATH], scrolling down to, or pressing ④ [Matrix] as seen in Figure 16. Scrolling down again to [J]Row ops. in Figure 17, and pressing the right arrow shows these four options.

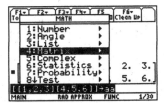
Figure 16: Matrix menu

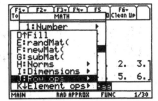
Figure 17: Row ops in matrix menu

To use the row operations efficiently it is convenient to store matrices in memory. The name of any matrix in your calculator memory can be inserted in place of aa; the letters r, m, and s stand for numbers you choose. Press ENTER after each command.

If you enter one of these commands, say, rowSwap(aa,1,3) then the new matrix, produced from aa, is stored in the matrix "ans(1)" (for "answer"). If instead you enter rowSwap(aa,1,3)→bb, then the answer is stored in a new matrix bb. If the new operation is mRowAdd(5,bb,1,2)→cc, then the result of changing bb is placed in cc, and so on.

The advantage of giving a new name to each new matrix is that you can easily go back a step if you don't like what you just did to a matrix. If, instead you enter mRowAdd(5,aa,1,2)→aa, then the result is placed back in aa and the "old" aa is lost. Of course the "reverse" operation mRowAdd(-5,aa,1,2)→aa, will bring back the old aa.

Note: For the simple problems in this section and the next, the multiple m you need in the command mRowAdd(m,aa,s,r) will usually be a small integer or fraction that you can compute in your head. In general, m may not be so easy to compute mentally. The next two paragraphs describe how to handle such a case.

The entry in row r and column c of a matrix aa is denoted by aa[r, c]. If the number stored in aa[r, c] is displayed with a decimal point, then the displayed values may be accurate to only about five digits. In this case, use the symbol aa[r, c] instead of the displayed value in calculations.

For instance, if you want to use the pivot entry aa[s, c] to change aa[r, c] to 0, enter the commands

    -aa[r,c]/aa[s,c]→m   Multiple of row s to be added to row r
    mRowAdd(m,aa,s,r)→aa   Adds m times row s to row r

or you can just use the command
    mRowAdd(-aa[r,c]/aa[s,c],aa,s,r)→aa

SECTION 1.3 Constructing a Matrix

TI-83+ The augment(operator concatenates two matrices or a matrix and a vector. For instance, if [A] is a matrix and [B] is a vector **b**, then the augmented matrix for the equation A**x**=**b** is given by augment([A],[B]). To construct this, press [2nd] [MATRX] [MATH] and select augment(. Fill in [A] and [B] separated by a comma and press [ENTER]. Matrices must have the same number of rows in order to augment them together. Given a vector equation such as $x_1\mathbf{a}_1 + x_2\mathbf{a}_2 + x_3\mathbf{a}_3 = \mathbf{b}$, you might store $\mathbf{a}_1$ in [A], $\mathbf{a}_2$ in [B], $\mathbf{a}_3$ in [C], and **b** in [D]. Then the augmented matrix for this system is created by the command

    augment(augment(augment([A],[B]),[C]),[D])→[E]

Excercises 11-14, 25-28, and 21 can be solved using the commands rowswap(, *row(, and row+(, described in notes for Section 1.1.

TI-86 The aug operator concatenates two matrices or a matrix and a vector. (see the second page of the [MATRX] [OPS] menu. Press [2nd] [MATRX] [F4] [MORE] to view the command, then [F1] to select it, Figure 6. Fill in with the two desired arguments and press [ENTER].) Matrices must have the same number of rows in order to augment them together. Given a vector equation such as $x_1a_1 + x_2a_2 + x_3a_3 = b$, the command aug(aug(a1,a2),aug(a3,b))→M creates the matrix $\begin{bmatrix} \mathbf{a}_1 & \mathbf{a}_2 & \mathbf{a}_3 & \mathbf{b} \end{bmatrix}$ and saves it as M; but, if you enter a1,a2,a3 and b as vectors, you must convert a1 and a3 to n × 1 matrices before using the aug command. (Use the [VtoM] program in the [PRGM] menu.) The same matrix is created by the command aug(A,b)→M—assuming that the matrix A and the vector **b**, have been created.

Exercises 11–14 and 25–28 and 21 can be solved using the commands mRAdd, rSwap, and multR, described in notes for Section 1.1.

TI-89   The augment(operator concatenates two matrices or a matrix and a vector. For instance, if aa is a matrix and bb is a vector **b**, then the augmented matrix for the equation Ax=**b** is given by augment(aa,bb). It can be found by pressing [2nd] [MATH], scrolling down to, or pressing [4] [Matrix]. Then scroll down to [7] [augment]. Fill in aa and bb separated by a comma and press [ENTER]. Matrices must have the same number of rows in order to augment them together. Given a vector equation such as $x_1\mathbf{a}_1 + x_2\mathbf{a}_2 + x_3\mathbf{a}_3 = \mathbf{b}$, you might store $\mathbf{a}_1$ in aa, $\mathbf{a}_2$ in bb, $\mathbf{a}_3$ in cc, and **b** in dd. Then the augmented matrix for this system is created by the command

   augment(augment(augment(aa,bb),cc),dd)→ee

Excercises 11-14, 25-28, and 21 can be solved using the commands mRowAdd(, rowSwap(, mRow(, described in notes for Section 1.1.

SECTION 1.4 Gauss and Bgauss

TI-83+   To solve Ax = **b**, row reduce the matrix C = [A **b**]. Matrix [C], or any other for that matter, can be constructed by the command

   augment([A],[B])→[C].

To speed up row reduction of [C], the program [GAUSS] (in the [ALINEAR] program) will use the leading entry in row r of [C] as a pivot, and use row replacements to create zeros in the pivot column below this pivot entry. The output of [GAUSS] is stored in the matrix [J]. If you wish, you can assign [J] to some other variable, such as [A] (the first matrix in the list).

   For the backward phase of row reduction, use the [BGAUSS] program, which selects the leading entry in row r as the pivot, and creates zeros in the column *above* the pivot. Use *row(to create leading 1's in the pivot positions. The output of [BGAUSS] is stored in the matrix [J]. If you wish, you can assign [J] to some other variable, such as [A] (the first matrix in the list).

   To run a program, see the notes on Data and Program Files at the beginning of the study guide notes. Both [GAUSS] and [BGAUSS] will prompt the user for the matrix and the row number.

TI-86   To solve Ax = **b**, row reduce the matrix C = [A **b**]. Matrix C, or any other for that matter, can be constructed by the command

   aug(A,B)→C.

To speed up row reduction of C, the program [GAUS] (in the [PRGM] menu) will use the leading entry in row r of M as a pivot, and use row replacements to create zeros in the pivot column below this pivot entry. The output of [GAUS] is stored in the matrix U. If you wish, you can assign U to some other variable, such as M itself.

   For the backward phase of row reduction, use the [BGAUS] program, which selects the leading entry in row r as the pivot, and creates zeros in the column *above* the pivot. Use multR to create leading 1's in the pivot positions. The output of [BGAUS] is stored in the matrix U. If you wish, you can assign U to some other variable, such as M itself.

   To run a program, see the notes on Data and Program Files at the beginning of the study guide notes. Both [GAUS] and [BGAUS] will prompt the user for the matrix and the row number.

TI-89   To solve Ax = **b**, row reduce the matrix [A **b**]. Matrix C, or any other for that matter, can be constructed by the command augment(aa,bb)→cc.

   To speed up row reduction of C, the program [gauss] (in the [alinear] program) will use the leading entry in row r of C as a pivot, and use row replacements to create zeros in the pivot column below this pivot entry. The output of [gauss] is stored in the matrix xx.

For the backward phase of row reduction, use the [bgauss] program, which selects the leading entry in row r as the pivot, and creates zeros in the column *above* the pivot. Use mRow(to create leading 1's in the pivot positions. The output of [bgauss] is stored in the matrix xx.

To run a program, see the notes on Data and Program Files at the beginning of the study guide notes. Both [gauss] and [bgauss] will prompt the user for the matrix and the row number.

SECTION 1.5 Zero Matrices

TI-83+ The command {m,n}→dim([A]) creates an m×n matrix and stores it in [A]. The command dim(is found by pressing [2nd] [MATRX] [MATH] and scrolling down. Using the command Fill(allows you to fill in a matrix with a particular value. Press [2nd] [MATRX] [MATH] and select Fill(. Then press zero and matrix [A] separated by a comma, and press [ENTER]. Matrix [A] should then be displayed with zeros as its entries.

TI-86   The command {m,n}→dim A creates an m×n matrix of zeros and stores it in A. The same procedure applies to vectors. The '{' and '}' symbols are found in the [LIST] menu, and the dim command is in the [MATRX] [OPS] menu When solving the equation A**x** = **0** create an augmented matrix aug(A,Z)→M, and assign it to the variable M, or execute the commands:

A→M

{m,n+1}→dim M   m and n are the number of rows and columns of A

to reformat the matrix A as an m×(n + 1) matrix with a column of zeros appended and save it as M. Then use [GAUS], [BGAUS] and multR to row reduce M completely.

TI-89   The command newmat(m,n)→aa creates an m×n matrix of zeros and stores it in aa. The function newmat(is found by pressing [2nd] [MATH], and scrolling down to, or pressing [4] [Matrix]. Then scroll down to newmat(.

SECTION 1.6 Rational Format

TI-83+ Chemical equation-balance problems are studied best using exact or symbolic arithmetic, because the balance variables must be whole numbers (with no round-off allowed). To convert floating-point numbers shown in matrices to rational answers, the command

[MATH] [Frac] [ENTER]

will give the last answer in fractions, if possible. This function will convert a number or a matrix into rational form. Once you find a rational solution of a chemical equation-balance problem, you can multiply the entries in the solution vector by a suitable integer to produce a solution that involves only whole numbers.

TI-86   Chemical equation-balance problems are studied best using exact or symbolic arithmetic, because the balance variables must be whole numbers (with no round-off allowed). To convert floating-point numbers shown in matrices to rational answers, the command

[2nd] [MATH] [MISC] [Frac] [ENTER]

will give the last answer in fractions, if possible. This function will convert a number or a matrix into rational form. Once you find a rational solution of a chemical equation-balance problem, you can multiply the entries in the solution vector by a suitable integer to produce a solution that involves only whole numbers.

TI-89   Chemical equation-balance problems are studied best using exact or symbolic arithmetic, because the balance variables must be whole numbers (with no round-off allowed). To convert floating-point numbers shown in matrices to rational answers, try setting the mode, page 2, to exact. This will give the answers in fractions, if possible. This setting will convert a number or a matrix into rational form. Once you find a rational solution of a chemical equation-balance problem, you can multiply the entries in the solution vector by a suitable integer to produce a solution that involves only whole numbers.

SECTION 1.10 Generating a Sequence

TI-83+ The data program for exercises 9-13 for Section 1.10 stores vectors into the matrix [B]. Use the command

   [B]*x [ENTER]

to generate the next value in the sequence. To continue generating the sequence press *x [ENTER] repeated times. This will compute the answer stored in memory times x.

TI-86   The data program for exercises 9-13 for Section 1.10 stores vectors into v1. Use the command

   v1*x [ENTER]

to generate the next value in the sequence. To continue generating the sequence press *x [ENTER] repeated times. This will compute the answer stored in memory times x.

TI-89   The data program for exercises 9-13 for Section 1.10 stores vectors into the matrix bb. Use the command

   bb*x [ENTER]

to generate the next value in the sequence. To continue generating the sequence press *x [ENTER] repeated times. This will compute the answer stored in memory times x.

SECTION 2.1 Matrix Notation and Operations

TI-83+ To create a matrix, begin with a square bracket, enter the data row-by-row, with a comma between entries, each row of the matrix must begin and end with square brackets. For instance, the command

   [[1,2,3][4,5,6]]→[A]

creates a 2×3 matrix [A]. If [A] is m×n, then dim([A]) (found in the [2nd] [MATRX] [MATH] menu) is the list {m n}. The (i, j)-entry in the matrix [A] is [A](i, j). To extract either the jth row or the jth column from a matrix, use program [JROW] or [JCOL] respectively. Each program will prompt for a matrix, and row or column number, and display the item in question.

   The TI-83+ uses the [+] , [−] , and [×] keys to denote matrix addition, subtraction and multiplication, respectively. (Multiplication shows on the screen and in these notes as *.) If [A] is square and k is a positive integer, [A] [^] k denotes the kth power of [A] ([A] [x²], is equivalent to [A] [^] 2). The transpose of [A] is [A]^T (the ^{T} operator is in the [2nd] [MATRX] [MATH] menu).

   If [A] and [B] are two column vectors of the same size, their inner product (or dot product) can be computed by [A]^T*[B] or [B]^T*[A].

   In this section you can experiment with random matrices. The command randM(returns a random matrix with integer entries in the interval [-9,9]. Go to the [2nd][MATRX] [MATH] menu to activate the command; it requires that you enter the two arguments—row size and column size—

then close the parenthesis and press ENTER . (To save the random matrix into memory press the STO◆ key and enter the name of the matrix before you press ENTER.)

TI-86   To create a matrix, begin with a square bracket, enter the data row-by-row, with a comma between entries, each row of the matrix must begin and end with square brackets. For instance, the command

        [[1,2,3][4,5,6]]→A

creates a 2×3 matrix A. If A is m×n, then dim A (found in the [MATRX] [OPS] menu) is the list {m n}. The (i, j)-entry in the matrix A is A(i, j). A(i) is the ith row of the matrix A, *given as a vector*. To extract the jth column of A, use the program [COL]. The program prompts for a matrix and a column number *j*, and returns the *j*th column as a vecto*r*.

        The TI-86 uses the ⊞ , ⊟ , and ⊠ keys to denote matrix addition, subtraction and multiplication, respectively. (Multiplication shows on the screen and in these notes as *.) If A is square and k is a positive integer, A ⌃ k denotes the kth power of A (A x^2, is equivalent to A ⌃ 2). The transpose of A is A^T (the ^{T} operator is in the [MATRX] [MATH] menu). Note: when A has complex entries, the (i, j)-entry of A^T is the complex conjugate of the (i, j)-entry of A.

        If **u** and **v** are two vectors of the same size, then dot(u,v) is their inner product (the dot operator is in the [VECTR] [MATH] menu). If the vectors are entered as n×1 matrices the inner product is v^T*u.

        In this section you can experiment with random matrices. The command randM returns a random matrix with integer entries in the interval [−9, 9]. Go to the [MATRX] [OPS] menu to activate the command; it requires that you enter the two arguments—row size and column size— then close the parenthesis and press ENTER . (To save the random matrix into memory press the STO◆ key and enter the name of the matrix before you press ENTER .)

TI-89   To create a matrix, begin with a square bracket, enter the data row-by-row, with a comma between entries, each row of the matrix must begin and end with square brackets. For instance, the command

        [[1,2,3][4,5,6]]→aa

creates a 2×3 matrix aa. If aa is m×n, then dim(aa) (found by pressing 2nd [MATH], scrolling down to, or pressing ④ [Matrix], scrolling down to [l]dimensions and pressing the right arrow.) is the list {m,n}. The [i, j]-entry in the matrix aa is aa[i, j]. aa[i] is the ith row of the matrix aa, *given as a vector*. To extract the jth column of aa, use the program [col], the program will prompt for a matrix, and column number, it will return the jth column *as a vector*.

        The TI-89 uses the ⊞, ⊟, and ⊠ keys to denote matrix addition, subtraction and multiplication, respectively. (Multiplication shows on the screen and in these notes as *.) If A is square and k is a positive integer, A ⌃ k denotes the kth power of A. The transpose of A is A^T (the ^{T} operator is found by pressing 2nd [MATH], and scrolling down to, or pressing ④[Matrix]).

        If **u** and **v** are two vectors of the same size, then dotP(u,v) is their inner product (the dotP(operator is found by pressing 2nd [MATH], and scrolling down to, or pressing ④ [Matrix]. Then scroll down to L:Vector ops and press the right arrow). The vectors can either be input as n×1 matrices or 1×n matrices when computing dot products.

        In this section you can experiment with random matrices. The command randMat(returns a random matrix with integer entries in the interval [−9, 9]. Press 2nd [MATH], scroll down to, or press ④ [Matrix], and scroll down to randMat(It requires that you enter the two arguments—row size and column size—then close the parenthesis and press ENTER . (To save the random matrix into memory press the [STO▸] key and enter the name of the matrix before you press [ENTER].)

SECTION 2.2 The Identity Matrix and A⁻¹

TI-83+ The n×n identity matrix is identity(n) (the identity(command is in the [2nd] [MATRX] [MATH] menu). If [A] is an n×n matrix, then the command
        augment([A],identity(n))→[B]
creates the augmented matrix [B] = [A   I]. Use the [GAUSS] and [BGAUSS] programs, and the *row(operation to reduce the augmented matrix [A I] completely.

There are other commands that can be used to row reduce matrices, invert matrices, and solve equations A**x** = **b**, but they are not discussed here because they will not help you to learn the concepts in this section.

TI-86 The n×n identity matrix is ident n (the ident command is in the [MATRX] [OPS] menu). If A is n×n , then the command
        aug(A,ident n)→M
creates the augmented matrix M = [A I]. Use the [GAUS] and [BGAUS] programs, and the multR operation to reduce [A I].

There are other commands that can be used to row reduce matrices, invert matrices, and solve equations A**x** = **b**, but they are not discussed here because they will not help you to learn the concepts in this section.

TI-89 To get an n×n identity matrix press identity(n) (the command is found by pressing [2nd] [MATH], and scrolling down to, or pressing [4] [Matrix]). If aa is an n×n matrix, then the command
        augment(aa,identity(n))→bb
creates the augmented matrix bb = [aa   I]. Use the [gauss] and [bgauss] programs, and the mRow(operation to reduce [aa   I].

There are other commands that can be used to row reduce matrices, invert matrices, and solve equations A**x** = **b**, but they are not discussed here because they will not help you to learn the concepts in this section.

SECTION 2.3 Characterizations of Invertible Matrices

TI-83+ Determining whether a specific numerical matrix is invertible is not always a simple matter. A fast and fairly reliable method is to enter the matrix [A] and press [A] [x⁻¹], which computes the inverse of [A]. If the matrix in question is not invertible, the TI-83+ will either display "SINGULAR MAT." or "ERR:INVALID DIM"

TI-86 Determining whether a specific numerical matrix is invertible is not always a simple matter. A fast and fairly reliable method is to enter the matrix A and press [2nd] A [x⁻¹], which computes the inverse of A. If the matrix in question is not invertible, the TI-86 will display the message "SINGULAR MAT". For exercises 38-41, the condition number, cond, is in the [MATRX] [MATH] menu.

TI-89 Determining whether a specific numerical matrix is invertible is not always a simple matter. A fast and fairly reliable method is to enter the matrix A and press [^] [(-)] [1], which computes the inverse of A. If the matrix in question is not invertible, the TI-89 will either display "Error: Dimension", or "Singular matrix".

SECTION 2.4 Partitioned Matrices

TI-83+ The TI-83+ uses partitioned matrix notation. For example, if [A], [B], [C], [D], [E], and [F] are matrices of appropriate sizes, then the command

augment(augment(augment([A],[B]),[C])$^\mathsf{T}$,augment(augment([D],[E]),[F])$^\mathsf{T}$)$^\mathsf{T}$➔[G]

creates a larger matrix of the form $[G] = \begin{bmatrix} A & B & C \\ D & E & F \end{bmatrix}$. Once [G] is formed, there is no record of the

partition that was used to create [G]. For instance, although [B] was the (1, 2)-block used to form [G], the number [G](1, 2) is the same as the (1, 2)-entry of [A].

TI-86 The TI-86 uses partitioned matrix notation. For example, if A, B, C, D, E, and F are matrices of appropriate sizes, then the command

aug(aug(aug(A,B),C)$^\mathsf{T}$,aug(aug(D,E),F)$^\mathsf{T}$)$^\mathsf{T}$➔M

creates a larger matrix of the form $M = \begin{bmatrix} A & B & C \\ D & E & F \end{bmatrix}$. Once M is formed, there is no record of the

partition that was used to create M. For instance, although B was the (1, 2)-block used to form M, the number M(1, 2) is the same as the (1, 2)-entry of A.

TI-89 The TI-89 uses partitioned matrix notation. For example, if aa, bb, cc, dd, ee, and ff are matrices of appropriate sizes, then the command

augment(augment(augment(aa,bb),cc)$^\mathsf{T}$,augment(augment(dd,ee),ff)$^\mathsf{T}$)$^\mathsf{T}$➔gg

creates a larger matrix of the form $gg = \begin{bmatrix} A & B & C \\ D & E & F \end{bmatrix}$. Once gg is formed, there is no record of the

partition that was used to create gg. For instance, although bb was the [1, 2]-block used to form gg, the number gg[1, 2] is the same as the [1, 2]-entry of aa.

SECTION 2.5 LU Factorization

TI-83+ Row reduction of A using the [GAUSS] program will produce the intermediate matrices needed for an LU factorization of A. You can try this on the matrix in Example 2. The matrices in (5) on page 145 in the text are produced by running the commands

      [GAUSS][A] 1   Returns a matrix [J] with 0's below the first pivot
      [GAUSS][J] 2   Returns a matrix [J] with 0's below pivots 1 and 2
      [GAUSS][J]3   Returns an echelon form, [J]

You can copy the information from your screen onto your paper, and divide by the pivot entries to produce L as in the text. For most text exercises, the pivots are integers and so are displayed accurately.

      The TI-83+ program [LU] produces U and an n×2n matrix J for some square matrices, but it does not handle the general case. From J, you can perform row swaps to make L and P. The left half of the matrix will form L and the right half will form P.

TI-86   Row reduction of A using the [GAUS] program will produce the intermediate matrices needed for an LU factorization of A. You can try this on the matrix in Example 2. The matrices in (5) on page 145 in the text are produced by running the commands

    [GAUS]A 1   Returns a matrix U with 0's below the first pivot
    [GAUS]U 2   Returns a matrix U with 0's below pivots 1 and 2
    [GAUS]U 3   Returns an echelon form, U

You can copy the information from your screen onto your paper, and divide by the pivot entries to produce L as in the text. For most text exercises, the pivots are integers and so are displayed accurately.

    The TI-86 command LU(A,L,U,P) produces a permuted LU factorization for some square matrices A, but it does not handle the general case.

TI-89   Row reduction of A using the [gauss] program will produce the intermediate matrices needed for an LU factorization of A. You can try this on the matrix in Example 2. The matrices in (5) on page 145 in the text are produced by running the commands

    [gauss]aa 1   Returns a matrix xx with 0's below the first pivot
    [gauss]xx 2   Returns a matrix xx with 0's below pivots 1 and 2
    [gauss]xx 3   Returns an echelon form, xx

You can copy the information from your screen onto your paper, and divide by the pivot entries to produce L as in the text. For most text exercises, the pivots are integers and so are displayed accurately.

    The TI-89 command LU aa,ll,uu,pp produces a permuted LU factorization for some square matrices aa, but it does not handle the general case.

SECTION 2.8 rref

TI-83+ The command rref(in the [2nd] [MATRX] [MATH] menu, produces the reduced row echelon form of a given matrix. This form gives you enough information to write down a basis for Col A, and the homogeneous equations that describe Nul A. (Don't forget that A is a coefficient matrix, not an augmented matrix.) The command rref(does not work if the number of columns in A is less than the number of rows in A. If this is the case, you can augment A with enough columns of zeros, and work with a square matrix. Use the command as usual, then ignore the columns of zeros when you interpret the output.

TI-86   The command rref A, in the [MATRX] [OPS] menu, produces the reduced row echelon form of A. This form gives you enough information to write down a basis for Col A, and the homogeneous equations that describe Nul A. (Don't forget that A is a coefficient matrix, not an augmented matrix.) The command rref does not work if the number of columns in A is less than the number of rows in A. If this is the case, you can augment A with enough columns of zeros, and work with a square matrix. Use the command as usual, then ignore the columns of zeros when you interpret the output.

TI-89   The command rref(, found by pressing [2nd] [MATH], and scrolling down to, or pressing [4] [Matrix], produces the reduced row echelon form of a matrix A. This form gives you enough information to write down a basis for Col A, and the homogeneous equations that describe Nul A. (Don't forget that A is a coefficient matrix, not an augmented matrix.) The command rref(does not work if the number of columns in A is less than the number of rows in A. If this is the case, you can augment A with enough columns of zeros, and work with a square matrix. Use the command as usual, then ignore the columns of zeros when you interpret the output.

SECTION 2.9 Rank

TI-83+ With a TI-83+, you should use an echelon form of A to determine the rank of A. You can row reduce A using [GAUSS], or you can use one of the commands rref(or ref(from the [2nd] [MATRX] [MATH] menu. The ref([A]) command produces a row echelon form of [A]. (Don't forget that A is a coefficient matrix, not an augmented matrix.) Both rref(and ref(only work on matrices with at least as many columns as rows. However, you can augment A with enough columns of zeros to make a square matrix, if necessary, and then ignore the columns of zeros when you interpret the output. With each of these three methods, roundoff error from an extremely small pivot entry can sometimes produce an incorrect echelon form.

TI-86 With a TI-86, you should use an echelon form of A to determine the rank of A. You can row reduce A using [GAUS], or you can use one of the commands rref or ref from the [MATRX] [OPS] menu. The ref A command produces a row echelon form of A. (Don't forget that A is a coefficient matrix, not an augmented matrix.) Both rref and ref only work on matrices with at least as many columns as rows. However, you can augment A with enough columns of zeros to make a square matrix, if necessary, and then ignore the columns of zeros when you interpret the output. With each of these three methods, roundoff error from an extremely small pivot entry can sometimes produce an incorrect echelon form.

TI-89 With a TI-89, you should use an echelon form of A to determine the rank of A. You can row reduce A using [gauss], or you can use one of the commands rref(or ref(found by pressing [2nd] [MATH], and scrolling down to, or pressing [4] [Matrix]. The ref(A) command produces a row echelon form of A. (Don't forget that A is a coefficient matrix, not an augmented matrix.) Both rref(and ref(only work on matrices with at least as many columns as rows. However, you can augment A with enough columns of zeros to make a square matrix, if necessary, and then ignore the columns of zeros when you interpret the output. With each of these three methods, roundoff error from an extremely small pivot entry can sometimes produce an incorrect echelon form.

SECTION 3.2 Computing Determinants

TI-83+ To compute det A, use the [GAUSS] program and rowswap(, repeatedly, to reduce A to a matrix U, which is an echelon form of A. Keep track of how many times you swap rows. Then except for a ±1, the determinant is obtained by using the program [MDIAV] to extract the diagonal of U, followed by the [PRDCT] program to compute the product of the diagonal. You can, of course, use the det(selection from the [2nd] [MATRX] [MATH] menu) to check your work, but the longer sequence of commands helps you to think about the *process* of computing det A.

TI-86 To compute det A, use the [GAUS] program and rSwap, repeatedly, to reduce A to a matrix U, which is an echelon form of A. Keep track of how many times you swap rows. Then, except possibly for a factor of ±1, the determinant is obtained by using the program [MdiaV] to extract the diagonal of U, followed by the [PRDCT] program to compute the product of the diagonal. You can, of course, use the det function (on the first page of the [MATRX] [MATH] menu) to check your work, but the longer sequence of commands helps you to think about the *process* of computing det A. For Exercise 46, the condition number, cond, is in the [MATRX] [MATH] menu.

TI-89   To compute det A, use the [gauss] program and rowSwap(, repeatedly, to reduce A to a matrix U, which is an echelon form of A. Keep track of how many times you swap rows. Then, except possibly for a factor of ±1, the determinant is obtained by using the program [mdiav] (in the [alinear] program) to extract the diagonal of U, followed by the [prdct] program to compute the product of the diagonal. You can, of course, use the det(command found by pressing [2nd] [MATH], and scrolling down to, or pressing [4] [Matrix], to check your work, but the longer sequence of commands helps you to think about the *process* of computing det A.

SECTION 4.3 ref and rref

TI-83+ With a TI-83+, you should use an echelon form of A to determine the rank of A. You can row reduce A using [GAUSS], or you can use one of the commands rref(or ref(from the [2nd] [MATRX] [MATH] menu. The ref([A]) command produces a <u>r</u>ow <u>e</u>chelon <u>f</u>orm of [A]. (Don't forget that A is a coefficient matrix, not an augmented matrix.) Both rref(and ref(only work on matrices with at least as many columns as rows. However, you can augment A with enough columns of zeros to make a square matrix, if necessary, and then ignore the columns of zeros when you interpret the output.   With each of these three methods, roundoff error from an extremely small pivot entry can sometimes produce an incorrect echelon form.

TI-86   With a TI-86, you should use an echelon form of A to determine the rank of A. You can row reduce A using [GAUS], or you can use one of the commands rref or ref from the [MATRX] [OPS] menu. The ref A command produces a <u>r</u>ow <u>e</u>chelon <u>f</u>orm of A. (Don't forget that A is a coefficient matrix, not an augmented matrix.) Both rref and ref only work on matrices with at least as many columns as rows. However, you can augment A with enough columns of zeros to make a square matrix, if necessary, and then ignore the columns of zeros when you interpret the output. With each of these three methods, roundoff error from an extremely small pivot entry can sometimes produce an incorrect echelon form.

TI-89   With a TI-89, you should use an echelon form of A to determine the rank of A. You can row reduce A using [gauss], or you can use one of the commands rref(or ref(found by pressing [2nd] [MATH], and scrolling down to, or pressing [4] [Matrix]. The ref(A) command produces a <u>r</u>ow <u>e</u>chelon <u>f</u>orm of A. (Don't forget that A is a coefficient matrix, not an augmented matrix.) Both rref(and ref(only work on matrices with at least as many columns as rows. However, you can augment A with enough columns of zeros to make a square matrix, if necessary, and then ignore the columns of zeros when you interpret the output. With each of these three methods, roundoff error from an extremely small pivot entry can sometimes produce an incorrect echelon form.

SECTION 4.4 The Inverse Operator $^{-1}$

TI-83+ If the equation Ax = **b** has a unique solution and A is a square matrix, the TI-83+ will automatically produce **x** if you use the command

[A]$^{-1}$*[B]→X

In this section, the equation will probably have the form **Pu** = **x**, with **u** the B-coordinate vector of **x**, and the command will be [P]⁻¹*[X]➔U

TI-86 If the equation A**x** = **b** has a unique solution and A is a square matrix, the TI-86 will automatically produce **x** if you use the command
        A⁻¹*b➔x
In this section, the equation will probably have the form **Pu** = **x**, with **u** the B-coordinate vector of **x**, and the command will be P⁻¹*x➔u.

TI-89 If the equation A**x** = **b** has a unique solution and A is a square matrix, the TI-89 will automatically produce **x** if you use the command
        aa⁻¹*bb➔x
In this section, the equation will probably have the form **Pu** = **x**, with **u** the B-coordinate vector of **x**, and the command will be pp⁻¹*xx➔uu.

SECTION 4.6 Rank and Random Matrices

TI-83+ With a TI-83+, you should use an echelon form of A to determine the rank of A. You can row reduce A using [GAUSS], or you can use one of the commands rref(or ref(from the [2nd] [MATRX] [MATH] menu.

    In this section you can experiment with random matrices. The command randM(returns a random matrix with integer entries in the interval [-9,9]. Go to the [2nd] [MATRX] [MATH] menu to activate the command; it requires that you enter the two arguments—row size and column size— then close the parenthesis and press [ENTER]. (To save the random matrix into memory press the [STO▶] key and enter the name of the matrix before you press [ENTER].)

TI-86 With a TI-86, you should use an echelon form of A to determine the rank of A. You can row reduce A using [GAUS] or you can use one of the commands rref or ref from the [MATRX] [OPS] menu.

    In this section you can experiment with random matrices. The command randM returns a random matrix with integer entries in the interval [−9, 9]. Go to the [MATRX] [OPS] menu to activate the command; it requires that you enter the two arguments—row size and column size— then close the parenthesis and press [ENTER]. (To save the random matrix into memory press the [STO▶] key and enter the name of the matrix before you press [ENTER].)

TI-89 With a TI-89, you should use an echelon form of A to determine the rank of A. You can row reduce A using [gauss], or you can use one of the commands rref(or ref(found by pressing [2nd] [MATH], and scrolling down to, or pressing [4] [Matrix].

    In this section you can experiment with random matrices. The command randMat(returns a random matrix with integer entries in the interval [−9, 9]. Press [2nd] [MATH], scroll down to, or press [4] [Matrix], and scroll down to randMat(It requires that you enter the two arguments—row size and column size—then close the parenthesis and press [ENTER]. (To save the random matrix into memory press the [STO▶] key and enter the name of the matrix before you press [ENTER].)

SECTION 4.7 Change of Coordinates Matrix

TI-83+ The rref(command will completely reduce a matrix $\begin{bmatrix} \mathbf{c}_1 & \mathbf{c}_2 & \mathbf{b}_1 & \mathbf{b}_2 \end{bmatrix}$ to the desired form. See the note for Section 1.3 to on how to construct this matrix.

TI-86 The rref command will completely reduce a matrix $\begin{bmatrix} \mathbf{c}_1 & \mathbf{c}_2 & \mathbf{b}_1 & \mathbf{b}_2 \end{bmatrix}$ to the desired form. See the note for Section 1.3 to on how to construct this matrix.

TI-89 The rref(command will completely reduce a matrix $\begin{bmatrix} \mathbf{c}_1 & \mathbf{c}_2 & \mathbf{b}_1 & \mathbf{b}_2 \end{bmatrix}$ to the desired form. See the note for Section 1.3 to on how to construct this matrix.

SECTION 4.8 Roots

TI-83+ To find the roots of a polynomial, use the solve(function. This can be found by pressing [2nd] [CATALOG] "s" and then scroll down. While looking in the catalog, the TI-83+ automatically puts the alpha lock on. The command should look similar to this:

        solve(polynomial,variable,guess,{lower bound,upper bound})
For additional help, consult your TI-83+ Guidebook.

TI-86 To find the roots of a polynomial, use the [2nd] [POLY] sequence. Input the order (degree) of the polynomial, and the coefficients as prompted (press [ENTER] after each entry), then press the [SOLVER] menu key. The roots of the polynomial are displayed on the screen. Complex roots appear as pairs of real numbers. Refer to your TI-86 Guidebook.

TI-89 To find the roots of a polynomial, use the solve(function found in the F2:Algebra menu. Press [ENTER] on the correct function, input the equation and the variable, separated by a comma, close the parentheses, and press [ENTER]. Refer to your TI-89 Guidebook for questions.

SECTION 4.9 Markov Chains

TI-83+ The notes for Section 1.10 contain information that is useful for the exercises in this section.

TI-86 The notes for Section 1.10 contain information that is useful for the exercises in this section.

TI-89 The notes for Section 1.10 contain information that is useful for the exercises in this section.

SECTION 5.1 Finding Eigenvectors

TI-83+ The program [NULB] (in the [ALINEAR] program) will simplify your homework by automatically producing a basis for an eigenspace. For example, if A is a 3×3 matrix with an eigenvalue 7, first input [A], then use the keystrokes

        [A]-7identity(3)→[C]
to produce the matrix [C] = [A − 7I]. Then run the [NULB] program; at the prompt for a matrix, enter [C]. The output is a matrix [J] whose columns form a basis for the eigenspace of A

corresponding to λ = 7. In general identity(k) (in the [2nd] [MATRX] [MATH] menu) produces the k×k identity matrix and [NULB] produces a matrix whose columns form a basis for Nul C (the same basis you would get if you started with rref([C]) and calculated by hand).
Remarks:
1.   The program [NULB] uses the command rref(and requires that the number of columns of the input matrix be greater than or equal to the number of rows.
2.   If the numbers in the basis matrix B are messy, try the sequence of keystrokes [MATH] [Frac] [ENTER] which shows the last answer in fractions, if possible.

TI-86   The program [NULB] (in the [PRGM] menu) will simplify your homework by automatically producing a basis for an eigenspace. For example, if A is a 3×3 matrix with an eigenvalue 7, first store A, then use the command
          A-7*ident 3 →C
to produce the matrix C = A − 7I. Run the [NULB] program. At the prompt for a matrix, enter C. The output is a matrix B whose columns form a basis for the eigenspace of A corresponding to λ = 7. In general ident k (in the [MATRX][OPS] menu) produces the k×k identity matrix and [NULB] produces a matrix whose columns form a basis for Nul C (the same basis you would get if you started with rref C and calculated by hand).
Remarks:
1.   The program [NULB] uses the command rref and requires that the number of columns of the input matix be greater than or equal to the number of rows.
2.   If the numbers in the basis matrix B are messy, try the sequence of keystrokes [2nd] [MATH] [MISC] [Frac] [ENTER], which shows the last answer in fractions, if possible.

TI-89   The program [nulb] (in the [alinear] program) will simplify your homework by automatically producing a basis for an eigenspace. For example, if aa is a 3×3 matrix with an eigenvalue 7, first store aa, then use the command
          aa-7identity(3)→cc
to produce the matrix C= A − 7I. Then run the [nulb] program; at the prompt for a matrix, enter cc. The output is a matrix xx whose columns form a basis for the eigenspace of A corresponding to λ = 7. In general identity(k) (found by pressing [2nd] [MATH], and scrolling down to, or pressing [4][Matrix]) produces the k×k identity matrix and [nulb] produces a matrix whose columns form a basis for Nul C (the same basis you would get if you started with rref(cc) and calculated by hand).
Remarks:
1.   The program [nulb] uses the command rref(and requires that the number of columns of the input matrix be greater than or equal to the number of rows.
2.   If the numbers in the basis matrix B are messy, try setting the mode, page 2, to exact.

SECTION 5.2 CHARA

TI-83+ You can use the [CHARA] program to check your answers in Exercises 9–14. Note that if A is n×n, running this program produces a vector listing the coefficients of the characteristic polynomial of A, in order of decreasing powers of λ, beginning with $λ^n$. If the polynomial is of odd degree, the coefficients are multiplied by −1, to make +1 the coefficient of $λ^n$; this corresponds to finding the determinant of $\lambda I - A$. ([CHARA] works for matrices up to size 3×3.)

TI-86 You can use the [CHAR] program to check your answers in Exercises 9–14. Note that if A is n×n, running this program produces a vector listing the coefficients of the characteristic polynomial of A, in order of decreasing powers of λ, beginning with λ^n. If the polynomial is of odd degree, the coefficients are multiplied by –1, to make +1 the coefficient of λ^n; this corresponds to finding the determinant of $\lambda I - A$. ([CHAR] works for matrices up to size 3×3.)

TI-89 You can use the [chara] program to check your answers in Exercises 9–14. Note that if A is n×n, running this program produces a vector listing the coefficients of the characteristic polynomial of A, in order of decreasing powers of λ, beginning with λ^n. If the polynomial is of odd degree, the coefficients are multiplied by –1, to make +1 the coefficient of λ^n; this corresponds to finding the determinant of $\lambda I - A$. ([chara] works for matrices up to size 3×3.)

SECTION 5.3 Diagonalization

TI-83+ To practice the diagonalization procedure in this section, you should use [NULB] to produce eigenvectors. For Exercises 33-36, you have to find the eigenvalues first. See notes from Section 5.1

The program [EIGEN] calculates eigenvalues for a given square matrix with less than four rows. For each eigenvalue, use [NULB] on $A - \lambda I$ to create a matrix whose column(s) give a basis for the eigenspace corresponding to the given eigenvalue.

TI-86 The command eigVl A→ev (eigVl is in the [MATRX] [MATH] menu) produces a list, of the eigenvalues of the matrix A. To learn the diagonalization procedure, you should use the method of Section 5.1 to produce the eigenvectors. For each eigenvalue, use [NULB] on $A - \lambda I$ to create a matrix whose column(s) give a basis for the eigenspace corresponding to the given eigenvalue.

In later work you can automate the diagonalization process. The command eigVc→P ([eigVc] is in the [MATRX] [MATH] menu) produces a matrix P such that AP = PD, where D is the diagonal matrix you can create from the list of eigenvalues, with the eigenvalues in the diagonal. (Use the command li▸vc, in the [LIST] menu, to change **ev** into a vector, then the program [VdiaM], with **ev** as input, to produce D.) If A happens to be diagonalizable, then P will be invertible. In any case, P is likely to be quite different from what you construct for your homework. The columns of P may be scaled. (The sequence [2nd][MATH][MISC][Frac][ENTER] shows the last answer in fractions, if possible.)

TI-89 The command eigVl(aa) (found by pressing [2nd] [MATH], and scrolling down to, or pressing [4] [Matrix]) produces a list of the eigenvalues of the matrix aa. To learn the diagonalization procedure, you should use the method of Section 5.1 to produce the eigenvectors. For each eigenvalue, use [nulb] on $A - \lambda I$ to create a matrix whose column(s) give a basis for the eigenspace corresponding to the given eigenvalue.

In later work you can automate the diagonalization process. The command eigVc→pp (found by pressing [2nd] [MATH], and scrolling down to, or pressing [4] [Matrix]) produces a matrix pp (or P) such that AP = PD, where D is the diagonal matrix you can create from the list of eigenvalues, with the eigenvalues in the diagonal. If A happens to be diagonalizable, then P will be invertible. In any case, P is likely to be quite different from what you construct for your homework. The columns of P may be scaled.

SECTION 5.5 Complex Eigenvalues

TI-83+ The 83+ does not compute complex eigenvalues and eigenvectors directly.

TI-86   The eigVl and eigVc functions (mentioned in Section 5.3) also work for matrices with complex eigenvalues. In this case the resulting list of eigenvalues or the matrix containing the eigenvectors as columns have some complex entries.

   For any matrix V the real and imag functions in the [MATRX] [CPLX] menu produce the real and imaginary parts of the entries in V, displayed as matrices of the same size as V.

TI-89   The eigVl(and eigVc(keys (mentioned in Section 5.3) also work for matrices with complex eigenvalues. In this case the resulting list of eigenvalues or the matrix containing the eigenvectors as columns have some complex entries. To view these complex values in the form $a+bi$, the Complex format in the Mode menu must be set to rectangular.

SECTION 5.6 Plotting Discrete Trajectories

TI-83+ Given a vector [B] (representing a point x_k) and a matrix [A] (the transition matrix) the command [A]*[B]→[B] will compute the "next" point on the trajectory. Use the [2nd] [ENTRY] keys to repeat the command over and over.

TI-86   Given a vector **x** (representing a point x_k) and a matrix A (the transition matrix) the command A*x→x displays x_{k+1}, the "next" point on the trajectory, and stores it in **x**. Use the [2nd] [ENTRY] keys to repeat the command over and over.

TI-89   Given a vector **x** (representing a point x_k) and a matrix A (the transition matrix) the command A*x→x will compute the "next" point on the trajectory. Use the [ENTER] key to repeat the command over and over.

SECTION 5.8 Power Method and Inverse Power Method

TI-83+ Set your calculator to display as many decimal places as possible. The algorithms below assume that A has a strictly dominant eigenvalue, and the initial vector is **x**, with largest entry 1 (in magnitude).

   The Power Method: When the following steps are executed over and over, the values of **x** (also known as matrix [B]) approach (in many cases) an eigenvector for a strictly dominant eigenvalue. (The program [VAMX] prompts for a vector and returns the entry in the vector with the maximum absolute value and stores it in M.)

[A]*[B]→[C]		(1)
[VAMX]	Returns M = estimate for the eigenvalue	(2)
[C]/M→[B]	Estimate for the eigenvector	(3)

As these commands are repeated, the numbers that appear are the μ_k that approach the dominant eigenvalue. You can program your TI-83+ to perform this sequence a certain number of times by using a loop structure. See the TI-83+ Guidebook for more information about programming with the TI-83+.

The Inverse Power Method: Store the initial estimate for the eigenvalue in the variable Z and enter the command [A]-Z*Identity(n)→[C], where n is the number of columns of A. Then enter the commands

[C]⁻¹*[B]→[D]	Solves the equation $(A - ZI)y = x$	(1)
[VAMX]		(2)
2nd [ANS] [x⁻¹]+Z→M	M = estimate for the eigenvalue	(3)
[C]/M→[B]	Estimate for the eigenvector	(4)

As these commands are repeated, lines (3) and (4) produce the sequences $\{v_k\}$ and $\{x_k\}$ described in the text.

TI-86   Set your calculator to display as many decimal places as possible. The algorithms below assume that A has a strictly dominant eigenvalue, and the initial vector is **x**, with largest entry 1 (in magnitude). (If your initial vector is called x_0, rename it by entering x_0→x.)

The Power Method: When the following steps are executed over and over, the values of **x** approach (in many cases) an eigenvector for a strictly dominant eigenvalue (the program [VAMX] prompts for a vector and returns the entry in the vector with the maximum absolute value and stores it in **mu**.)

A*x→y		(1)
[VAMX]	**mu** = estimate for the eigenvalue	(2)
y/mu→x	Estimate for the eigenvector	(3)

As these commands are repeated, the numbers that appear are the μ_k that approach the dominant eigenvalue. You can program your TI-86 to perform this sequence a certain number of times by using a loop structure. See the TI-86 Guidebook for more information about programming with the TI-86.

The Inverse Power Method: Store the initial estimate for the eigenvalue in the variable **a**, and enter the command A-a*ident n→C, where n is the number of columns of A. Then enter the commands

C⁻¹*x→y	Solves the equation $(A - aI)y = x$	(1)
[VAMX]		(2)
2nd [ANS] 2nd [x⁻¹]+a→nu **nu** = estimate for the eigenvalue		(3)
y/nu→x	Estimate for the eigenvector	(4)

As these commands are repeated, lines (3) and (4) produce the sequences $\{v_k\}$ and $\{x_k\}$ described in the text.

TI-89 Set your calculator to display as many decimal places as possible. The algorithms below assume that A has a strictly dominant eigenvalue, and the initial vector is **x**, with largest entry 1 (in magnitude).

The Power Method: When the following steps are executed over and over, the values of **x** approach (in many cases) an eigenvector for a strictly dominant eigenvalue. The program [vamx] prompts for a vector and returns the entry in the vector with the maximum absolute value and stores it in m.

aa*xx→yy		(1)
[vamx]	m = estimate for the eigenvalue	(2)
yy/m→xx	Estimate for the eigenvector	(3)

As these commands are repeated, the numbers that appear are the μ_k that approach the dominant eigenvalue. You can program your TI-89 to perform this sequence a certain number of times by using a loop structure. See the TI-89 Guidebook for more information about programming with the TI-89.

The Inverse Power Method: Store the initial estimate for the eigenvalue in the variable z and enter the command aa-z*identity(n)→yy, where n is the number of column of aa. Then enter the commands

yy ⌃ (-) 1 *xx→zz	Solves the equation $(A - ZI)\mathbf{y} = \mathbf{x}$	(1)
[vamx]		(2)
uu ⌃ (-) 1 +z→m	m = estimate for the eigenvalue	(3)
yy/m→xx	Estimate for the eigenvector	(4)

As these commands are repeated, lines (3) and (4) produce the sequences $\{v_k\}$ and $\{x_k\}$ described in the text.

SECTION 6.1 Inner Product and Norm

TI-83+ If [A] and [B] are two column vectors of the same size, their inner product (or dot product) can be computed by $[A]^T*[B]$ or $[B]^T*[A]$. To calculate the norm of a vector, you need only take the square root of the dot product of a vector with itself. The keystrokes needed to find the length of the vector stored in [A] are:

$[A]^T*[A]$ Computes the dot product of [A] with itself

[2nd] [√] [2nd] [ANS] (1,1) Computes the square root of the dot product

See the note for Section 2.1.

TI-86   The inner product of two vectors is found with the dot command in the [VECTR] [MATH] menu. The norm command in the same menu produces the length of a vector. See the note for Section 2.1.

TI-89   The inner product of two vectors is found with the dotP(command (found by pressing [2nd] [MATH], and scrolling down to, or pressing 4 [Matrix], then choosing L:Vector ops). The norm(command, in the same matrix menu but under H:Norms, produces the length of a vector. See the note for Section 2.1.

SECTION 6.2 Orthogonality

TI-83+ In Exercises 1–9 and 17–22, the fastest way (counting keystrokes) with the TI-83+ to test a set such as $\{\mathbf{u}_1, \mathbf{u}_2, \mathbf{u}_3\}$ for orthogonality is to use a matrix $[A] = [\mathbf{u}_1 \ \mathbf{u}_2 \ \mathbf{u}_3]$. See the proof of Theorem 6.

For vectors **y** and **u**, the orthogonal projection of **y** onto **u**, (called [B] and [C] respectively) is:

$([B]^T*[C]/[C]^T*[C])*[C]$

TI-86   In Exercises 1–9 and 17–22, the fastest way (counting keystrokes) with the TI-86 to test a set such as $\{\mathbf{u}_1, \mathbf{u}_2, \mathbf{u}_3\}$ for orthogonality is to use a matrix $U = [\mathbf{u}_1 \ \mathbf{u}_2 \ \mathbf{u}_3]$. See the proof of Theorem 6. For vectors **y** and **u**, the orthogonal projection of **y** onto **u**, is:

(dot(y,u)/dot(u,u))*u

TI-89   In Exercises 1–9 and 17–22, the fastest way (counting keystrokes) with the TI-89 to test a set such as $\{\mathbf{u}_1, \mathbf{u}_2, \mathbf{u}_3\}$ for orthogonality is to use a matrix $U = [\mathbf{u}_1 \ \ \mathbf{u}_2 \ \ \mathbf{u}_3]$. See the proof of Theorem 6. For vectors **y** and **u,** the orthogonal projection of **y** onto **u,** denoted yy and uu, is:

(dotP(yy,uu)/dotP(uu,uu))*uu

SECTION 6.3 Orthogonal Projections

TI-83+ The orthogonal projection of **y** onto a single vector was described in the TI-83+ note for Section 6.2. The orthogonal projection onto the set spanned by an orthogonal set of nonzero vectors is the sum of the one-dimensional projections. Another way to construct this projection is to normalize the orthogonal vectors, (use the [UNITV] program) place them in the columns of a matrix [A], and use Theorem 10. For instance if {[B], [C], [D]} is an orthogonal set of nonzero vectors, they can be normalized by running the [UNITV] program for each vector. Then, to construct [A], input the following commands:

   augment([B],[C])→[A]   Augments [B] and [C] together
   augment([A],[D])→[A]   Augments [B] and [C] and [D] together

The resulting matrix [A] has orthonormal columns, and $[A]*[A]^T*y$ produces the orthogonal projection of y onto the subspace spanned by {[B], [C], [D]}.

TI-86 The orthogonal projection of **y** onto a single vector was described in the TI-86 note for Section 6.2. The orthogonal projection onto the set spanned by an orthogonal set of nonzero vectors is the sum of the one-dimensional projections. Another way to construct this projection is to normalize the orthogonal vectors, (use the unitV function in the [VECTR] [MATH] menu) place them in the columns of a matrix U, and use Theorem 10. For instance if $\{\mathbf{y}_1, \mathbf{y}_2, \mathbf{y}_3\}$ is an

orthogonal set of nonzero vectors, then the matrix $U = \left[\dfrac{y_1}{\text{norm}(y_1)} \quad \dfrac{y_2}{\text{norm}(y_2)} \quad \dfrac{y_3}{\text{norm}(y_3)} \right]$

can be created with the sequence:

  unitV y1→u1   **u1** is a unit vector in the direction of **y1**
  [VtoM] u1→U   Convert **u1** to an n×1 matrix and store in U
  unitV y2→u2   **u2** is a unit vector in the direction of **y2**
  aug(U,u2)→U   Append u2 to U
  unitV y3→u3   **u3** is a unit vector in the direction of **y3**
  aug(U,u3)→U   Append **u3** to U.

The resulting matrix U has orthonormal columns, and $U*U^T*y$ produces the orthogonal projection of y onto the subspace spanned by $\{\mathbf{y}_1, \mathbf{y}_2, \mathbf{y}_3\}$.

TI-89   The orthogonal projection of **y** onto a single vector was described in the TI-89 note for Section 6.2. The orthogonal projection onto the set spanned by an orthogonal set of nonzero vectors is the sum of the one-dimensional projections. Another way to construct this projection is to normalize the orthogonal vectors, (use the unitV(command by pressing [2nd] [MATH], and scrolling down to, or pressing [4] [Matrix] and then choosing L:Vector Ops) place them in the columns of a matrix U, and use Theorem 10. For instance if {aa, bb, cc} is an orthogonal set of nonzero vectors, they can be normalized by the unitV(command for each vector. Then to construct U, input the following commands:

   augment(aa,bb)→uu   Augments aa and bb together

```
augment(uu,cc)→uu
```
Augments aa and bb and cc together

The resulting matrix U has orthonormal columns, and U*U^T*y produces the orthogonal projection of y onto the subspace spanned by {aa, bb, cc}.

SECTION 6.4 The Gram-Schmidt Process

TI-83+ The program [PROJ] prompts for two vectors. It produces the orthogonal projection of one onto the other. If you wish to compute the projection of a vector onto the subspace spanned by a set of vectors given as columns of a matrix, [PROJV] will prove to be more useful.

TI-86 The program [PROJ] prompts for the vectors **v2** and **v1**; it produces the orthogonal projection of **v2** onto **v1**. If A has three columns, then add the commands:
```
B(3)→x3
x3-(dot(x3,v1)/dot(v1,v1))*v1-(dot(x3,v2)/dot(v2,v2))*v2→v3
```
You can continue to use the program [PROJ], although in this situation the [PROJV] program—which computes the projection of a vector **x** onto the subspace spanned by a set of vectors given as columns of a matrix V—will prove to be more useful. For example,

TI-89 The program [proj] prompts for two vectors. It produces the orthogonal projection of one onto the other. If you wish to compute the projection of a vector onto the subspace spanned by a set of vectors given as columns of a matrix, [projv] will prove to be more useful.

SECTION 6.5 The Inverse Operator $^{-1}$

TI-83+ The least squares solution to A**x** = **b** is the solution of $A^TAx=A^Tb$. The reduced row echelon form for the augmented matrix for the system is displayed by the command
```
rref augment([A]ᵀ[A],[A]ᵀ[B]),
```
from which the set of solutions can be determined by the method described in Section 1.2 of the text.

If the matrix A is square and invertible, then Theorem 14 tells us that the unique least squares solution to A**x** = **b** is given by $(A^TA)^{-1}A^Tb$. The solution in this case is displayed explicitly as a nx1 matrix by the command
```
([A]ᵀ*[A])⁻¹*[A]ᵀ*[B]
```
For Exercises 15 and 16, see the Numerical Note on page 410 in the text. For Exercise 26, the command
```
augment([A]ᵀ,[B]ᵀ)ᵀ
```
creates a (partitioned) matrix whose top block is [A] and the bottom block is [B]. This command works as long as [A] and [B] have the same number of columns.

TI-86 The least squares solution to A**x** = **b** is the solution of $A^TAx=A^Tb$. The reduced row echelon form for the augmented matrix for the system is displayed by the command
```
rref aug(AᵀA,Aᵀb),
```
from which the set of solutions can be determined by the method described in Section 1.2 of the text.

If the matrix A is square and invertible, then Theorem 14 tells us that the unique least squares solution to A**x** = **b** is given by $(A^TA)^{-1}A^Tb$. The solution in this case is displayed explicitly as a nx1 matrix by the command
```
(Aᵀ*A)⁻¹*Aᵀ*b
```

For Exercises 15 and 16, see the Numerical Note on page 410 in the text. For Exercise 26, the command

        aug(A1^T,A2^T)^T

creates a (partitioned) matrix whose top block is A1 and the bottom block is A2. This command works as long as A1 and A2 have the same number of columns.

TI-89   The least squares solution to A**x** = **b** is the solution of $A^TAx=A^Tb.$ The reduced row echelon form for the augmented matrix for the system is displayed by the command

        rref augment(aa^{T}aa,aa^{T}bb),

from which the set of solutions can be determined by the method described in Section 1.2 of the text.

   If the matrix A is square and invertible, then Theorem 14 tells us that the unique least squares solution to A**x** = **b** is given by $(A^TA)^{-1}A^Tb$. The solution in this case is displayed explicitly as a nx1 matrix by the command

        (aa^T*aa)$^{-1}$*aa^T*bb

For Exercises 15 and 16, see the Numerical Note on page 410 in the text. For Exercise 26, the command

        augment(aa^T,bb^T)^T

creates a (partitioned) matrix whose top block is aa and the bottom block is bb. This command works as long as aa and bb have the same number of columns.

SECTION 6.6 Least-Squares Solutions, Functions of Vectors

TI-83+ Once you create the design matrix X and the observation vector **y**, your computations for least-square solutions here are the same as those described in notes for Section 6.5. Here [A] and [B] are replaced by X and **y**, respectively. The command

        ref augment([A]^T[A],[A]^T[B]),

leads to the general description of all least-squares solutions. When X has linearly independent columns, the command

        ([A]^{T*}[A])$^{-1*}$[A]^{T*}[B]

creates the least-squares solution. In other courses, you may choose simply to use [A]$^{-1}$*[B], which also produces a least-squares solution, except when X is square and singular (or nearly singular).

TI-86   Once you create the design matrix X and the observation vector **y**, your computations for least-square solutions here are the same as those described in notes for Section 6.5. Here [A] and [B] are replaced by X and **y**, respectively. The command

        ref aug(A^{T}A,A^{T}b),

leads to the general description of all least-squares solutions. When X has linearly independent columns, the command

        (A^T*A)$^{-1}$*A^T*b

creates the least-squares solution. In other courses, you may choose simply to use [A]$^{-1}$*[B], which also produces a least-squares solution, except when X is square and singular (or nearly singular).

TI-89   Once you create the design matrix X and the observation vector **y**, your computations for least-square solutions here are the same as those described in notes for Section 6.5. Here [A] and [B] are replaced by X and **y**, respectively. The command

        ref augment(aa^{T}aa,aa^{T}bb),

leads to the general description of all least-squares solutions. When X has linearly independent columns, the command

(aa^T*aa)$^{-1}$*aa^T*bb

creates the least-squares solution. In other courses, you may choose simply to use [A]$^{-1}$*[B], which also produces a least-squares solution, except when X is square and singular (or nearly singular).

SECTION 7.1 Orthogonal Diagonalization

TI-83+ The program [EIGEN] calculates eigenvalues for a given square matrix with less than four rows. Once the eigenvalues have been found, use the [NULB] program to obtain eigenvectors, as in Section 5.3.

TI-86   The eigVl and eigVc functions orthogonally diagonalize any symmetric matrix A, but you miss the opportunity to learn the procedure of this section. Use eigVl to find the eigenvalues of the matrix and the [NULB] program to obtain eigenvectors, as in Section 5.3.

TI-89   The eigVl(and eigVc(keys orthogonally diagonalize any symmetric matrix A, but you miss the opportunity to learn the procedure of this section. Use eigVl(to find the eigenvalues of the matrix and the [nulb] program to obtain eigenvectors, as in Section 5.3.

SECTION 7.4 The Singular Value Decomposition

TI-83+ The TI-83+ does not have the capabilities to compute eigenvectors for this exercise.

TI-86   The commands eigVl A^{T}A→EV and eigVc A^{T}A→P produce the list of eigenvalues EV and an orthogonal matrix P of eigenvectors of the matrix A^{T}A, but the eigenvalues may not be in decreasing order. In such a case, you will have to rearrange things to form V and $\sum$.

TI-89   The commands eigVl(aa^T*aa)→e and eigVc(aa^T*aa)→pp produce the list of eigenvalues e and an orthogonal matrix pp of eigenvectors of the matrix A^{T}A, but the eigenvalues may not be in decreasing order. In such a case, you will have to rearrange things to form V and $\sum$.

Index of TI-83+ Commands

Index of TI-86 Commands

Symbols

Addition key ⊞ TI-12
Alpha key [ALPHA] TI-3
Answer key [ANS] TI-7,23
Bracket keys [[ ] [ ]] TI-2
Exponent key [∧] TI-12
Enter key [ENTER] TI-2,3,5,7,8,10-12,18,19,21
Entry key [ENTRY] TI-22
Exit key [EXIT] TI-3
Inverse key [x^{-1}] TI-13,18,23,26,27
List menu [LIST] TI-10,21
Math menu [MATH] TI-10,20,21
Matrix menu [MATRX] TI-2,3,6,8,10,12,13,15-18,20-22
Memory key [MEM] TI-5
More key [MORE] TI-6,18
Multiplication key [×] TI-11,12,22-27
Poly key [POLY] TI-19
Program menu [PRGM] TI-5,8,9,20
Second key [2nd] TI-2,6,8,10,13,19-23
Solver menu [SOLVER] TI-19
Squaring key [x^2] TI-12
Store key [STO▶] TI-2,7-10,12-14,18,20-23,25,26
Subtraction key [−] TI-7,12,26
Vector menu [VECTR] TI-2,3,12,24,25

Supplemental Programs

BGAUS TI-9,10,13
CHAR TI-21
GAUS TI-9,10,13,15-18
MdiaV TI-16
MtoV TI-3
NULB TI-20,21,28
PRDCT TI-16
PROJ TI-26
PROJV TI-26
VAMX TI-23
VdiaM TI-21
VtoM TI-3,8,25

Built in Commands

augment TI-8-10,13,14,25-27
cond TI-13,16
det TI-16
dim TI-10,12
dot TI-12,24,26
eigVc TI-21,22,28
eigVl TI-21,22,28
Frac TI-10,20,21
ident TI-13,20,23
imag TI-22
li▶vc TI-21
LU TI-15
mRadd TI-6-8
multR TI-6,8,9,13
norm TI-24
randM TI-12,18
real TI-22
ref TI-16-18,27
rref TI-15-20,26
rSwap TI-6-8,16
T TI-12,14,25-28
unitV TI-25

Index of TI-89 Commands

NOTES FOR THE HP-48G CALCULATOR

GETTING STARTED WITH AN HP-48G CALCULATOR

Using the HP-48G with *Linear Algebra ind its Applications*

All of the commands contained in these notes are either standard functions on the HP-48G or are functions in the $\boxed{\text{LALG}}$ directory which may be downloaded from www.laylinalgebra.com. To download this directory onto the HP-48G, first download it onto a computer. By using the Serial Interface Cable and a data transfer protocol such as Kermit or Xmodem, software may be downloaded onto the calculator. Freeware may be used to handle the data transfer; see www.hpcalc.org and the HP-48G User's Guide for more information. The easiest way to transfer data from a PC to a HP-48G is to use the HP Serial Interface Kit, which is available from HP distributors (see www.hpcalc.org for a complete list with a price comparison). This package includes the Serial Interface Cable and connectivity software which allows immediate sharing of information between a PC and the HP-48G. The connectivity software is also available by a free download from www.hp.com. Once the directory is on a HP-48, it may be transferred to another calculator via the infrared interface.

Within the $\boxed{\text{LALG}}$ directory there are two subdirectories: $\boxed{\text{LT}}$ and $\boxed{\text{TBOX}}$. See the HP-48G Manual for information on the $\boxed{\text{LT}}$ directory. The $\boxed{\text{TBOX}}$ directory contains a variety of linear algebra-related programs which are used at various points in the text, and are referred to in the Study Guide Boxes which follow. The $\boxed{\text{TBOX}}$ directory should be loaded onto the calculator for the duration of the linear algebra course.

Also available for downloading from www.laylinalgebra.com is the data for many of the homework exercises from the text. The data is organized first into files which contain the data for a chapter ar a half-chapter. Within these files there is a directory for each section of the text, and then subdirectories for each exercise for which there is data. The menu keys in these subdirectories provide the student with the appropriate data.

Creating Matrices with the HP-48G

There are several ways to create matrices on the HP-48G. The most direct way is to use the calculator's command line. For example, to enter the matrix $\begin{bmatrix} 1 & 2 & 3 \\ 4 & 5 & 6 \end{bmatrix}$, the following sequence of commands is used:

1. Press $\boxed{\text{[ ]}}$ (purple $\boxed{\times}$) key twice to open the delimiters for the matrix and for the first row.

2. Key in the first row: [1] [SPC] [2] [SPC] [3] .

3. Press the [▶] key to move the cursor past the first] delimiter.

4. Key in the rest of matrix – further brackets are unnecessary: [4] [SPC] [5] [SPC] [6] .

5. Press [ENTER] .

   Another way to enter a matrix is to assemble it from its column vectors; see the HP-48G Note for Section 1.3 below for details on how to do this. Finally, the HP-48G has a MatrixWriter application which can be used to enter and edit matrices. See Chapter 14 of the HP-48G User's Guide for directions on how to use this application.

STUDY GUIDE NOTES

SECTION 1.1   Row Operations

Row operations on a matrix A are performed by the following keys, which are found in the [MATH] [MATR] [ROW] menu: [RSWP] swaps rows, [RCI] scales a row by a non-zero constant, and [RCIJ] performs a row replacement operation. The [RSWP] key is on the second page of the menu. These keys are used as follows; you may also find the User's Guide (p. 14-19) helpful.

[RSWP]:   With the matrix on level 1, enter the numbers of the rows you want swapped. For example, 1 [ENTER] 2 [ENTER] [RSWP] will swap rows 1 and 2.

[RCI]:   With the matrix on level 1, enter the constant c by which you want to multiply row i. Next enter the row number i. For example, 5 [ENTER] 1 [ENTER] [RCI] will multiply row 1 by 5.

[RCIJ]:   With the matrix on level 1, first enter the constant c by which you want to multiply row number i. Next enter the row number i of the row you wish to multiply, then enter the row number j of the row to which you want to add c times row i. For example, 5 [ENTER] 1 [ENTER] 2 [ENTER] [RCIJ] will add 5 times row 1 to row 2.

   The new matrix will now be on level 1 of the stack; if you wish to keep a copy of it for later use, simply press [ENTER] ; a copy of it will then appear on level 2. If you then perform a row operation that you don't like for some reason, simply use [DROP] (the purple backspace key) to remove it from the stack. The old matrix on level 2 will now move down to level 1, ready for your next operation. Make sure to recopy it using [ENTER] before proceeding. More permanent storage can be achieved using the [STO] key; see the User's Guide (p. 5-11) for more information.

Note: For the simple problems in this section and the next, the multiple c you will need in the $\boxed{\text{RCI}}$ and $\boxed{\text{RCIJ}}$ commands will usually be a small integer or fraction that you can compute in your head. In general, c may not be so easy to compute mentally. The paragraphs that follow describe a simple way to write c in terms of the entries in A.

The (i, j) entry in A is denoted by the algebraic expression `'A(i,j)'`. To use this expression in calculations, it must be surrounded by single quotes, and the matrix A must be stored as a variable in your current directory. You can use this expression to help you row reduce matrices.

For instance, if you want to scale row i of A to change the value of A(i,k) to 1, you can enter A, then `'A(i,k)'`, then press $\boxed{1/x}$ and $\boxed{\text{EVAL}}$. The proper scaling factor should now be on level 1. Finally, enter the row number i and press $\boxed{\text{RCI}}$.

If you want to use a pivot entry A(i,j) to change A(k,j) to 0, you can enter A, then `'A(k,j)'` `'A(i,j)'` $\boxed{\div}$ $\boxed{+/-}$ $\boxed{\text{EVAL}}$ to produce the proper factor. Then enter i, then k, then press $\boxed{\text{RCIJ}}$.

SECTION 1.3   Constructing a Matrix

To create the matrix $A = \begin{bmatrix} \mathbf{a_1} & \mathbf{a_2} & \mathbf{a_3} & \mathbf{b} \end{bmatrix}$, do the following steps. First enter $\mathbf{a_1}$ *as a vector*. This can be confusing – you must enter it as a row. Open one pair of brackets with the $\boxed{[\,]}$ (purple $\boxed{\times}$) key, then enter the entries of the vector from top to bottom as the cursor proceeds from left to right. Separate the entries with a $\boxed{\text{SPC}}$; when you have completed typing the entries, press $\boxed{\text{ENTER}}$. Enter $\mathbf{a_2}$, $\mathbf{a_3}$, and $\mathbf{b}$ onto the stack in similar fashion. You then enter the number of columns (4 in this case), and press $\boxed{\text{COL}\rightarrow}$, which is found in the $\boxed{\text{MTH}}$ $\boxed{\text{MATR}}$ $\boxed{\text{COL}}$ menu.

To append the column vector $\mathbf{b}$ to the matrix A, thus forming $\begin{bmatrix} A & \mathbf{b} \end{bmatrix}$, first place A on the stack, then enter $\mathbf{b}$ as described above. You then enter the number of the column in A which you wish $\mathbf{b}$ to become, and press the $\boxed{\text{COL+}}$ key in the $\boxed{\text{MTH}}$ $\boxed{\text{MATR}}$ $\boxed{\text{COL}}$ menu. Consult your User's Guide (pp.14-3, 14-5) for more details.

Exercises 11-14, 25-28, and 31 can be solved using the $\boxed{\text{RSWP}}$, $\boxed{\text{RCI}}$, and $\boxed{\text{RCIJ}}$ keys described in the HP-48G Note for Section 1.1.

SECTION 1.4 $\boxed{\text{GAUS}}$ and $\boxed{\text{BGAU}}$

To solve $Ax = b$, row reduce the matrix $\begin{bmatrix} A & \mathbf{b} \end{bmatrix}$, which you can create by the method outlined in the HP-48G Note for Section 1.3. Recall that you enter column vectors as rows; thus the vector

$$x = \begin{bmatrix} 1 \\ 2 \\ 3 \end{bmatrix} \text{ is entered as } \begin{bmatrix} 1 & 2 & 3 \end{bmatrix}.$$

To multiply a matrix A by a vector x place A on the stack, then place x on the stack and press $\boxed{\times}$. The number of entries in x must match the number of columns in A. You should interpret the result as a column vector.

To speed up row reduction of the augmented matrix $M = [\ A\ |\ b\ ]$, the $\boxed{\text{GAUS}}$ key in your $\boxed{\text{TBOX}}$ directory may be used. Place the matrix M on the stack, then enter the number of the row you wish to use. The $\boxed{\text{GAUS}}$ program will now use the leading entry in the given row of M as a pivot, and use row replacements to create zeroes in the pivot column below this pivot entry. The result is returned to the stack. For the backward phase of row reduction, use the $\boxed{\text{BGAU}}$ key which is also in your $\boxed{\text{TBOX}}$ directory. The key works exactly as the $\boxed{\text{GAUS}}$ key, except that the program creates zeroes in the pivot column *above* the pivot entry. You may then use the $\boxed{\text{RCI}}$ key to create 1's in the pivot positions. The $\boxed{\text{TBOX}}$ directory which contains the $\boxed{\text{GAUS}}$ and $\boxed{\text{BGAU}}$ programs is a subdirectory of the directory $\boxed{\text{LALG}}$. The $\boxed{\text{LALG}}$ directory may be downloaded from www.laylinalgebra.com, as is described in the above section "Getting Started with an HP-48G Calculator."

SECTION 1.5   Zero Matrices

To create an $m \times n$ matrix of zeroes, enter the list $\{\ \texttt{m}\ \texttt{n}\ \}$, then 0, then press the $\boxed{\text{CON}}$ key in the $\boxed{\text{MTH}}$ $\boxed{\text{MATR}}$ $\boxed{\text{MAKE}}$ menu. To create a vector containing m zeroes, proceed as above, except use the list $\{\ \texttt{m}\ \}$. When solving the equation $A\mathbf{x} = \mathbf{0}$, where A is an $m \times n$ matrix, you can create the matrix $\begin{bmatrix} A & \mathbf{0} \end{bmatrix}$ by entering A, then creating a vector of m zeroes by the above method. Finally use the $\boxed{\text{COL+}}$ key in the $\boxed{\text{MTH}}$ $\boxed{\text{MATR}}$ $\boxed{\text{COL}}$ menu (described in the HP-48G Note for Section 1.3) to append the vector onto the matrix. You can then use the $\boxed{\text{GAUS}}$, $\boxed{\text{BGAU}}$ and $\boxed{\text{RCI}}$ keys to row reduce $\begin{bmatrix} A & \mathbf{0} \end{bmatrix}$ completely.

SECTION 1.6   Rational Format

Chemical equation-balance problems are studied best using exact or symbolic arithmetic, because the balance variables must be whole numbers (with no round-off allowed). The $\boxed{\rightarrow Q}$ key will take a floating point number at level 1 of the stack and return a rational approximation for this floating number. To find the $\boxed{\rightarrow Q}$ key, press the left arrow (purple) key, then the $\boxed{9}$ key, then the $\boxed{\text{NXT}}$ key. The $\boxed{\rightarrow Q}$ key should now be one of the menu options. Note that this key will not work on an entire matrix at one time. You must either retype the entry you want to convert, or reference the entry in the $\texttt{'A(i,j)'}$ format mentioned in the HP-48G Note for Section 1.1, then use the $\boxed{\rightarrow Q}$ key on that single floating point number.

Once you find a rational solution of a chemical equation-balance problem, you can multiply the entries in the solution vector by a suitable integer to produce a solution that involves only whole numbers.

SECTION 1.10   Generating a Sequence

To generate the sequence $x_1, x_2, \ldots$, enter and store the matrix M. You can then enter the vector x_0 onto the stack, and press $\boxed{\text{ENTER}}$ to copy it. Press $\boxed{\text{VAR}}$ if necessary to produce a list of your variables, then press the menu key labelled $\boxed{\text{M}}$. The series of commands $\boxed{\text{SWAP}}$ $\boxed{\times}$ $\boxed{\text{ENTER}}$ will compute x_1 and copy it onto the stack. Repeating this process will yield the sequence of vectors in order on your stack; the final vector in the sequence will appear twice. The stack will extend upwards as long as the calculator has memory to hold it; you shouldn't worry about exhausting your calculator's memory with the exercises in this section.

Numbers are entered into the HP-48G without commas. The number 6,000,000,000,000 in HP-48G scientific notation is 6.E12. A small number such as .0000000000012 is 1.2E-12.

SECTION 2.1   Matrix Notation and Operations

To create a matrix, you may use the MatrixWriter; see Chapter 8 of your User's Guide for more information. You may also use the command line to enter a matrix. For example, the keystrokes

$\boxed{[\ ]}$ $\boxed{[\ ]}$ 1 $\boxed{\text{SPC}}$ 2 $\boxed{\text{SPC}}$ 3 $\boxed{\blacktriangleright}$ 4 $\boxed{\text{SPC}}$ 5 $\boxed{\text{SPC}}$ 6 $\boxed{\text{ENTER}}$

will create the 2×3 matrix

$$\begin{bmatrix} 1 & 2 & 3 \\ 4 & 5 & 6 \end{bmatrix}.$$

If A is an $m \times n$ matrix, you can find its size by placing A on level 1 of the stack and pressing the $\boxed{\text{SIZE}}$ key in the $\boxed{\text{PRG}}$ $\boxed{\text{LIST}}$ $\boxed{\text{ELEM}}$ menu. The list { m n } will be returned to the stack. As was noted in Section 1.5, the (i, j) element in the matrix A is 'A(i,j)'.

The HP-48G uses the $\boxed{+}$, $\boxed{-}$, and $\boxed{\times}$ keys to denote matrix addition, subtraction, and multiplication, respectively. Note that the $\boxed{y^x}$ key will not operate on matrices, but the $\boxed{x^2}$ key will. You can produce the transpose of a matrix by using the $\boxed{\text{TRN}}$ key in the $\boxed{\text{MTH}}$ $\boxed{\text{MATR}}$ $\boxed{\text{MAKE}}$ menu. To compute the inner product of two vectors, place them at levels 1 and 2 of the stack and use the $\boxed{\text{DOT}}$ key in $\boxed{\text{MTH}}$ $\boxed{\text{VECTR}}$ menu. In order to take the outer product uv^T of two vectors, you must enter them as $n \times 1$ matrices, and use the matrix commands $\boxed{\text{TRN}}$ and $\boxed{\times}$. For complex vectors or matrices, consult your User's Guide.

The $\boxed{\text{MTH}}$ $\boxed{\text{MATR}}$ and $\boxed{\text{MTH}}$ $\boxed{\text{MATR}}$ $\boxed{\text{MAKE}}$ menus also contain commands which will help you construct many special matrices. For example, the following sequences of commands yield the following matrices; more information on these commands is available in the User's Guide.

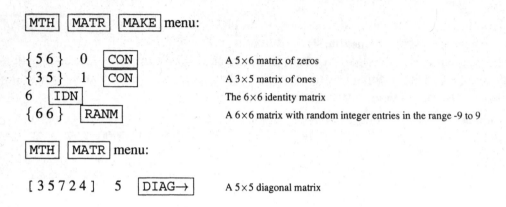

MTH MATR MAKE menu:

$\{5\ 6\}$ 0   CON   A 5×6 matrix of zeros
$\{3\ 5\}$ 1   CON   A 3×5 matrix of ones
6   IDN   The 6×6 identity matrix
$\{6\ 6\}$ RANM   A 6×6 matrix with random integer entries in the range -9 to 9

MTH MATR menu:

$[\ 3\ 5\ 7\ 2\ 4\ ]$ 5   DIAG→   A 5×5 diagonal matrix

SECTION 2.2   Constructing A^{-1}

To produce the $n \times n$ identity matrix, enter n and press the IDN key in the MTH MATR MAKE menu. You may augment the matrix A with an identity matrix by means of the COL+ command mentioned in the HP-48G Note for Section 1.3. Use the GAUS , BGAU and RCI keys to row reduce $\begin{bmatrix} A & I \end{bmatrix}$.

There are other keys that can be used to row reduce matrices, invert matrices, and solve equations $A\mathbf{x} = \mathbf{b}$. They will be discussed later. after you have studied the concepts and algorithms in this section.

SECTION 2.3 $1/x$, COND , and Hilbert matrices

Determining whether a specific numerical matrix is invertible is not always a simple matter. A fast and fairly reliable method is to enter the matrix onto the stack and press $1/x$, which computes the inverse of the matrix. An error message is given if the calculator finds that the matrix is not invertible.

For Exercises 41-44, the COND key in the MTH MATR NORM computes a type of condition number for the matrix on level 1 of the stack. Since this condition number is not the same type of condition number as used in the text, your answers will not match those in the back of the text. However, the note preceding Exercise 42 is still valid for the HP-48G's condition number.

To perform the experiment described in Exercise 42, store the matrix A on the calculator. To produce x, use the commands $\{\ 4\ 1\ \}$ RANM , and store the result as X. The sequence of keystrokes

A   X   ×   A $1/x$ SWAP   ×

will produce $\mathbf{x}_1$. With this vector on level 1 of the stack, the keystrokes

X   SWAP   −

will compute $x - x_1$. Displaying the value of $x - x_1$ is the best way to compare x and x_1. To repeat the process, create and store a new x.

The data from www.laylinalgebra.com which accompanies Exercises 44 and 45 is actually a program which produces, respectively, a 5×5 Hilbert matrix and a 12×12 Hilbert matrix. These programs may be edited to produce Hilbert matrices of any size.

SECTION 2.4   Partitioned Matrices

You may use the COL+ key in the MTH MATR COL menu and the ROW+ key in the MTH MATR ROW menu to append matrices to each other, thus creating partitioned matrices. Consult your User's Guide (p. 14-5) for more details.

SECTION 2.5   LU Factorization and the ÷ Key

Row reduction of A using the GAUS key described in the HP-48G Note for Section 2.2 will produce the intermediate matrices needed for an LU factorization of A. You can try this on the matrix in Example 2 of Section 2.5. The matrices in equation (5) on page 145 of the text are produced by placing A on the stack and keying

1	GAUS	This produces a matrix with 0's below the first pivot
2	GAUS	This produces a matrix with 0's below pivots 1 and 2
3	GAUS	This produces the echelon form of the matrix

You can copy the information from your screen onto your paper, and divide by the pivot entries to produce L as in the text. (For most text exercises, the pivots are integers and so are displayed accurately.) The LU key in the MTH MATR FACTR menu produces the ingredients for a permuted LU factorization of a square matrix A, but does not handle the general case. The calculator produces three matrices on the stack. On level 1 you will find a matrix P, on level 2 an upper triangular matrix U, and on level 3 a lower triangular matrix L. Notice that in this case the ones lie on the diagonal of U, not L. These three matrices satisfy the identity $PA = LU$, or $A = P^{-1}LU$. The matrix $P^{-1}L$ is a permuted lower triangular matrix, so the factorization $A = (P^{-1}L)U$ is a permuted LU factorization of A. As the algorithm used by the calculator differs from that in your text, you should not expect your permuted LU factorization to agree with that of the calculator.

When A is invertible, the best way to solve $Ax = b$ is to use the ÷ key. Enter b onto the stack (as a vector), then enter A. Pressing ÷ now will cause x to be produced, again as a vector. The HP-48G performs a permuted LU factorization on A and uses the matrices P, L, and U to find $A^{-1} = U^{-1}L^{-1}P$, then $x = A^{-1}b$. The ÷ operation uses 15-digit internal precision, which provides for a more accurate answer than would be obtained by calculating A^{-1} by using the 1/x operation. The $\boxed{1/x}$ operation also uses a permuted LU decomposition, but does not carry as great an internal precision.

SECTION 2.8 $\boxed{\text{RREF}}$

By now, the row reduction algorithm should be second nature, so now it's time to cut to the chase. Applying the $\boxed{\text{RREF}}$ key in the $\boxed{\text{MTH}}$ $\boxed{\text{MATR}}$ $\boxed{\text{FACTR}}$ menu to a matrix A produces the reduced row echelon form of A. From that you will immediately be able to write a basis for Col A and to write the homogeneous equations that describe Nul A. Don't forget that A is a coefficient matrix, not an augmented matrix.

SECTION 2.9 $\boxed{\text{RANK}}$

You can use the $\boxed{\text{RREF}}$ key to check the rank of A, but roundoff error or small pivot entries can produce an incorrect reduced row echelon form. A more reliable strategy is to use $\boxed{\text{RANK}}$. The $\boxed{\text{RANK}}$ key is located on the second page of the $\boxed{\text{MTH}}$ $\boxed{\text{MATR}}$ $\boxed{\text{NORM}}$ menu. By default the HP-48G sets all "tiny" elements in a matrix to 0. This helps avoid problems with roundoff error in calculations, but can also generate unexpected results. See the User's Guide (pp. 14-9, 14-20, D-5) for more information.

SECTION 3.2   Computing Determinants

To compute det A, place A on the stack and then repeatedly use the $\boxed{\text{GAUS}}$ and $\boxed{\text{RSWP}}$ keys as needed to reduce A to a matrix U which is in echelon form. (See the HP-48G Note for Section 2.2.) Keep track of how many times you swap rows. Then except for a ±1, the determinant of A can be found by placing U on the stack and executing the keystrokes $\boxed{\rightarrow\text{DIAG}}$ $\boxed{\text{PROD}}$. The $\boxed{\rightarrow\text{DIAG}}$ key is found on the second page of the $\boxed{\text{MTH}}$ $\boxed{\text{MATR}}$ menu; the $\boxed{\text{PROD}}$ key is found in the $\boxed{\text{TBOX}}$ directory. The $\boxed{\rightarrow\text{DIAG}}$ key extracts the diagonal entries from U and places them in a vector, and the $\boxed{\text{PROD}}$ key computes the product of those entries. You can, of course, use the $\boxed{\text{DET}}$ key (found on the second page of the $\boxed{\text{MTH}}$ $\boxed{\text{MATR}}$ $\boxed{\text{NORM}}$ menu) to check your work, but the longer sequence of commands helps you to think about the *process* of computing det A.

SECTION 4.1   Graphing Functions

The following procedure will graph the function f in Exercise 37. Press the $\boxed{\text{PLOT}}$ (aqua 8) key, which will open an input form. You fill in the different pieces of data on this form, using the arrow keys ($\boxed{\blacktriangle}$, $\boxed{\blacktriangleleft}$, $\boxed{\blacktriangledown}$, $\boxed{\blacktriangleright}$) to maneuver bewteen input boxes. For example, if you highlight the "TYPE" box and press the menu key labelled $\boxed{\text{CHOOSE}}$, a list of types of graphs will appear. Use the up and down arrows to highlight the "Function" option, then press $\boxed{\text{OK}}$. To enter the function f, highlight the "EQ" box and enter 1-8*COS(T)^2+8*COS(T)^4 then press $\boxed{\text{OK}}$. Similarly you enter T as INDEP (the independent variable) and 0 and 6.28 as the H-VIEW limits. You may choose to have the calculator scale the vertical dimension of the graph automatically by placing a check in the AUTOSCALE area (press $\boxed{\checkmark\text{CHK}}$ to do this), or you may enter V-VIEW limits on your own. After all of this is entered, press $\boxed{\text{ERASE}}$ to erase any previous graph, then $\boxed{\text{DRAW}}$ to create the graph. See Chapter 24 of your User's Guide for more information on plotting.

SECTION 4.3 $\boxed{\text{RREF}}$ and Applying Functions to Lists

Applying the $\boxed{\text{RREF}}$ key in the $\boxed{\text{MTH}}$ $\boxed{\text{MATR}}$ $\boxed{\text{FACTR}}$ menu to a matrix A produces the reduced row echelon form of A. From that you will immediately be able to write a basis for $\text{Col}A$ and to write the homogeneous equations that describe $\text{Nul}A$. Don't forget that A is a coefficient matrix, not an augmented matrix.

For Exercise 38: To form the necessary coefficient matrix in this case, you can first produce each column then use the $\boxed{\text{COL}\rightarrow}$ key (See the HP-48G Note for Section 1.3 or the User's Guide p. 14-3). To produce each column, you will want to apply the function $\cos^k(t)$ to each element in a vector. To do this, enter the vector on the stack **enclosed not by brackets, but by set braces ({ and }).** This is an example of what the HP-48G calls a list. You may operate on lists just as you do on numbers, so pressing $\boxed{\text{COS}}$ k $\boxed{y^x}$ will apply the function $\cos^k(t)$ to each element in the list. To change this list into a vector, press $\boxed{\text{OBJ}\rightarrow}$ $\boxed{\rightarrow\text{ARR}}$. These keys are found in the $\boxed{\text{PRG}}$ $\boxed{\text{TYPE}}$ menu. For more on lists, see Chapter 17 of the User's Guide.

SECTION 4.4 The Division Key $\boxed{\div}$

If the equation $A\mathbf{x} = \mathbf{b}$ has a unique solution and A is a square matrix, you may calculate the solution $\mathbf{x}$ by entering first $\mathbf{b}$ then A and pressing $\boxed{\div}$. In this section, the equation will probably have the form $P\mathbf{u} = \mathbf{x}$.

This command actually computes $A^{-1}\mathbf{b}$, but uses 15-digit internal precision. This provides a more precise result than computing $A^{-1}\mathbf{b}$ by inverting A and multiplying. If A is not invertible, the HP-48G will give you an "Infinite Result" error.

SECTION 4.6 $\boxed{\text{RREF}}$, $\boxed{\text{RANK}}$, and $\boxed{\text{RANM}}$

In this course you may use either the $\boxed{\text{RREF}}$ or the $\boxed{\text{RANK}}$ key to check the rank of a matrix. In practical work the $\boxed{\text{RANK}}$ key should be used, since this key uses a more reliable algorithm. This algorithm is based on the singular value decomposition (see Section 7.4).

The $\boxed{\text{RANM}}$ key in the $\boxed{\text{MTH}}$ $\boxed{\text{MATR}}$ $\boxed{\text{MAKE}}$ menu produces a matrix with random integer entries between -9 and 9; see your User's Guide for details.

SECTION 4.7   Change-of-Coordinates Matrix

Data for Exercises 7-10 and 17-19 may be downloaded from www.laylinalgebra.com. The RREF key will completely row reduce the matrix $\begin{bmatrix} c_1 & c_2 & b_1 & b_2 \end{bmatrix}$ to the desired form.

SECTION 4.8   Finding the Roots of a Polynomial

To find the roots of a polynomial, begin by entering a vector which contains the coefficients of the polynomial in decreasing powers of the variable. For example, to find the roots of $t^2 - 5t + 6$, enter the vector [1 -5 6]. Press the purple (left) arrow then the 7 key, which sends you to an alternative version of the SOLVE application. Press the POLY menu key, then the PROOT menu key. A vector of roots will be returned to the stack.

SECTION 4.9   Generating a Sequence

The HP-48G Note for Section 1.10 contains information that is useful for homework here.

SECTION 5.1   Finding Eigenvectors

When you know an eigenvalue, your TBOX directory has a program NULB that will simplify your homework by helping to produce a basis for an eigenspace. For example, if A is a 5×5 matrix with an eigenvalue 7, first place A on level 1 of the stack, then use the keystrokes

$$5 \;\; \boxed{\text{IDN}} \;\; 7 \;\; \boxed{\times} \;\; \boxed{-}$$

to produce the matrix $A - 7I$. Then pressing the NULB key will produce a set of vectors on the stack which forms a basis for the eigenspace for A corresponding to $\lambda = 7$. In general, k IDN produces the $k \times k$ identity matrix, and C NULB produces a set of vectors which is a basis for Nul C.

  For Exercises 37-40, you need the EGVL program, which is located on the second page of the MTH MATR menu. This program produces a vector containing the eigenvalues of the matrix on level 1 of the stack, which we stored as A. For example, the keystrokes

$$\boxed{A} \;\; \boxed{\text{ENTER}} \;\; \boxed{\text{EGVL}} \;\; \text{'E'} \;\; \boxed{\text{STO}} \;\; 5 \;\; \boxed{\text{IDN}} \;\; \text{'E(2)'} \;\; \boxed{\text{EVAL}} \;\; \boxed{\times} \;\; \boxed{-} \;\; \boxed{\text{NULB}}$$

compute a basis for the eigenspace corresponding to the second eigenvalue listed in the vector E. It is dangerous to use EGVL and simply "look" at the list of eigenvalues to use in the NULB command. You make a mistake when you type an eigenvalue, particularly when the HP-48G does not show all of the nonzero digits of them.

SECTION 5.2   CHAR

You can use the CHAR key in the TBOX directory to check your answers in Exercises 9-14. Note that if A is $n \times n$, pressing this key with A at level 1 produces a vector listing the coefficients of the characteristic polynomial of A, in order of decreasing powers of λ, beginning with λ^n. If the polynomial is of odd degree, the coefficients are multiplied by -1, to make $+1$ the coefficient of λ^n. This corresponds to finding the determinant of $\lambda I - A$.

For Exercises 28 and 29, use the RANM key to create a 4×4 matrix with random integer entries. See the HP-48G Note for Section 2.1 on how to use this key. For Exercise 29, use the GAUS key and perhaps the RSWP key to create the echelon form without row scaling. See the HP-48G Note for Section 1.4.

SECTION 5.3   Diagonalization and EGV

To practice the diagonalization procedure, you should use NULB to produce the eigenvectors. For Exercises 33-36, you should use the EGVL program to produce the eigenvalues. See the HP-48G Note for Section 5.1.

In later work, you can automate the diagonalization process. The EGV key in the MTH MATR menu produces a vector containing the eigenvalues of A on level 1 of the stack, and a matrix P on level 2 of the stack. To convert the vector containing the eigenvalues into the diagonal matrix D, begin with that vector at level 1 of the stack. Enter the number of rows (or columns) of A onto the stack, and press the DIAG→ key, which is on the second page of the MTH MATR menu. The matrix D will now be at level 1 of the stack. The matrix P which is also produced by EGV satifies the equation $AP = PD$, where D is the diagonal matrix created from the vector of eigenvalues. If P is invertible, then A is diagonalizable. Check whether $PDP^{-1} - A$ is the zero matrix. In any case the P matrix generated by EGV is likely to be quite different from what you construct for your homework.

SECTION 5.5   Complex Eigenvalues

The EGVL and EGV keys also work for matrices with complex eigenvalues.

For a matrix on level 1 of the stack, the RE and IM keys in the CMPL menu produce the real and the imaginary parts of the entries in a matrix, displayed as matrices the same size as the given matrix. The CMPL menu is located on the second page of the MTH menu.

SECTION 5.6   Plotting Discrete Trajectories

Given a vector x, you may compute the product Ax by placing x on the stack, then placing A on the stack, pressing SWAP, then ×. The product vector will be left on the stack, and you may repeat the above procedure over and over if you wish.

The following program creates a "trajectory" matrix whose rows are the points $\mathbf{x}$, $A\mathbf{x}$, $A^2\mathbf{x}$, ..., $A^{15}\mathbf{x}$. (Change 15 to any number you wish.) This program asssumes that the matrix A is on level 2 of the stack and the vector $\mathbf{v}$ is on level 3 of the stack when the program is run from level 1 of the stack.

```
« → V A                          Inputs data.
    « V DUP                      Places v on stack.
        1 15 START               This loop repeats the next line 15 times.
            A SWAP * DUP         Computes the next point on the trajectory.
        NEXT                     End of the loop.
        DROP DEPTH ROW→          Assembles points into the matrix.
    »
»
```

If you place this program on level 1 of the stack and press $\boxed{\text{ENTER}}$, the result is the trajectory matrix. If you intend to use the program more than once, store it as a variable. If you want to plot the points in the output matrix, store the matrix under some name and enter the PLOT utility. Choose the plot type "Scatter" in the window labelled TYPE and choose your matrix in the window labelled ΣDAT. You can also elect to change the corners of your viewing window at this point, if you so desire. Pressing the ERASE and DRAW menu keys will produce a graph of the trajectory. If you have the data for another trajectory stored in another matrix, you can plot both trajectories on the same graph by plotting first one and then the other. Do not press ERASE in between your plots. See Chapters 22, 23, and 29 in your User's Guide for more information about programming and plotting with the HP-48G.

For Exercise 17, you will first need to generate a appropriate vector of values y_i for which you wish to plot the points $(1, y_1)$, $(2, y_2)$, ..., $(8, y_8)$. Place this vector on level 1 of the stack, then do the following keystrokes (the $\boxed{\text{OBJ}\rightarrow}$ and the $\boxed{\rightarrow\text{ARR}}$ keys are found in the $\boxed{\text{PRG}}$ $\boxed{\text{TYPE}}$ menu).

$\boxed{\text{OBJ}\rightarrow}$ $\boxed{\text{DROP}}$

1 $\boxed{\text{SPC}}$ 2 $\boxed{\text{SPC}}$ 3 $\boxed{\text{SPC}}$ 4 $\boxed{\text{SPC}}$ 5 $\boxed{\text{SPC}}$ 6 $\boxed{\text{SPC}}$ 7 $\boxed{\text{SPC}}$ 8 $\boxed{\text{SPC}}$ $\boxed{\text{ENTER}}$

{ 2 8 } $\boxed{\text{ENTER}}$ $\boxed{\rightarrow\text{ARR}}$ 1 2 $\boxed{\text{RSWP}}$ $\boxed{\text{TRN}}$ 'A' $\boxed{\text{STO}}$

The matrix A now holds the data you wish to plot. Use the above directions to do a scatter plot of the entries in A.

SECTION 5.7   Solutions of Differential Equations

For the eigenvalues of A, use the $\boxed{\text{EGVL}}$ key and store the resulting vector of eigenvalues as E. If the eigenvalue $E(1)$ is complex, then the corresponding eigenvector $\mathbf{v}$ will also be complex. The $\boxed{\text{RE}}$ and $\boxed{\text{IM}}$ keys in the $\boxed{\text{CMPL}}$ menu produce the real and the imaginary parts of $\mathbf{v}$.

SECTION 5.8   Power Method and Inverse Power Method

The algorithms below assume that A has a strictly dominant eigenvalue, and the initial vector is $\mathbf{x}$, with largest entry 1 (in magnitude). Store A, then place A and your initial vector $\mathbf{x}$ on the stack. You proceed as follows, using the $\boxed{\text{VAMX}}$ key in your $\boxed{\text{TBOX}}$ directory.

The Power Method When the following keystrokes are repeated over and over, the resulting vectors approach (in many cases) an eigenvector for a strictly dominant eigenvalue:

$\boxed{\times}$ (1)

$\boxed{\text{VAMX}}$ estimate for eigenvalue at level 1   (2)

$\boxed{\div}$ estimate for the eigenvector   (3)

In (2), the program $\boxed{\text{VAMX}}$ finds the entry of largest absolute value in the vector $A\mathbf{x}$. To repeat the process, recall A to the stack and press $\boxed{\text{SWAP}}$. As these commands are repeated, the numbers that appear at level 1 after you press $\boxed{\text{VAMX}}$ are the μ_k that approach the dominant eigenvalue. You could program your HP-48G to perform this algorithm a certain number of times by using a loop structure (see the HP-48G Note for Section 5.6).

The Inverse Power Method Store the initial estimate of the eigenvalue in the variable B, then perform the following keystrokes, where n is the number of columns in A:

$$\boxed{A}\ \boxed{B}\ n\ \boxed{\text{IDN}}\ \boxed{\times}\ \boxed{-}$$

Store the resulting matrix as C. Place your initial vector $\mathbf{x}$ on the stack, followed by C. You then enter the keystrokes

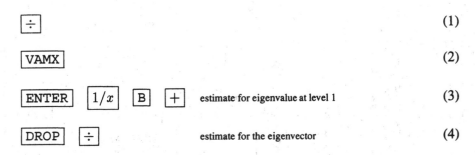

$\boxed{\div}$ (1)

$\boxed{\text{VAMX}}$ (2)

$\boxed{\text{ENTER}}\ \boxed{1/x}\ \boxed{B}\ \boxed{+}$ estimate for eigenvalue at level 1   (3)

$\boxed{\text{DROP}}\ \boxed{\div}$ estimate for the eigenvector   (4)

You may now enter C onto the stack and repeat the keystrokes. As you repeat these keystrokes, the numbers at level 1 after step 3 form the sequence referred to as $\{\nu_k\}$ in the text; the vectors at level 1 after step 4 form the sequence $\{x_k\}$.

SECTION 6.1 Inner Product and Norm

The inner product of two vectors may be found using the $\boxed{\text{DOT}}$ key in the $\boxed{\text{MTH}}$ $\boxed{\text{VECTR}}$ menu. The $\boxed{\text{ABS}}$ key in the same menu produces the length of a vector. See the HP-48G Note for Section 2.1.

SECTION 6.2 Orthogonality

In Exercises 1-9 and 17-22, the quickest way to test a set such as $\{u_1, u_2, u_3\}$ for orthogonality is to create the matrix $U = \begin{bmatrix} u_1 & u_2 & u_3 \end{bmatrix}$ whose columns are the vectors in the set, and test whether $U^T U$ is a diagonal matrix. See the proof of Theorem 6.

To find the orthogonal projection of y onto u, compute the inner product of y and u and divide by the inner product of u with itself. Take this number and multiply it by u. This process is easily done using the stack; you should enter 3 copies of u and one copy of y, then use the following keystrokes:

$$\boxed{\text{DOT}} \quad \boxed{\text{SWAP}} \quad \boxed{\text{ABS}} \quad \boxed{x^2} \quad \boxed{\div} \quad \boxed{\times}.$$

SECTION 6.3 Orthogonal Projections

The orthogonal projection of y onto a single vector was described in the HP-48G Note for Section 6.2. The orthogonal projection onto the set spanned by an orthogonal set of vectors is the sum of the one-dimensional projections. Another way to construct this projection is to normalize the orthogonal vectors, place them in the columns of a matrix U, and use Theorem 10. That is, the desired projection is $UU^T y$.

SECTION 6.4 The Gram-Schmidt Process, $\boxed{\text{PROJ}}$, $\boxed{\text{GS}}$, and $\boxed{\text{GS.O}}$

If A has only two columns, then the Gram-Schmidt process can be implemented using the following keystrokes. The $\boxed{\text{GCOL}}$ key in your $\boxed{\text{TBOX}}$ directory will be used to get columns from A. Store the matrix under the variable A.

$\boxed{\text{A}}$ 1 $\boxed{\text{GCOL}}$ 'V1' $\boxed{\text{STO}}$

$\boxed{\text{A}}$ 2 $\boxed{\text{GCOL}}$ $\boxed{\text{ENTER}}$ V1 $\boxed{\text{DOT}}$ V1 $\boxed{\text{ENTER}}$ $\boxed{\text{DOT}}$ $\boxed{\div}$ V1 $\boxed{\times}$ $\boxed{-}$

'V2' $\boxed{\text{STO}}$

If A has three columns, add the keystrokes

[A] 3 [GCOL] [ENTER] [ENTER] [V1] [DOT] [V1] [ENTER] [DOT] [÷] [V1]

[×] [SWAP] [V2] [DOT] [V2] [ENTER] [DOT] [÷] [V2] [×] [+] [−]

'V3' [STO]

You should use these keystrokes for awhile, to learn the general procedure. After that, you can use the [PROJ] key in your [TBOX] directory, which computes the projection of a vector x onto the subspace spanned by a set of vectors. To use the [PROJ] program, place x on the stack, then enter the set of vectors one by one onto the stack. If your set of vectors is the set of columns of a matrix, you may enter the matrix then press [→COL] [DROP] to enter the vectors. Pressing the [PROJ] key will produce the projection of the first vector entered onto the span of the remaining vectors.

To implement the Gram-Schmidt process on a matrix A with three columns using [PROJ], you would use the following keystrokes:

[A] 1 [GCOL] 'V1' [STO]

[A] 2 [GCOL] [V1] [PROJ] [A] 2 [GCOL] [SWAP] [−] 'V2' [STO]

[A] 3 [GCOL] [V1] [V2] [PROJ] [A] 3 [GCOL] [SWAP] [−] 'V3' [STO]

The set of vectors need not be orthogonal for the [PROJ] program to work, but if they are, the resulting vector will usually agree with those computed via Theorem 10 in Section 6.3 to ten or more decimal places.

To check your work or save time, you can use the [GS] and [GS.O] keys in the [TBOX] directory to perform the Gram-Schmidt process on the columns of a given matrix. The [GS] key will produce a matrix whose columns are orthogonal, while the [GS.O] key will produce a matrix with orthonormal columns. Thus to find Q for a matrix A, use the [GS.O] key.

Your calculator computes a permuted QR factorization of a matrix A with the [QR] key in the [MTH] [MATR] [FACTR] menu. This command will produce matrices Q, R, and P at levels 3, 2, and 1 respectively. The matrix Q is orthogonal, R is upper triangular, and P is again a permutation matrix (See the HP-48G Note for Section 2.5) such that $AP = QR$.

SECTION 6.5   The [÷] and [LSQ] Keys

When A has linearly dependent columns, you can write down the general description of all least-squares solutions on paper after you row reduce the augmented matrix for the normal equations: $[A^T A \quad A^T \mathbf{b}]$. When A has linearly independent columns, enter the matrix $A^T A$ and the vector

$A^T\mathbf{b}$ onto the stack and press the $\boxed{\div}$ key to solve the system. You could also use calculate $(A^T A)^{-1} A^T \mathbf{b}$ directly by using the $\boxed{1/x}$ key, or by row reducing the augmented matrix as above. For Exercises 15 and 16, see the Numerical Note in Section 6.5 of the text. In these exercises, you find the least-squares solution to $A\mathbf{x} = \mathbf{b}$ using the QR factorization: solve the system $R\hat{\mathbf{x}} = Q^T\mathbf{b}$ by using the $\boxed{\div}$ key as above.

Yet another way to produce a least-squares solution to $A\mathbf{x} = \mathbf{b}$ is to use the $\boxed{\text{LSQ}}$ key. This key is located in the $\boxed{\text{MTH}}$ $\boxed{\text{MATR}}$ menu. To use it, enter $\mathbf{b}$ then A onto the stack directly and press $\boxed{\text{LSQ}}$. While this operation is very easy, you should use the normal equations or the QR factorization for computations here instead of the $\boxed{\text{LSQ}}$ key. This will give you a solid conceptual background for applying least-squares techniques later in your career.

For Exercise 26, you can create a (partitioned) matrix whose top block is $A1$ and bottom block is $A2$ by placing $A1$ and $A2$ on the stack, then entering the number of rows in $A1$ plus 1. Now pressing the $\boxed{\text{ROW+}}$ key in the $\boxed{\text{MTH}}$ $\boxed{\text{MATR}}$ $\boxed{\text{ROW}}$ menu will produce the desired matrix.

SECTION 6.6   Least-Squares Solutions

Once you create the design matrix X and the observation vector $\mathbf{y}$, your computations for least-squares solutions here are the same as those described in the HP-48G Note for Section 6.5. Here, A and $\mathbf{b}$ are replaced by X and $\mathbf{y}$, respectively.

Row reducing the augmented matrix $[\ X^T X \quad X^T\mathbf{y}\ ]$ for the normal equations keads to a general description of all least-squares solutions. When X has linearly independent columns, enter the matrix $X^T X$ and the vector $X^T\mathbf{y}$ onto the stack and press the $\boxed{\div}$ key to solve the system. In subsequent courses, you may choose simply to to use the $\boxed{\text{LSQ}}$ key, as described in the HP-48G Note for Section 6.5. This key also produces a least-squares solution.

SECTION 6.8   Graphing Functions

Once you find an nth order Fourier approximation to a function f by hand computation, you can plot the result using the advice in the HP-48G Note for Section 4.1.

SECTION 7.1   Orthogonal Diagonalization

You can use the $\boxed{\text{EGVL}}$ and $\boxed{\text{NULB}}$ keys to orthogonally diagonalize a matrix by the procedure of this section. You can obtain eigenvectors as in Section 5.3; if you encounter a two-dimensional eigenspace with a basis $\{\mathbf{u}_1, \mathbf{u}_2\}$, replace $\mathbf{u}_2$ with a new eigenvector $\mathbf{v}_2$ orthogonal to $\mathbf{u}_1$. You can use the $\boxed{\text{PROJ}}$ key introduced in the HP-48G Note for Section 6.4 to help with this calculation. After you normalize these vectors and create P, you can check that P is indeed an orthogonal matrix by confirming that $P^T P = I$.

SECTION 7.4   The Singular Value Decomposition

Applying the $\boxed{\text{EGV}}$ operation on the matrix $A^T A$ will produce a matrix of eigenvectors and a vector of eigenvalues for $A^T A$, but there are two problems. First, the matrix of eigenvectors may not be orthogonal. The procedure in the HP-48G Note for Section 7.1 can help you to produce an orthogonal matrix of eigenvectors. Second, the eigenvalues in the vector may not be in decreasing order. In such a case you will have to rearrange things things to form V and Σ. The $\boxed{\text{CSWP}}$ key in the $\boxed{\text{MTH}}$ $\boxed{\text{MATR}}$ $\boxed{\text{COL}}$ menu allows you to swap columns just as $\boxed{\text{RSWP}}$ allows you to swap rows. To form U for the singular value decomposition, normalize the nonzero columns of AV. If U needs more columns, use the method of Example 4.

After you thoroughly understand the singular value decomposition, you will want to use the much faster and more numerically reliable $\boxed{\text{SVD}}$ key. This key is found in the $\boxed{\text{MTH}}$ $\boxed{\text{MATR}}$ $\boxed{\text{FACTR}}$ menu. If you place the matrix A on level 1 of the stack and press $\boxed{\text{SVD}}$, the result will be the matrix U on level 3, the matrix V on level 2, and a vector of singular values on level 1. You can create the matrix Σ from this vector by using the $\boxed{\text{DIAG}\rightarrow}$ key (described in the HP-48G Notes for Sections 2.1 and 5.3), then appending rows and/or columns of zeros if necessary.

SECTION 7.5   Computing Principal Components

The RMEAN menu key (abbreviated $\boxed{\text{RMEA}}$) in the $\boxed{\text{TBOX}}$ directory takes a matrix X on level 1 of the stack and produces a vector whose entries list the averages of the rows of X. With this vector you may create a diagonal matrix whose diagonal entries are the row averages of X by using the $\boxed{\text{DIAG}\rightarrow}$ key in the $\boxed{\text{MTH}}$ $\boxed{\text{MATR}}$ menu. See the HP-48G Note for Section 2.6 for more information. Finally multiplying this diagonal matrix on the right by a matrix of all ones creates a matrix A which is the size of X, whose columns are all the same: each column of A lists the row averages of X. To create a matrix of all ones of the appropriate size, place a copy of X on the stack, enter a 1, then press the $\boxed{\text{CON}}$ key in the $\boxed{\text{MTH}}$ $\boxed{\text{MATR}}$ $\boxed{\text{MAKE}}$ menu.

To convert the data in X into mean-deviation form, find the matrix A above, then use

$$B = X - A.$$

The sample covariance matrix S is produced by the formula

$$S = \frac{1}{N-1} BB^T,$$

where N is the number of columns in B.

The principal component data you need is produced by using the $\boxed{\text{SVD}}$ key in the $\boxed{\text{MTH}}$ $\boxed{\text{MATR}}$ $\boxed{\text{FACTR}}$ menu. If you place the matrix

$$\frac{B^T}{\sqrt{N-1}}$$

on level 1 of the stack and press $\boxed{\text{SVD}}$, the result will be the matrix U on level 3, the matrix V on level 2, and a vector of singular values on level 1. The columns of V are the principal components of the data, and the squares of the singular values list the variances of the new variates.

INDEX OF HP-48G COMMANDS